Communications
in Computer and Information Science　384

Lazaros Iliadis   Harris Papadopoulos
Chrisina Jayne (Eds.)

Engineering Applications of Neural Networks

14th International Conference, EANN 2013
Halkidiki, Greece, September 13-16, 2013
Proceedings, Part II

 Springer

Volume Editors

Lazaros Iliadis
Democritus University of Thrace, Orestiada, Greece
E-mail: liliadis@fmenr.duth.gr

Harris Papadopoulos
Frederick University of Cyprus, Nicosia, Cyprus
E-mail: harris.papadopoulos@gmail.com

Chrisina Jayne
Coventry University, UK
E-mail: ab1527@coventry.ac.uk

ISSN 1865-0929   e-ISSN 1865-0937
ISBN 978-3-642-41015-4   e-ISBN 978-3-642-41016-1
DOI 10.1007/978-3-642-41016-1
Springer Heidelberg New York Dordrecht London

Library of Congress Control Number: Applied for

CR Subject Classification (1998): I.2.6, I.5.1, H.2.8, J.2, J.1, J.3, F.1.1, I.5, I.2, C.2

Typesetting: Camera-ready by author, data conversion by Scientific Publishing Services, Chennai, India

Printed on acid-free paper

Springer is part of Springer Science+Business Media (www.springer.com)

Preface

Artificial Intelligence is a branch of computer science, continuously and rapidly evolving. It is a fact that more and more sophisticated modeling techniques are published in the literature all the time, capable of tackling complicated and challenging problems. Artificial Neural Networks (ANN) and other Soft Computing approaches seek inspiration from the world of biology to enable the development of real world intelligent systems.

EANN is a well established event with a very long and successful history. Eighteen years have passed since the first organization in Otaniemi, Finland, in 1995. For the following years it has a continuous and dynamic presence as a major European scientific event. An important milestone is year 2009, when its guidance by a steering committee of the INNS (*EANN Special Interest Group*) was initiated. Thus, from that moment the conference has been continuously supported technically, by the International Neural Network Society (INNS).

This volume contains the papers that were accepted for oral presentation at the 14th EANN conference and its satellite workshops. This volume belongs to the CCIS Springer Series. The event was held during September 13–16, 2013 at the "Athina Pallas" Resort and Conference Center in Halkidiki, Greece, and was supported by the Aristotle University of Thessaloniki and the Democritus University of Thrace.

Three workshops on timely AI subjects were organized successfully and co-located with EANN'2013:

1. The Second Mining Humanistic Data (MHD) Workshop supported by the Ionian University and the University of Patras. We wish to express our gratitude to Spyros Sioutas and Christos Makris for their common effort towards the organization of the Second MHD Workshop. Also we would like to thank Vassilios Verykios of the Hellenic Open University, Greece, and Evaggelia Pitoura of the University of Ioannina, Greece, for their keynote lectures in the MHD workshop

2. The Third Computational Intelligence Applications in Bioinformatics (CIAB) Workshop supported by the University of Patras. We are grateful to Spyros Likothanasis for his kind efforts towards the management of the CIAB Workshop and for his keynote lecture in the frame of this event.

3. The First Innovative European Policies and Applied Measures for Developing Smart Cities (IPMSC) Workshop, supported by the Hellenic Telecommunications Organization. The IPMSC was driven by the hard work of Ioannis P. Chochliouros and Ioannis M. Stephanakis Hellenic Telecommunications Organization - OTE, Greece.

Three keynote speakers were invited and they gave lectures in timely aspects of AI and ANN. Finally, a highly interesting tutorial entitled "Neural Networks for Digital Media Analysis and Description" was delivered by Anastasios Tefas of the Aristotle University of Thessaloniki, Greece. We wish to express our sincere thanks to the invited keynote speakers and to Anastasios Tefas.

The diverse nature of papers presented, demonstrates the vitality of neural computing and related soft computing approaches and proves the very wide range of ANN applications as well. On the other hand, this volume contains basic research papers, presenting variations and extensions of several approaches.

The Organizing Committee was delighted by the overwhelming response to the call for papers. All papers have passed through a peer review process by at least 2 independent academic referees. Where needed a third referee was consulted to resolve any conflicts. Overall 40% of the submitted manuscripts (totally 91) were accepted to be presented in the EANN and in the three satellite workshops. The accepted papers of the 8th AIAI conference are related to the following thematic topics:

evolutionary algorithms, adaptive algorithms, control approaches, soft computing applications, ANN, ensembles, bioinformatics, classification, pattern recognition, medical applications of AI, fuzzy Iinference, filtering, SOM, RBF, image – video analysis, learning, social media applications, community based governance

The authors came from 28 different countries from all over Europe (e.g. Austria, Bulgaria, Cyprus, Czech Republic, Finland, France, Germany, Greece, Holland, Italy, Poland, Portugal, Slovakia, Slovenia, Spain, UK, Ukraine, Russia, Romania, Serbia), Americas (e.g. Brazil, USA, Mexico), Asia (e.g., China, India, Iran, Pakistan,), Africa (e.g. Egypt, Tunisia, Algeria) and Oceania (New Zealand).

September 2013   Lazaros Iliadis
                                                               Harris Papadopoulos
                                                                  Chrisina Jayne

Organization

General Chair

Konstantinos Margaritis   University of Macedonia, Greece

Advisory chairs

Nikola Kasabov   KEDRI Auckland University of Technology,
                                  New Zealand
Vera Kurkova   Czech Academy of Sciences, Czech Republic
Mikko Kolehmainen   University of Eastern Finland, Finland

Honorary Chair

Dominic Palmer Brown   Dean London Metropolitan University, UK

Program Committee Co-chairs

Lazaros Iliadis   Democritus University of Thrace, Greece
Chrisina Jayne   University of Coventry, UK
Haris Papadopoulos   Frederick University, Cyprus

Workshop chair

Spyros Sioutas   Ionian University, Greece
Christos Makris   University of Patras, Greece

Organizing chair

Yannis Manolopoulos   Aristotle University of Thessaloniki, Greece

Web chair

Ioannis Karydis   Ionian University, Greece

Program Committee Members

Athanasios Alexiou   Ionian University, Greece
Luciano Alonso Renteria   Universidad de Cantabria, Spain

Georgios Anastasopoulos   Democritus University of Thrace, Greece
Ioannis Andreadis   Democritus University of Thrace, Greece
Andreas Andreou   University of Cyprus, Cyprus
Costin Badica   University of Craiova, Romania
Zorana Bankovic   Technical University of MAdrid, Spain
Kostas Berberidis   University of Patras, Greece
Nick Bessis   University of Derby, UK
Monica Bianchini   University of Siena, Italy
Ivo Bukovsky   Czech Technical University in Prague,
                                Czech Republic
George Caridakis   National Technical University of Athens,
                                Greece
Aristotelis Chatziioannou   Institute of Biological Research &
                                Biotechnology, NHRF, Greece
Javier Fernandez De
  Canete Rodriguez   University of Malaga, Spain
Ruggero Donida Labati   University of Milano, Italy
Anestis Fachantidis   Aristotle University of Thessaloniki, Greece
Maurizio Fiasche   Politecnico di Milano, Italy
Ignazio Gallo   University of Insubria, Italy
Francisco Garcia   University of Oviedo, Spain
Christos Georgiadis   University of Macedonia, Greece
Efstratios F. Georgopoulos   Technological Educational Institute of
                                Kalamata, Greece
Giorgio Gnecco   University of Genova, Italy
Petr Hajek   University of Pardubice, Czech Republic
Ioannis Hatzilygeroudis   University of Patras, Greece
Emmanouil Hourdakis   Forthnet, Greece
Raul Jimenez Naharro   Universidad de Huelva, Spain
Jacek Kabzinski   Lodz University of Technology, Poland
Antonios Kalampakas   Democritus University of Thrace, Greece
Ryotaro Kamimura   Hiratsuka Kanagawa, Japan
Kostas Karatzas   Aristotle University of Thessaloniki, Greece
Kostas Karpouzis   National Technical University of Athens,
                                Greece
Ioannis Karydis   Ionian University, Greece
Katia Kermanidis   Ionian University, Greece
Kyriaki Kitikidou   Democritus University of Thrace, Greece
Petia Koprinkova-Hristova   Bulgarian Academy of Sciences, Bulgaria
Konstantinos Koutroumbas   National Observatory of Athens, Greece
Paul Krause   University of Surrey, UK
Pekka Kumpulainen   Tampere University of Technology, Finland
Efthyvoulos Kyriacou   Frederick University, Cyprus
Sin Wee Lee   University of East London, UK
Spyros Likothanasis   University of Patras, Greece

Ilias Maglogiannis	University of Piraeus, Greece
George Magoulas	University of London, UK
Mario Malcangi	University of Milano, Italy
Francesco Marcelloni	University of Pisa, Italy
Avlonitis Markos	Ionian University, Greece
Marisa Masvoula	University of Athens, Greece
Nikolaos Mitianoudis	Democritus University of Thrace, Greece
Haris Mouratidis	University of East London, UK
Phivos Mylonas	National Technical University of Athens, Greece
Nicoletta Nicolaou	University of Cyprus, Cyprus
Vladimir Olej	University of Pardubice, Czech Republic
Mihaela Oprea	Universitatea Petrol-Gaze din Ploiesti, Romania
Ioannis Partalas	Aristotle University of Thessaloniki, Greece
Daniel Perez	University of Oviedo, Spain
Elias Pimenidis	University of East London, UK
Jefferson Rodrigo de Souza	Universidade de Sao Paulo, Brazil
Nick Ryman-Tubb	University of Surrey, UK
Marcello Sanguineti	University of Genova, Italy
Thomas Schack	Bielefeld University, Germany
Christos Schizas	University of Cyprus
Abe Shigeo	Kobe University, Japan
Alexandros Sideridis	Agricultural University of Athens, Greece
Luis Silva	University of Aveiro, Portugal
Spyros Sioutas	Ionian University, Greece
Stephanos Spartalis	Democritus University of Thrace, Greece
Ioannis Stamelos	Aristotle University of Thessaloniki, Greece
Kathleen Steinhofel	King's College, UK
Ioannis Stephanakis	Organization of Telecommunications, Greece
Tatiana Tambouratzis	University of Piraeus, Greece
Panos Trahanias	Forthent, Greece
Thanos Tsadiras	Aristotle University of Thessaloniki, Greece
Nicolas Tsapatsoulis	Technical University of Cyprus, Cyprus
George Tsekouras	University of Aegean, Greece
Aristeidis Tsitiridis	University of Swansea, UK
Grigorios Tsoumakas	Aristotle University of Thessaloniki, Greece
Nikolaos Vasilas	TEI of Athens, Greece
Panayiotis Vlamos	Ionian University, Greece
George Vouros	University of Piraeus, Greece
Peter Weller	City University, UK
Shigang Yue	University of Lincoln, UK
Achilleas Zapranis	University of Macedonia, Greece
Rodolfo Zunino	University of Genova, Italy

Keynotes

Nikola Kasabov: Founding Director and Chief Scientist of the Knowledge Engineering and Discovery Research Institute (KEDRI), Auckland. Chair of Knowledge Engineering at the School of Computing and Mathematical Sciences at Auckland University of Technology. Fellow of the Royal Society of New Zealand, Fellow of the New Zealand Computer Society and a Senior Member of IEEE.

Keynote Presentation Subject: "Neurocomputing for Spatio/Spectro-Temporal Pattern Recognition and Early Event Prediction: Methods, Systems, Applications"

Neurocomputing for Spatio/Spectro-Temporal Pattern Recognition and Early Event Prediction: Methods, Systems, Applications

Nikola Kasabov, Fellow IEEE, Fellow RSNZ
Director, Knowledge Engineering and Discovery Research Institute - KEDRI,
Auckland University of Technology, NZ
nkasabov@aut.ac.nz, www.kedri.aut.ac.nz

Abstract. The talk presents a brief overview of contemporary methods for neurocomputation, including: evolving connections systems (ECOS) and evolving neuro-fuzzy systems [1]; evolving spiking neural networks (eSNN) [2-5]; evolutionary and neurogenetic systems [6]; quantum inspired evolutionary computation [7,8]; rule extraction from eSNN [9]. These methods are suitable for incremental adaptive, on-line learning from spatio-temporal data and for data mining. But the main focus of the talk is how they can learn to predict early the outcome of an input spatio-temporal pattern, before the whole pattern is entered in a system. This is demonstrated on several applications in bioinformatics, such as stroke occurrence prediction, and brain data modeling for brain-computer interfaces [10], on ecological and environmental modeling [11]. eSNN have proved superior for spatio-and spectro-temporal data analysis, modeling, pattern recognition and early event prediction as outcome of recognized patterns when partially presented.

Future directions are discussed. Materials related to the lecture, such as papers, data and software systems can be found from www.kedri.aut.ac.nz and also from: www.theneucom.com and http://ncs.ethz.ch/projects/evospike/.

References

[1] Kasabov, N.: Evolving Connectionist Systems: The Knowledge Engineering Approach. Springer, London (2007), http://www.springer.de (first edition published in 2002)

[2] Wysoski, S., Benuskova, L., Kasabov, N.: Evolving Spiking Neural Networks for Audio-Visual Information Processing. Neural Networks 23(7), 819–835 (2010)

[3] Kasabov, N.: To spike or not to spike: A probabilistic spiking neural model. Neural Networks 23(1), 16–19 (2010)

[4] Mohemmed, A., Schliebs, S., Kasabov, N.: SPAN: Spike Pattern Association Neuron for Learning Spatio-Temporal Sequences. Int. J. Neural Systems (2011, 2012)

[5] Kasabov, N., Dhoble, K., Nuntalid, N., Indiveri, G.: Dynamic Evolving Spiking Neural Networks for On-line Spatio- and Spectro-Temporal Pattern Recognition. Neural Networks 41, 188–201 (2013)

[6] Benuskova, L., Kasabov, N.: Computaional Neurogenetic Modelling. Springer (2007)

[7] Defoin-Platel, M., Schliebs, S., Kasabov, N.: Quantum-inspired Evolutionary Algorithm: A multi-model EDA. IEEE Trans. Evolutionary Computation 13(6), 1218–1232 (2009)

[8] Nuzly, H., Kasabov, N., Shamsuddin, S.: Probabilistic Evolving Spiking Neural Network Optimization Using Dynamic Quantum Inspired Particle Swarm Optimization. In: Wong, K.W., Mendis, B.S.U., Bouzerdoum, A. (eds.) ICONIP 2010, Part I. LNCS, vol. 6443. Springer, Heidelberg (2010)

[9] Soltic, S., Kasabov, N.: Knowledge extraction from evolving spiking neural networks with a rank order population coding. Int. J. Neural Systems 20(6), 437–445 (2010)

[10] Kasabov, N. (ed.): The Springer Handbook of Bio- and Neuroinformatics. Springer (2013)

[11] Schliebs, S., Platel, M.D., Worner, S., Kasabov, N.: Integrated Feature and Parameter Optimization for Evolving Spiking Neural Networks: Exploring Heterogeneous Probabilistic Models. Neural Networks 22, 623–632 (2009)

Erkki Oja: Professor with the Aalto University, Finland, Recipient of the 2006 IEEE Computational Intelligence Society Neural Networks Pioneer Award, Director of the Adaptive Informatics Research Centre, Chairman of the Finnish Research Council for Natural Sciences and Engineering, Visiting Professor at the Tokyo Institute of Technology, Japan, Member of the Finnish Academy of Sciences, IEEE Fellow, Founding Fellow of the International Association of Pattern Recognition (IAPR), Past President of the European Neural Network Society (ENNS), Fellow of the International Neural Network Society (INNS). Author of the scientific books:

o "Subspace Methods of Pattern Recognition", New York: Research Studies Press and Wiley, 1983, translated into Chinese and Japanese,

o "Kohonen Maps", Elsevier, 1999,

o "Independent Component Analysis", Wiley, 2001 translated in Chinese and Japanese.

Machine Learning for Big Data Analytics

Erkki Oja

Aalto University, Finland

`erkki.oja@aalto.fi`

Abstract. During the past 30 years, the amount of stored digital data has roughly doubled every 40 months. Today, about 2.5 quintillion bytes are created very day. This data comes from sensor networks, cameras, microphones, mobile devices, software logs etc. Part of it is scientific data especially in particle physics, astronomy and genomics, part of it comes from other sectors of society such as internet text and documents, web logs, medical records, military surveillance, photo and video archives and e-commerce. This data poses a unique challenge in data mining: finding meaningful things out of the data masses. Central algorithmic techniques to process and mine the data are classification, clustering, neural networks, pattern recognition, regression, visualization etc. Many of these fall under the term machine learning. In the author's research group at Aalto University, Finland, machine learning techniques are developed and applied to many of the above problems together with other research institutes and industry. The talk will cover some recent algorithmic discoveries and illustrate the problem area with case studies in speech recognition and synthesis, video recognition, brain imaging, and large-scale climate research.

Marios Polycarpou is a Fellow of the IEEE and currently serves as the President of the IEEE Computational Intelligence Society. He has served as the Editor-in-Chief of the *IEEE Transactions on Neural Networks and Learning Systems* from 2004 until 2010. He participated in more than 60 research projects/grants, funded by several agencies and industry in Europe and the United States. In 2011, Dr. Polycarpou was awarded the prestigious European Research Council (ERC) Advanced Grant.

Distributed Sensor Fault Diagnosis in Big Data Environments

Marios Polycarpou

University of Cyprus

`mpolycar@ucy.ac.cy`

Abstract. The emergence of networked embedded systems and sensor/actuator networks has given rise to advanced monitoring and control applications, where a large amount of sensor data is collected and processed in real-time in order to achieve smooth and efficient operation of the underlying system. The current trend is towards larger and larger sensor data sets, leading to so called big data environments. However, in situations where faults arise in one or more of the sensing devices, this may lead to a serious degradation in performance or even to an overall system failure. The goal of this presentation is to motivate the need for fault diagnosis in complex distributed dynamical systems and to provide a methodology for detecting and isolating multiple sensor faults in a class of non-linear dynamical systems. The detection of faults in sensor groups is conducted using robust analytical redundancy relations, formulated by structured residuals and adaptive thresholds. Various estimation algorithms will be presented and illustrated, and directions for future research will be discussed.

We hope that these proceedings will help researchers worldwide to understand and to be aware of new ANN aspects. We do believe that they will be of major interest for scientists over the globe and that they will stimulate further research in the domain of Artificial Neural Networks and AI in general.

Table of Contents – Part II

Evaluating Sentiment in Annual Reports for Financial Distress
Prediction Using Neural Networks and Support Vector Machines   1
    Petr Hájek and Vladimír Olej

Identification of All Exact and Approximate Inverted Repeats in
Regular and Weighted Sequences   11
    Carl Barton, Costas S. Iliopoulos, Nicola Mulder, and Bruce Watson

Query Expansion with a Little Help from Twitter   20
    Ioannis Anagnostopoulos, Gerasimos Razis, Phivos Mylonas, and
    Christos-Nikolaos Anagnostopoulos

Recognizing Emotion Presence in Natural Language Sentences   30
    Isidoros Perikos and Ioannis Hatzilygeroudis

Classification of Event Related Potentials of Error-Related Observations
Using Support Vector Machines   40
    Pantelis Asvestas, Errikos M. Ventouras, Irene Karanasiou, and
    George K. Matsopoulos

A Novel Hierarchical Approach to Ranking-Based Collaborative
Filtering   50
    Athanasios N. Nikolakopoulos, Marianna Kouneli, and
    John Garofalakis

Mimicking Real Users' Interactions on Web Videos through a Controlled
Experiment   60
    Antonia Spiridonidou, Ioannis Karydis, and Markos Avlonitis

Mining Student Learning Behavior and Self-assessment for Adaptive
Learning Management System   70
    Konstantina Moutafi, Paraskevi Vergeti, Christos Alexakos,
    Christos Dimitrakopoulos, Konstantinos Giotopoulos,
    Hera Antonopoulou, and Spiros Likothanassis

Exploiting Fuzzy Expert Systems in Cardiology   80
    Efrosini Sourla, Vasileios Syrimpeis,
    Konstantina-Maria Stamatopoulou, Georgios Merekoulias,
    Athanasios Tsakalidis, and Giannis Tzimas

The Strength of Negative Opinions 90
Thanos Papaoikonomou, Mania Kardara,
Konstantinos Tserpes, and Theodora Varvarigou

Extracting Knowledge from Web Search Engine Using Wikipedia 100
Andreas Kanavos, Christos Makris, Yannis Plegas, and
Evangelos Theodoridis

AppendicitisScan Tool: A New Tool for the Efficient Classification
of Childhood Abdominal Pain Clinical Cases Using Machine Learning
Tools 110
Athanasios Mitroulias, Theofilatos Konstantinos,
Spiros Likothanassis, and Mavroudi Seferina

Mining the Conceptual Model of Open Source CMS Using a Reverse
Engineering Approach 119
Vassiliki Gkantouna, Spyros Sioutas, Georgia Sourla,
Athanasios Tsakalidis, and Giannis Tzimas

Representation of Possessive Pronouns in Universal Networking
Language 129
Velislava Stoykova

Sleep Spindle Detection in EEG Signals Combining HMMs
and SVMs 138
Iosif Mporas, Panagiotis Korvesis, Evangelia I. Zacharaki, and
Vasilis Megalooikonomou

Classifying Ductal Trees Using Geometrical Features and Ensemble
Learning Techniques 146
Angeliki Skoura, Tatyana Nuzhnaya, Predrag R. Bakic, and
Vasilis Megalooikonomou

Medical Decision Making via Artificial Neural Networks: A Smart
Phone-Embedded Application Addressing Pulmonary Diseases'
Diagnosis 156
George-Peter K. Economou and Vaios Papaioannou

A Simulator for Privacy Preserving Record Linkage 164
Alexandros Karakasidis and Vassilios S. Verykios

Development of a Clinical Decision Support System Using AI, Medical
Data Mining and Web Applications 174
Dimitrios Tsolis, Kallirroi Paschali, Anna Tsakona,
Zafeiria-Marina Ioannou, Spiros Likothanassis,
Athanasios Tsakalidis, Theodore Alexandrides, and
Athanasios Tsamandas

Supporting and Consulting Infrastructure for Educators during
Distance Learning Process: The Case of Russian Verbs of Motion 185
  *Oksana Kalita, Alexander Gartsov, Georgios Pavlidis, and
  Photis Nanopoulos*

Classification Models for Alzheimer's Disease Detection 193
  *Christos-Nikolaos Anagnostopoulos, Ioannis Giannoukos,
  Christian Spenger, Andrew Simmons, Patrizia Mecocci,
  Hikka Soininen, Iwona Kłoszewska, Bruno Vellas,
  Simon Lovestone, and Magda Tsolaki*

Combined Classification of Risk Factors for Appendicitis Prediction in
Childhood 203
  *Theodoros Iliou, Christos-Nikolaos Anagnostopoulos,
  Ioannis M. Stephanakis, and George Anastassopoulos*

Analysis of DNA Barcode Sequences Using Neural Gas and Spectral
Representation 212
  *Antonino Fiannaca, Massimo La Rosa, Riccardo Rizzo, and
  Alfonso Urso*

A Genetic Algorithm for Pancreatic Cancer Diagnosis 222
  *Charalampos Moschopoulos, Dusan Popovic, Alejandro Sifrim,
  Grigorios Beligiannis, Bart De Moor, and Yves Moreau*

Enhanced Weighted Restricted Neighborhood Search Clustering:
A Novel Algorithm for Detecting Human Protein Complexes from
Weighted Protein-Protein Interaction Graphs 231
  *Christos Dimitrakopoulos, Konstantinos Theofilatos,
  Andreas Pegkas, Spiros Likothanassis, and Seferina Mavroudi*

A Hybrid Approach to Feature Ranking for Microarray Data
Classification 241
  *Dusan Popovic, Alejandro Sifrim, Charalampos Moschopoulos,
  Yves Moreau, and Bart De Moor*

Derivation of Cancer Related Biomarkers from DNA Methylation Data
from an Epidemiological Cohort 249
  *Ioannis Valavanis, Emmanouil G. Sifakis, Panagiotis Georgiadis,
  Soterios Kyrtopoulos, and Aristotelis A. Chatziioannou*

A Particle Swarm Optimization (PSO) Model for Scheduling Nonlinear
Multimedia Services in Multicommodity Fat-Tree Cloud Networks...... 257
  *Ioannis M. Stephanakis, Ioannis P. Chochliouros,
  George Caridakis, and Stefanos Kollias*

Intelligent and Adaptive Pervasive Future Internet: Smart Cities
for the Citizens 269
    George Caridakis, Georgios Siolas, Phivos Mylonas,
    Stefanos Kollias, and Andreas Stafylopatis

Creative Rings for Smart Cities 282
    Simon Delaere, Pieter Ballon, Peter Mechant, Giorgio Parladori,
    Dirk Osstyn, Merce Lopez, Fabio Antonelli, Sven Maltha,
    Makis Stamatelatos, Ana Garcia, and Artur Serra

Energy Efficient E-Band Transceiver for Future Networking 292
    Evangelia M. Georgiadou, Mario Giovanni Frecassetti,
    Ioannis P. Chochliouros, Evangelos Sfakianakis, and
    Ioannis M. Stephanakis

Social and Smart: Towards an Instance of Subconscious Social
Intelligence 302
    M. Graña, B. Apolloni, M. Fiasché, G. Galliani, C. Zizzo,
    G. Caridakis, G. Siolas, S. Kollias,
    F. Barrientos, and S. San Jose

Living Labs in Smart Cities as Critical Enablers for Making Real the
Modern Future Internet 312
    Ioannis P. Chochliouros, Anastasia S. Spiliopoulou,
    Evangelos Sfakianakis, Evangelia M. Georgiadou, and
    Eleni Rethimiotaki

Author Index 323

Table of Contents – Part I

Invited

Neural Networks for Digital Media Analysis and Description . . . .　　1
　　Anastasios Tefas, Alexandros Iosifidis, and Ioannis Pitas

Evolutionary Algorithms

Temperature Forecasting in the Concept of Weather Derivatives:
A Comparison between Wavelet Networks and Genetic Programming . . .　　12
　　Antonios K. Alexandiris and Michael Kampouridis

MPEG-4 Internet Traffic Estimation Using Recurrent CGPANN . . . .　　22
　　Gul Muhammad Khan, Fahad Ullah, and Sahibzada Ali Mahmud

SCH-EGA: An Efficient Hybrid Algorithm for the Frequency
Assignment Problem . . . .　　32
　　Shaohui Wu, Gang Yang, Jieping Xu, and Xirong Li

Improving the RACAI Neural Network MSD Tagger . . . .　　42
　　Tiberiu Boroş and Stefan Daniel Dumitrescu

Adaptive Algorithms - Control Approaches

Neural Network Simulation of Photosynthetic Production . . . .　　52
　　Tibor Kmet and Maria Kmetova

A Novel Artificial Neural Network Based Space Vector Modulated DTC
and Its Comparison with Other Artificial Intelligence (AI) Control
Techniques . . . .　　61
　　Sadhana V. Jadhav and B.N. Chaudhari

General Aspects of AI Evolution

Thinking Machines versus Thinking Organisms . . . .　　71
　　Petro Gopych

Soft Computing Applications

Study of Influence of Parameter Grouping on the Error of Neural
Network Solution of the Inverse Problem of Electrical Prospecting . . . . . .　　81
　　*Sergey Dolenko, Igor Isaev, Eugeny Obornev, Igor Persiantsev, and
　　Mikhail Shimelevich*

Prediction of Foreign Currency Exchange Rates Using CGPANN   91
  Durre Nayab, Gul Muhammad Khan, and Sahibzada Ali Mahmud

Coastal Hurricane Inundation Prediction for Emergency Response
Using Artificial Neural Networks   102
  Bernard Hsieh and Jay Ratcliff

Crossroad Detection Using Artificial Neural Networks   112
  Alberto Hata, Danilo Habermann, Denis Wolf, and Fernando Osório

Application of Particle Swarm Optimization Algorithm to Neural
Network Training Process in the Localization of the Mobile Terminal . . .   122
  Jan Karwowski, Michał Okulewicz, and Jarosław Legierski

Modeling Spatiotemporal Wild Fire Data with Support Vector
Machines and Artificial Neural Networks   132
  *Georgios Karapilafis, Lazaros Iliadis, Stefanos Spartalis,
  S. Katsavounis, and Elias Pimenidis*

Prediction of Surface Texture Characteristics in Turning of FRPs Using
ANN   144
  *Stefanos Karagiannis, Vassilis Iakovakis, John Kechagias,
  Nikos Fountas, and Nikolaos Vaxevanidis*

ANN Ensembles

A State Space Approach and Hurst Exponent for Ensemble
Predictors   154
  Ryszard Szupiluk and Tomasz Ząbkowski

Bioinformatics

3D Molecular Modelling of the Helicase Enzyme of the Endemic,
Zoonotic Greek Goat Encephalitis Virus   165
  Dimitrios Vlachakis, Georgia Tsiliki, and Sophia Kossida

Classification - Pattern Recognition

Feature Comparison and Feature Fusion for Traditional Dances
Recognition   172
  *Ioannis Kapsouras, Stylianos Karanikolos, Nikolaos Nikolaidis, and
  Anastasios Tefas*

Intelligent Chair Sensor: Classification of Sitting Posture   182
  *Leonardo Martins, Rui Lucena, João Belo, Marcelo Santos,
  Cláudia Quaresma, Adelaide P. Jesus, and Pedro Vieira*

Hierarchical Object Recognition Model of Increased Invariance   192
  Aristeidis Tsitiridis, Ben Mora, and Mark Richardson

Detection of Damage in Composite Materials Using Classification and
Novelty Detection Methods   203
  Ramin Amali and Bradley J. Hughes

Impact of Sampling on Neural Network Classification Performance in
the Context of Repeat Movie Viewing   213
  Elena Fitkov-Norris and Sakinat Oluwabukonla Folorunso

Discovery of Weather Forecast Web Resources Based on Ontology and
Content-Driven Hierarchical Classification   223
  *Anastasia Moumtzidou, Stefanos Vrochidis, and
  Ioannis Kompatsiaris*

Towards a Wearable Coach: Classifying Sports Activities with Reservoir
Computing   233
  Stefan Schliebs, Nikola Kasabov, Dave Parry, and Doug Hunt

Real-Time Psychophysiological Emotional State Estimation in Digital
Gameplay Scenarios   243
  Pedro A. Nogueira, Rui Rodrigues, and Eugénio Oliveira

Medical Applications of AI

Probabilistic Prediction for the Detection of Vesicoureteral Reflux   253
  Harris Papadopoulos and George Anastassopoulos

Application of a Neural Network to Improve the Automatic
Measurement of Blood Pressure   263
  *Juan Luis Salazar Mendiola, José Luis Vargas Luna,
  José Luis González Guerra, and Jorge Armando Cortés Ramírez*

An Immune-Inspired Approach for Breast Cancer Classification   273
  Rima Daoudi, Khalifa Djemal, and Abdelkader Benyettou

Classification of Arrhythmia Types Using Cartesian Genetic
Programming Evolved Artificial Neural Networks   282
  *Arbab Masood Ahmad, Gul Muhammad Khan, and
  Sahibzada Ali Mahmud*

Artificial Neural Networks and Principal Components Analysis for
Detection of Idiopathic Pulmonary Fibrosis in Microscopy Images   292
  *Spiros V. Georgakopoulos, Sotiris K. Tasoulis,
  Vassilis P. Plagianakos, and Ilias Maglogiannis*

Fuzzy Inference - Filtering

Prediction of Air Quality Indices by Neural Networks and Fuzzy
Inference Systems – The Case of Pardubice Microregion................   302
    Petr Hájek and Vladimír Olej

Novel Neural Architecture for Air Data Angle Estimation   313
    Manuela Battipede, Mario Cassaro, Piero Gili, and Angelo Lerro

Audio Data Fuzzy Fusion for Source Localization   323
    Mario Malcangi

Boosting Simplified Fuzzy Neural Networks   330
    Alexey Natekin and Alois Knoll

SOM-RBF

A Parallel and Hierarchical Markovian RBF Neural Network:
Preliminary Performance Evaluation...............................   340
    Yiannis Kokkinos and Konstantinos Margaritis

Data Mining and Modelling for Wave Power Applications Using Hybrid
SOM-NG Algorithm...   350
    *Mario J. Crespo-Ramos, Iván Machón-Gonzílez,
    Hilario López-García, and Jose Luis Calvo-Rolle*

Automatic Detection of Different Harvesting Stages in Lettuce Plants
by Using Chlorophyll Fluorescence Kinetics and Supervised Self
Organizing Maps (SOMs)   360
    *Xanthoula Eirini Pantazi, Dimitrios Moshou, Dimitrios Kasampalis,
    Pavlos Tsouvaltzis, and Dimitrios Kateris*

Analysis of Heating Systems in Buildings Using Self-Organizing
Maps...   370
    *Pablo Barrientos, Carlos J. del Canto, Antonio Morán,
    Serafín Alonso, Miguel A. Prada, Juan J. Fuertes, and
    Manuel Domínguez*

Image-Video Analysis

IMMI: Interactive Segmentation Toolkit   380
    Jan Masek, Radim Burget, and Vaclav Uher

Local Binary Patterns and Neural Networks for No-Reference Image
and Video Quality Assessment....................................   388
    *Marko Panić, Dubravko Ćulibrk, Srdjan Sladojević, and
    Vladimir Crnojević*

Learning Accurate Active Contours 396
    *Adas Gelzinis, Antanas Verikas, Marija Bacauskiene, and
    Evaldas Vaiciukynas*

Pattern Recognition in Thermal Images of Plants Pine Using Artificial
Neural Networks... 406
    *Adimara Bentivoglio Colturato, André Benjamin Gomes,
    Daniel Fernando Pigatto, Danielle Bentivoglio Colturato,
    Alex Sandro Roschildt Pinto, Luiz Henrique Castelo Branco,
    Edson Luiz Furtado, and
    Kalinka Regina Lucas Jaquie Castelo Branco*

Direct Multi-label Linear Discriminant Analysis...................... 414
    Maria Oikonomou and Anastasios Tefas

Image Restoration Method by Total Variation Minimization Using
Multilayer Neural Networks Approach 424
    Mohammed Debakla, Khalifa Djemal, and Mohamed Benyettou

Learning

Algorithmic Problem Solving Using Interactive Virtual Environment:
A Case Study 433
    Plerou P. Antonia and Panayiotis M. Vlamos

No-Prop-*fast* - A High-Speed Multilayer Neural Network Learning
Algorithm: MNIST Benchmark and Eye-Tracking Data Classification ... 446
    André Frank Krause, Kai Essig, Martina Piefke, and Thomas Schack

CPL Criterion Functions and Learning Algorithms Linked to the Linear
Separability Concept 456
    Leon Bobrowski

Learning Errors of Environmental Mathematical Models 466
    Dimitri Solomatine, Vadim Kuzmin, and Durga Lal Shrestha

Social Media – Community Based Governance Applications

On Mining Opinions from Social Media 474
    Vicky Politopoulou and Manolis Maragoudakis

Automata on Directed Graphs for the Recognition of Assembly Lines ... 485
    Antonios Kalampakas, Stefanos Spartalis, and Lazaros Iliadis

On the Quantification of Missing Value Impact on Voting Advice
Applications.. 496
   Marilena Agathokleous, Nicolas Tsapatsoulis, and Ioannis Katakis

Author Index.. 507

Evaluating Sentiment in Annual Reports for Financial Distress Prediction Using Neural Networks and Support Vector Machines

Petr Hájek and Vladimír Olej

Institute of System Engineering and Informatics
Faculty of Economics and Administration, University of Pardubice, Studentská 84
532 10 Pardubice, Czech Republic
{petr.hajek,vladimir.olej}@upce.cz

Abstract. Sentiment in annual reports is recognized as being an important determinant of future financial performance. The aim of this study is to examine the effect of the sentiment on future financial distress. We evaluated the sentiment in the annual reports of U.S. companies using word categorization (rule-based) approach. We used six categories of sentiment, together with financial indicators, as the inputs of neural networks and support vector machines. The results indicate that the sentiment information significantly improves the accuracy of the used classifiers.

Keywords: Sentiment analysis, annual reports, financial distress, neural networks, support vector machines.

1   Introduction

Understanding how financial and non-financial determinants affect corporate financial distress is obviously of great importance to the stakeholders of companies. Corporate financial distress has a major effect on creditor-debtor relationships and corporate capital structure. The most serious forms of corporate financial distress are represented by corporate default on creditor claims and filing for bankruptcy. Thus, the attention of the researchers has been attracted to two related branches, i.e. corporate credit rating prediction and bankruptcy prediction, for reviews see [1], [2] and [3]. Much research has been aimed at elucidating the mechanisms of corporate financial distress prediction using financial determinants such as profitability, liquidity, and debt ratios. These approaches have evolved from the use of univariate and multivariate statistical models [4] to recent use of artificial intelligence (AI) methods such as neural networks (NNs) [5,6,7], support vector machines (SVMs) [6,7], fuzzy rule-based systems [8,9], or evolutionary algorithms [10].

One of the AI approaches commonly used in the literature for financial distress prediction is the application of NNs and related support vector machines SVMs. The fact that these methods significantly outperform traditional statistical methods in financial distress prediction problems has been demonstrated in many previous studies, see e.g. [1,2]. Even though various AI methods have been used in previous research, they have not been capable to fully explain the described complex relations

L. Iliadis, H. Papadopoulos, and C. Jayne (Eds.): EANN 2013, Part II, CCIS 384, pp. 1–10, 2013.
© Springer-Verlag Berlin Heidelberg 2013

yet. We assume that this fact is caused by omitting important qualitative determinants which can be substantially extracted from corporate annual reports.

A growing body of literature has recently employed textual analysis of annual reports, newspaper articles and other documents to forecast financial performance [11] and stock prices trend [12,13]. The results have indicated that the tone of the documents (sentiment) is significantly correlated with profitability, trading volume, unexpected earnings, etc. [14], Sentiment analysis has been also successfully used to predict fraud detection [14]. However, far too little attention has been paid to the use of sentiment analysis in financial distress prediction. The objective of our work is to explore the consequences of using both quantitative (i.e. financial ratios) and qualitative (sentiment) information in predicting corporate financial distress. We hypothesize that the use of sentiment results in significantly more accurate corporate financial distress prediction models. We use NNs and SVMs to test this hypothesis.

The rest of this paper is organized as follows. First, we briefly review previous literature on sentiment analysis of corporate annual reports. Next, we present the methods used for the sentiment analysis and for the prediction of financial distress. In section 4, dataset is described. Finally, the results of experiments are provided and analyzed.

2   Previous Literature on Sentiment Analysis of Annual Reports

Annual reports are one of the most important external documents that reflect the organizations' financial performance and strategy. They are an important vehicle for organizations to communicate with their stakeholders. They are usually composed by two types of data, quantitative data (accounting and financial data drawn from financial statements) and qualitative data (narrative texts). Our assumption on the applicability of annual reports in corporate financial distress prediction problems is supported by the following literature. Annual reports are not only the best possible description of a company, but are also a description of a company's managerial priorities. Thus, communication strategies hidden in annual reports differ in terms of the subjects emphasized when the company's performance worsens [15]. There are two general approaches to sentiment analysis of text documents in the literature, word categorization (bag of words) method (also known as a rule-based approach) and statistical methods. Word categorization requires available dictionary of terms and their categorization according to their sentiment (pre-defined rules). The main problem of this approach is that such a dictionary is context sensitive. Thus, specific dictionary has to be developed for each domain. Statistical approaches do not require a dictionary but, on the other hand, the likelihood ratios have to be estimated based on difficult to replicate and subjective classification of texts' tone [12]. The study by [16] uses word classification scheme into positive and negative categories to measure the tone change in the management discussion and analysis section of corporate annual reports. The results indicate that stock market reactions are significantly associated with the tone change of the annual reports. An alternative statistical approach (Naïve Bayes classifier) is employed also by [17,18] to demonstrated the same result, i.e. that the stock market reacts to the sentiment of annual reports and that the prediction of future stock return can be significantly improved using the sentiment.

The quarterly reports of three global telecommunication companies were analyzed by [19]. The results showed that the reports, besides giving information about past performance, showed some indications about future performance. The changes in the writing style of the reports tended to approach the financial performance for the following period. When a company performed well, and expected to continue doing so, the tone of the report was positive with extensive use of optimistic vocabulary, active verbs and clause constructions. On the contrary, if a company anticipated worsening of its financial performance in the next quarter, the tone of financial report became less optimistic and more conservative, using words and short sentence construction with particularly negative financial connotation. The company avoided directly stating the accomplished results in its quarterly reports, instead shifting the subjects of emphasis in the report.

In [14], the dictionary for financial domain is proposed as an alternative to traditional Harvard's General Inquirer. Thus, the context specific tone of annual reports can be captured. The comparative advantage of the financial dictionary has been demonstrated on the predictions of returns, trading volume, return volatility, fraud, material weakness, and unexpected earnings. In the study by [20], qualitative textual content in annual reports is examined to predict fraud. Using the χ^2 test, the results suggest that six categories of linguistic cues are associated with fraudulent financial reporting, concretely the use of: (1) complex sentential structures, (2) low readability, (3) positive tone, (4) passive voice, (5) uncertainty markers and (6) adverbs. The management discussion and analysis section of corporate annual reports are used by [11] to calculate a concept score and, thus, to develop a financial ontology. This ontology is further applied to discriminate between bankrupt / non-bankrupt and fraudulent / non-fraudulent firms. However, only relatively low prediction accuracy (less than 84%) has been achieved in the case of bankruptcy prediction problem mainly due to a small dataset. Apart from annual reports, the sentiment of other financial text documents has been studied recently such as news stories [12,21], IPO prospectuses [22] or earnings press releases [23]. Statistical approaches such as Naïve Bayes classifier, vector distance classifier, discriminant-based classifier, and adjective-adverb phrase classifier are used by [24] to analyze the sentiment of stock message boards. The sentiment analysis proves to be a significant determinant of stock index levels, trading volumes and volatility. In general, previous findings support the hypothesis that qualitative verbal communication by managers is, together with quantitative information, important determinant of future corporate financial performance and stock returns.

3   Research Methodology

Our research methodology is depicted in Fig. 1. First, data were collected both from annual reports (10-Ks) and financial statements. The linguistic pre-processing step includes tokenization and lemmatization [25]. The sequence of tagged lemmas represents potential term candidates which were compared with the financial dictionary developed by [14] in order to obtain word categorization according to the sentiment. In the next stage, the *tf.idf* term weighting scheme was applied to obtain the importance of terms in the corpus of annual reports. An average weight was then calculated for each sentiment category. The quantitative data pre-processing included the treatment of missing values and data standardization. We used median values to replace missing values. Then data were standardized using the Z-score to prevent problems with different scales. Both sentiment and financial indicators were used as inputs to the NNs and SVMs classifiers. In this section, we present the basic notions of the used methods, i.e. sentiment analysis (the construction of the term weighting scheme) and classification methods (the prediction of financial distress).

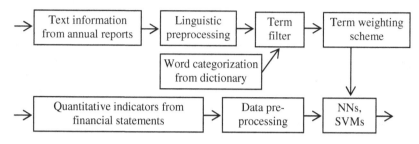

Fig. 1. Research methodology

3.1   Sentiment Analysis

Extracting the sentiment from text is a complex problem because the categorization of words is not always unambiguous. The ambiguity can be resolved using context knowledge. The main issue with the sentiment analysis of textual documents is the right choice of positive (neutral, negative, etc.) terms. This word categorization (bag of words) is difficult since words may have various meanings and, moreover, the tone of the words may be different in individual domains. Thus, specific word categorizations have been proposed for financial domain [14]. We used the most common *tf.idf* (term frequency-inverse document frequency) term weighting scheme, where weights are defined as follows

$$
w_{i,j} = \begin{cases} \dfrac{(1 + \log(tf_{i,j}))}{(1 + \log(a))} \log \dfrac{N}{df_i} & \text{if } tf_{i,j} \geq 1 \\ 0 & \text{otherwise} \end{cases}, \tag{1}
$$

where N represents the total number of documents in the sample, df_i denotes the number of documents with at least one occurrence of the i-th term, $tf_{i,j}$ is the frequency of the i-th term in the j-th document, and a denotes the average term count in the document.

Following [14], the following categories of terms were considered in our study:

a.   Negative (e.g. loss, bankruptcy, problem, suffer, unable, weak), $wf = 2349$,
b.   Positive (e.g. achieve, effective, gain, progress, strong, succeed), $wf = 354$,
c.   Uncertainty (e.g. ambiguity, assume, risk, unknown, variable), $wf = 291$,
d.   Litigious (e.g. allege, amend, bail, contract, indict, legal, sue), $wf = 871$,
e.   Modal strong (e.g. always, definitely, strongly, undoubtedly), $wf = 19$,
f.   Modal weak (e.g. nearly, seldom, sometimes, suggest), $wf = 27$,

where wf denotes the frequency of terms in the word categories listed in the financial dictionary. The frequency of net positive words was calculated as the positive term count minus the count for negation (positive terms are easily qualified or compromised) [14].

3.2   Neural Networks and Support Vector Machines

The j-th output $f_j(\mathbf{x},d,\mathbf{w})$ of perceptron type feed-forward neural networks (FFNNs) [26] can be expressed for example as follows

$$f_j(\mathbf{x},d,\mathbf{w}) = \sum_{k=1}^{K} \mathbf{v}_k (d(\sum_{j=1}^{J} \mathbf{w}_{j,k}\mathbf{x}_{j,k})), \qquad (2)$$

where $\mathbf{v}_k$ is the vector of synapses' weights among neurons in hidden layer and output neuron, $\mathbf{w}_{j,k}$ is the vector of synapses' weights among input neurons and neurons in hidden layer, k is the index of neuron in hidden layer, K is the number of neurons in hidden layer, d is the activation function, j is the index of the input neuron, J is the number of the input neurons per one neuron in hidden layer, and $\mathbf{x}_{j,k}$ is the input vector of FFNN. The j-th output $f_j(\mathbf{x},H,\mathbf{w})$ radial basis radial basis function (RBF) NN [26] can be defined this way

$$f_j(\mathbf{x},H,\mathbf{w}) = \sum_{i=1}^{q} \mathbf{w}_{j,i} h_i(\mathbf{x}), \qquad (3)$$

where $H=\{h_1(x),h_2(x), \ldots ,h_i(x), \ldots ,h_q(x)\}$ is a set of activation functions RBF of neurons in hidden layer and $\mathbf{w}_{j,i}$ are synapse weights. Each of m components of vector $\mathbf{x}=(x_1,x_2, \ldots ,x_k, \ldots ,x_m)$ is an input value for q activation functions $h_i(x)$ (usually Gaussian). The j-th output $f_j(\mathbf{x},H,\mathbf{w})$ of RBF NN represents a linear combination of outputs from q RBF neurons and corresponding synapse weights $\mathbf{w}$.

The output $f(\mathbf{x}_t)$ of SVMs is defined this way

$$f(\mathbf{x}_t) = \sum_{i=1}^{N} \alpha_i y_i k(\mathbf{x}_i,\mathbf{x}_t) + b, \qquad (4)$$

where $\mathbf{x}_t$ is the evaluated pattern, N is the number of support vectors, $\mathbf{x}_i$ are support vectors, α_i are Lagrange multipliers determined in the optimization process, k is the actual kernel function $k(\mathbf{x},\mathbf{x}_i)$. Given some training data D, a set of n points of the form $D=\{(\mathbf{x}_i,y_i)| \mathbf{x}_i \in R^p, y_i \in \{-1,1\}\}_{i=1}^{n}$ where the y_i is either 1 or -1, indicating the class to which the point $\mathbf{x}_i$ belongs. For standard SVM problem, the smallest possible optimization involves two Lagrange multipliers because the Lagrange multipliers must obey a linear equality constraint. At every step, the SVM trained by the sequential minimal optimization (SMO) [27] chooses two Lagrange multipliers to jointly optimize, finds the optimal values for these multipliers, and updates the SVM to reflect new optimal values. The SVM trained by stochastic gradient descent (SGD) method [28] are based on non-differentiable loss functions, where we optimize

$$\lambda /2\|\mathbf{w}\|^2 + \sum[1-(y\mathbf{x}\mathbf{w} + b)]_+, \qquad (5)$$

where $\mathbf{w}$ is the weight vector, b the bias, λ the regularization parameter and the class labels y are from $\{-1,1\}$.

4   Dataset

In our study, the prediction of financial distress was realized as a two-class problem. The classes were represented by investment grade (IG) and non-investment grade (NG) assigned by a highly regarded Standard & Poor's rating agency. IG bonds expose investors to a low default risk because their issuers are considered able to meet their obligations. On the contrary, NG bonds are considered high risk for bondholders.

Input variables (determinants of financial distress) used in our study for describing companies can be divided into two main groups: financial indicators and sentiment indicators, see Table 1.

Table 1. Input and output variables describing the dataset

	Variable		Variable
x_1	Enterprise value	x_{11}	Dividend yield
x_2	Cash	x_{12}	Payout ratio
x_3	Revenues	x_{13}	Standard deviation of stock price
x_4	Earnings per share	x_{14}	Frequency of negative terms
x_5	Return on equity	x_{15}	Frequency of positive terms
x_6	Price to book value	x_{16}	Frequency of uncertainty terms
x_7	Enterprise value / earnings	x_{17}	Frequency of litigious terms
x_8	Price to earnings per share	x_{18}	Frequency of strong modal terms
x_9	Market debt / total capital	x_{19}	Frequency of weak modal terms
x_{10}	High to low stock price	class	{IG, NG}

We used several subgroups of financial indicators such as size, profitability ratios, liquidity ratios, leverage ratios, and market value ratios. Sentiment indicators are strongly related to financial indicators as demonstrated in previous studies. In addition, sentiment analysis also refers to the business position of a company. This position involves business diversification, business risk, character (reputation), organizational problems, management evaluation, accounting quality, etc. These parameters are difficult to measure quentitatively since their evaluation requires expert knowledge. Therefore, we employed sentiment analysis in order to cover the business position and, thus, to improve the performance of classifiers.

Quantitative financial indicators were drawn from the Value Line database, while sentiment indicators were drawn from annual reports available at U.S. Securities and Exchange Commission EDGAR System. Both types of input variables were collected for U.S. companies in the year 2010, while the output classes assigned by Standard & Poor's rating agency were obtained for the year 2011. We excluded the companies from mining and financial industries to prevent problems with industry-specific input variables [9]. As a result, we were able to collect data for 520 U.S. companies, 195 of them classified into IG and 325 classified into NG category.

The financially distressed companies (NG class) showed the following sentiment characteristic: a more negative, less positive, more uncertain, less litigious and more modal tone (Table 2).

Further, we tested the merits of input attributes using the Relief algorithm [29] with 10-fold cross-validation (Table 3). The leverage ratio x_{14} and market value ratios seem to be the most important predictors of financial distress. However, the sentiment in annual reports (especially positive, litigious and weak modal) also appears to be a relevant determinant of financial distress.

Table 2. Descriptive statistics on sentiment indicators

	IG				NG			
	Mean	Std.Dev.	Min	Max	Mean	Std.Dev.	Min	Max
x_{14}	1.975	0.052	1.779	2.145	1.982	0.066	1.804	2.196
x_{15}	1.803	0.048	1.639	1.980	1.787	0.041	1.616	1.890
x_{16}	1.910	0.046	1.783	2.034	1.914	0.037	1.676	2.002
x_{17}	1.929	0.116	1.663	2.200	1.908	0.119	1.623	2.210
x_{18}	1.715	0.069	1.493	1.862	1.727	0.058	1.384	1.854
x_{19}	1.574	0.098	1.302	1.913	1.576	0.080	1.323	1.833

Table 3. The ranking of input variables using the Relief algorithm

	Merit	Rank		Merit	Rank
x_9	0.049±0.003	1.2±0.4	x_6	0.004±0.001	11.9±1.3
x_{10}	0.044±0.002	1.8±0.4	x_2	0.004±0.000	12.2±1.4
x_{13}	0.037±0.002	3.0±0.0	x_{18}	0.003±0.001	13.6±1.7
x_{11}	0.018±0.001	4.0±0.0	x_{16}	0.003±0.001	13.6±2.0
x_1	0.014±0.001	4.0±0.0	x_{12}	0.003±0.001	13.7±1.7
x_{15}	0.013±0.002	5.8±0.6	x_5	0.002±0.001	16.3±0.8
x_{17}	0.008±0.001	7.8±1.3	x_{14}	0.001±0.001	16.4±0.9
x_{19}	0.008±0.001	8.4±1.0	x_7	0.001±0.000	17.6±0.5
x_3	0.008±0.002	8.6±0.8	x_4	-0.002±0.000	19.0±0.0
x_8	0.006±0.002	9.8±2.1			

5   Financial Distress Prediction Using NNs and SVMs

The dataset was divided into training and testing data in relation 1:1. This division was realized five times. Since we used a two-class dataset we recorded classification accuracy (Acc), F-measure (a combination of precision and recall) and the Matthews correlation coefficient (MCC) (a correlation coefficient between the observed and predicted binary classifications, also known as the phi coefficient) on testing data as the measures of classification performance. In order to avoid over-fitting, we carried out experiments with different values of NNs and SVMs' parameters using 10-fold cross-validation on the training data.

We used the FFNN, RBF NN, SVM trained by the SMO, and SVM trained by the SGD method for financial distress prediction. In addition, we employed logistic regression (LR) as a representative of traditional statistical methods. The FFNN was trained using the backpropagation algorithm with momentum. The following parameters of the FFNN were set and examined: the number of neurons in the hidden layer = {5,10, ... ,30}, learning rate = 0.1, momentum = 0.2, and the number of epochs = 2000. The

RBF NN was trained with the BFGS (Broyden-Fletcher-Goldfarb-Shanno) method. The initial centres for the Gaussian RBFs were found using k-means algorithm. The initial sigma values were set to the maximum distance between any center and its nearest neighbour in the set of centers. We tested various numbers of the RBF functions from the set $\{2^1, 2^2, \ldots, 2^5\}$. The SVM trained by the SMO was tested for linear, polynomial (exponent = 2) and RBF kernel functions (gamma = 0.01) with the complexity $C = \{2^0, 2^1, \ldots, 2^{10}\}$. Finally, the SVM trained using the SGD was examined for the number of epochs = 100 and learning rate = $\{0.01, 0.05\}$.

First, we examined the performance of classifiers using only financial indicators (Table 4). In this case, the FFNN outperformed other methods with the accuracy of 91.84%. This also holds true for the F-measure and MCC measures indicating that the FFNN worked well for both classes, IG and NG. The second set of experiments made use of sentiment indicators, too. As shown in Table 5, the significant improvement of the classification performance was achieved for RBF NNs and SVMs using the sentiment information. On the other hand, the LR and FFNN classifiers' performance improved but not significantly.

Table 4. Classification performance using only financial indicators

	Acc [%]	F-measure	MCC
LR	89.60±0.00	0.895±0.000	0.777±0.000
FFNN	91.84±0.61	0.918±0.006	0.826±0.014
RBF	88.68±0.02	0.886±0.019	0.757±0.039
SVM_{SMO}	89.80±1.79	0.896±0.019	0.781±0.037
SVM_{SGD}	90.78±1.91	0.907±0.021	0.803±0.041

Table 5. Classification performance using financial and sentiment indicators

	Acc [%]	F-measure	MCC
LR	90.52±1.15	0.905±0.012	0.797±0.025
FFNN	92.22±0.73	0.923±0.007	0.835±0.007
RBF	91.86±0.54**	0.918±0.005**	0.826±0.012**
SVM_{SMO}	91.90±0.25*	0.919±0.003*	0.827±0.005*
SVM_{SGD}	92.40±1.05*	0.924±0.011*	0.838±0.023*

Legend: we tested if the improvement over the non-sentiment models was statistically significant using the paired t-test at $p = 0.01$ ***, $p = 0.05$ ** and $p = 0.1$*.

6   Conclusion

A strong relationship between qualitative (textual) information and corporate financial performance and stock returns has been reported in the literature. Prior studies that have noted the importance of sentiment in annual reports examined the impact of the sentiment of financial market ratios such as stock prices, trading volumes or volatility.

The present study was designed to determine the effect of sentiment in annual reports on financial distress represented by IG and NG classes. Returning to the hypothesis posed at the beginning of this study, it is now possible to state that the use of sentiment indicators results in significantly more accurate corporate financial distress prediction models. Another important finding is that it seems to be the frequency of positive terms that has the most important impact on financial distress prediction. The study has gone some way towards enhancing our understanding of financial distress prediction. Our findings have broader important implications for related issues such as the prediction of the start of financial crises since corporate financial distress prediction models are used as early warning indicators.

A high complexity of the investigated problem was documented by the necessity to use high values of the complexity parameter in SVMs ($C = 2^9$ or $C = 2^{10}$). SVMs performed better than NNs when employing the sentiment information. This finding suggests that the SVMs better coped with the growing dimensionality of the data induced by using additional inputs. The growing dimensionality increased computational complexity in the case of the NNs but not in the case of SVMs. Thus, NNs became more prone to both over-fitting and local minima. A number of caveats need to be noted regarding the present study. The most important limitation lies in the fact that the words in the word categories may have various importance for financial distress prediction and, thus, the accuracy improvement of the models is limited. A further study with more focus on statistical approach is therefore suggested. In addition, we suggest to make use of the unlabelled data in future research [2,30].

The experiments in this study were carried out in Statistica 10 (linguistic pre-processing) and Weka 3.7.5 (NNs and SVMs) in MS Windows 7 operation system.

Acknowledgments. This work was supported by the scientific research project of the Czech Sciences Foundation Grant No: 13-10331S.

References

1. Ravi Kumar, P., Ravi, V.: Bankruptcy Prediction in Banks and Firms via Statistical and Intelligent Techniques - A Review. Europ. J. of Operational Research 180(1), 1–28 (2007)
2. Hajek, P., Olej, V.: Credit Rating Modelling by Kernel-Based Approaches with Supervised and Semi-Supervised Learning. Neural Computing and Applications 20(6), 761–773 (2011)
3. Kirkos, E.: Assessing Methodologies for Intelligent Bankruptcy Prediction. Artificial Intelligence Review 11 (2012)
4. Altman, E.I.: Financial Ratios, Discriminant Analysis and the Prediction of Corporate Bankruptcy. Journal of Finance 23(4), 589–609 (1968)
5. Wilson, R.L., Sharda, R.: Bankruptcy Prediction using Neural Networks. Decision Support Systems 11(5), 545–557 (1994)
6. Huang, Z., Chen, H., et al.: Credit Rating Analysis with Support Vector Machines and Neural Networks: A Market Comparative Study. Decision Support Systems 37(4), 543–558 (2004)
7. Hajek, P.: Municipal Credit Rating Modelling by Neural Networks. Decision Support Systems 51(1), 108–118 (2011)
8. Chen, H.L., Yang, B., et al.: A Novel Bankruptcy Prediction Model based on an Adaptive Fuzzy k-nearest Neighbour Method. Knowledge-Based Systems 24(8), 1348–1359 (2011)
9. Hajek, P.: Credit Rating Analysis using Adaptive Fuzzy Rule-Based Systems: An Industry-Specific Approach. Central European Journal of Operations Research 20(3), 421–434 (2012)
10. Varetto, F.: Genetic Algorithms Applications in the Analysis of Insolvency Risk. Journal of Banking and Finance 22(10-11), 1421–1439 (1998)

11. Cecchini, M., Aytug, H., et al.: Making Words Work: Using Financial Text as a Predictor of Financial Events. Decision Support Systems 50(1), 164–175 (2010)
12. Tetlock, P.C.: Giving Content to Investor Sentiment: The Role of Media in the Stock Market. Journal of Finance 62, 1139–1168 (2007)
13. Loughran, T., McDonald, B.: When Is a Liability Not a Liability? Textual Analysis, Dictionaries, and 10-Ks. The Journal of Finance 66(1), 35–65 (2011)
14. Hajek, P., Olej, V., Myskova, R.: Forecasting Stock Prices using Sentiment Information in Annual Reports - A Neural Network and Support Vector Regression Approach. WSEAS Transactions on Systems (in press, 2013)
15. Kohut, G.F., Segars, A.H.: The President's Letter to Stockholders: An Examination of Corporate Communication Strategy. Journal of Business Communication 29(1), 7–21 (1992)
16. Feldman, R., Govindaraj, S., et al.: Management's Tone Change, Post Earnings Announcement Drift and Accruals. Review of Accounting Studies 15, 915–953 (2010)
17. Li, F.: The Information Content of Forward-Looking Statements in Corporate Filings - A Naïve Bayesian Machine Learning Approach. Journal of Accounting Research 48(5), 1049–1102 (2010)
18. Huang, A., Zang, A., et al.: Informativeness of Text in Analyst Reports: A Naïve Bayesian Machine Learning Approach. Working Paper. The Hong Kong University of Science and Technology (2010)
19. Magnusson, C., Arppe, A., et al.: The Language of Quarterly Reports as an Indicator of Change in the Company's Financial Status. Information & Management 42(4), 561–574 (2005)
20. Goel, S., Gangolly, J.: Beyond the Numbers: Mining the Annual Reports for Hidden Cues Indicative of Financial Statement Fraud. Intelligent Systems in Accounting, Finance and Management 19(2), 75–89 (2012)
21. Lu, Y.C., Shen, C.H., et al.: Revisiting Early Warning Signals of Corporate Credit Default using Linguistic Analysis. Pacifin-Basin Finance Journal 24, 1–21 (2013)
22. Hanley, K.W., Hoberg, G.: The Information Content of IPO Prospectuses. Review of Financial Studies 23(7), 2821–2864 (2010)
23. Demers, E.A., Vega, C.: Soft Information in Earnings Announcements: News or Noise? Working paper. INSEAD (2010)
24. Das, S., Chen, M.: Yahoo! for Amazon: Opinion Extraction from Small Talk on the Web. Working paper. Santa Clara University (2001)
25. Feldman, R., Fresko, M., et al.: Text Mining at the Term Level. In: Żytkow, J.M. (ed.) PKDD 1998. LNCS, vol. 1510, pp. 65–73. Springer, Heidelberg (1998)
26. Haykin, S.: Neural Networks: A Comprehensive Foundation, 2nd edn. Prentice-Hall Inc., New Jersey (1999)
27. Platt, J.C.: Fast Training of Support Vector Machines using Sequential Minimal Optimization. In: Schoelkopf, B., Burges, C., Smola, A. (eds.) Advances in Kernel Methods - Support Vector Learning. MIT Press (1998)
28. Bifet, A., Frank, E.: Sentiment Knowledge Discovery in Twitter Streaming Data. In: Pfahringer, B., Holmes, G., Hoffmann, A. (eds.) DS 2010. LNCS, vol. 6332, pp. 1–15. Springer, Heidelberg (2010)
29. Kira, K., Rendell, L.A.: A Practical Approach to Feature Selection. In: Proc. of the 9th International Workshop on Machine Learning, pp. 249–256 (1992)
30. Hajek, P., Olej, V.: Municipal Creditworthiness Modelling by Kernel-Based Approaches with Supervised and Semi-supervised Learning. In: Palmer-Brown, D., Draganova, C., Pimenidis, E., Mouratidis, H. (eds.) EANN 2009. CCIS, vol. 43, pp. 35–44. Springer, Heidelberg (2009)

Identification of All Exact and Approximate Inverted Repeats in Regular and Weighted Sequences

Carl Barton[1], Costas S. Iliopoulos[1,3], Nicola Mulder[2], and Bruce Watson[3]

[1] King's College London, Dept. of Informatics, London WC2R 2LS, UK
carl.barton@kcl.ac.uk
[2] Computational Biology Group, University of Cape Town, Cape Town
nicola.mulder@uct.ac.za
[3] Fastar Group, University of Pretoria, Pretoria
bwatson@fastar.org

Abstract. The detection of various types of repeats is a fundamental and well studied problem in stringology. In this paper we present extensions to this problem with applications to bioinformatics. In this paper we consider the detection of all exact and approximate inverted repeats, as well as all exact and approximate weighted inverted repeats and give efficient algorithms for their computation.

1 Introduction

Next generation sequencing technologies produce DNA/RNA sequences as a huge number of short fragments. These fragments are then reconstructed into a single string through various assembly methods. The reconstructed DNA/RNA sequence is one dimensional and does not take into account any type of secondary structures which may be present in the original sequence. Deriving the secondary structure is an important step in the analysis of sequences, in the case of RNA, the secondary structure can have a great affect on the resulting proteins function. Inverted repeats are one such secondary structure and represent areas where the DNA/RNA can fold back on itself to create the well known double helix structure. From a stringological perspective, inverted repeats or 'hairpins' in DNA or RNA sequences are factors of the sequence that occur with a reverse and complemented version of it in close proximity, where the complement is defined by the Watson-Crick pairing. It is known that determining the optimal secondary structure of RNA is NP-Hard under some models of computation and the only polynomial algorithms, for other models of computation, are far from practical with a complexity of $\mathcal{O}(n^{80})$ [15]. However, the same is not true for detecting inverted repeats. Existing work on detecting inverted repeats, from a biological context, tends to focus on probabilistic methods for their detection, such as the the method proposed in [12] which makes use of local alignments as well probabilistic techniques. Additionally there exists a well know statistical tool called the 'Inverted Repeats Finder'[1] which uses a stochastic model of

---

[1] Tandem repeat finder can be found here http://tandem.bu.edu/trf/trf.html

L. Iliadis, H. Papadopoulos, and C. Jayne (Eds.): EANN 2013, Part II, CCIS 384, pp. 11–19, 2013.
© Springer-Verlag Berlin Heidelberg 2013

repeats to analyse DNA. Inverted Repeats Finder is a popular tool which gives good results, however, none of the approaches mentioned above provide an exact solution to the problem of finding inverted repeats.

Pseudoknots are another secondary that can occur as part of the secondary structure of RNA and are known to be an important part for certain cell activity [4], as well as playing an important role in viral infiltration [19]. A pseudoknot occurs when two or more inverted repeats 'overlap', in the sense that one inverted repeat forms on part of an existing inverted repeat. Although there exists polynomial time algorithms for computing *some* pseudoknots, the problem is in general know to be NP-Hard and as such, a number of heuristics were proposed. Most heuristics for this task are based on dynamic programming [1] and struggle to detect the complex interactions of the non-nested base pairs that form pseudoknots, even the newer methods of stochastic context free grammars [2,13] struggle to detect this. Clearly inverted repeats are the building blocks for pseudoknots, so efficient algorithms for the detection of inverted repeats may lead to more accurate and biologically meaningful heuristics for detecting pseudoknots.

2 Background

The detection of inverted repeats can be considered as the detection of palindromes in strings. Within the stringology community there has been a great deal of work on the detection of palindromes. There exists linear time algorithms for the detection of the longest palindrome [17] in a string and for the detection of the maximal palindromes in a string [7]. Additionally there is work on approximate detection of palindromes, [20] detects maximal palindromes under edit distance in time $\mathcal{O}(k^2n)$, later [9] did the same but in time $\mathcal{O}(kn)$. These techniques make use of global alignments and incremental string comparison techniques to compute the maximal palindromes. These solutions essentially compute maximal inverted repeats, however, it is not clear that the maximal inverted repeat is more biologically meaningful than any other. Therefore it is an important problem to be able to identify all of the inverted repeats and not just maximal inverted repeats.

A related problem is the detection of all or just the distinct squares in a string. Detecting squares is a well studied problem in stringology and is fundamental in the detection of repetitions. Early work on detecting squares lead to linear time algorithms to test squarefreeness of a string and $\mathcal{O}(n \log n)$ algorithms for computing all the squares in a string [16]. Later it was shown that simple algorithms based on suffix tree traversals can also lead to an $\mathcal{O}(n \log n)$ algorithm [21] or a suffix tree can be 'marked', in linear time, in such a way that all of the squares are implicitly represented in the tree. This trick allows Gusfield et al [8] to avoid reporting all the squares and get around the $\mathcal{O}(n \log n)$ barrier implied by works such as [5] and instead they report the distinct squares in linear time, as the number of distinct squares is $\mathcal{O}(n)^2$. It was later shown by Kolpakov and Kucherov [14] that the number of maximal periodicities, also

---

[2] See [8] for details on this bound.

known as runs, was actually linear in the length of the string and they gave an algorithm for their computation.

In this paper we first show that, for the exact case, it is possible to report all inverted repeats in linear time, we then propose extensions to approximate inverted repeats and the case of weighted strings.

The paper is structured as follows, Section 1 gives an introduction, Section 2 provides background information on previous works, Section 3 gives preliminaries and problem definitions, Section 4 presents our algorithms, and finally Section 5 some concluding remarks.

3 Preliminaries

An *alphabet* Σ is a finite non-empty set, of size σ, whose elements are called *letters*. A *string* on an alphabet Σ is a finite, possibly empty, sequence of elements of Σ. The zero-letter sequence is called the *empty string*, and is denoted by ε. The *length* of a string x is defined as the length of the sequence associated with the string x, and is denoted by $|x|$. We denote by $x[i]$, for all $0 \le i < |x|$, the letter at index i of x. Each index i, for all $0 \le i < |x|$, is a position in x when $x \ne \varepsilon$. It follows that the ith letter of x is the letter at position i in x, and that

$$x = x[0 .. |x| - 1].$$

A string x is a *factor* of a string y if there exist two strings u and v, such that $y = uxv$. Let the strings $x, y, u,$ and v, such that $y = uxv$. If $u = \varepsilon$, then x is a *prefix* of y. If $v = \varepsilon$, then x is a *suffix* of y.

Let x be a non-empty string and y be a string. We say that there exists an *occurrence* of x in y, or, more simply, that x *occurs in* y, when x is a factor of y. Every occurrence of x can be characterised by a position in y. Thus we say that x occurs at the *starting position* i in y when $y[i .. i + |x| - 1] = x$. It is sometimes more suitable to consider the *ending position* $i + |x| - 1$. For succinctness we sometimes refer to $x[i]$ as i_x. We denote by x^R the reverse of x.

A *weighted sequence* x is a sequence of position, where each position $x[i]$ consists of a set of ordered pairs. Each pair has the form $(\sigma, \pi_i(\sigma))$, where $\pi_i(\sigma)$ is the probability of having the character, σ, at position i. For every position $x[i]$, $0 \le i < n$, $\Sigma_{A_\sigma} \pi_i(\sigma) = 1$.

A factor of a weighted sequence starting at position i and of length j is defined is a sequence $u[0]u[1] .. u[j-1]$ s.t,$u[k] \in x[k+i]$ for $0 \le k < i-j$. The probability of a factor occurring is $\prod_{i=0}^{j-1} \pi_i(u[i])$

The *Hamming distance* between strings u and v, both of length n, is the number of positions i, $0 \le i < n$, such that $u[i] \ne v[i]$. Given a nonnegative integer k, we write $u \equiv_k v$ if the Hamming distance between $\mathbf{u}$ and v is at most k.

For symbols the following are equivalent under complement, denoted $\dot{=}$.

- $A \dot{=} T$
- $C \dot{=} G$

- $G\dot{=}C$
- $T\dot{=}A$

Given 2 non empty strings of equal length u, v we say $u\dot{=}v$ if and only if $u[i]\dot{=}v[i]$ for $0 \leq i < n$ and we denoted the complement of a factor u as $\bar{u}$.

In the case of a weighted sequence the complement takes the probabilities of its complementary symbol.

Problem 1. Given a text x find all inverted repeats. Where a inverted repeats is a factor of the text of the form $u\bar{u}^R$ or $ua\bar{u}^R$ and $a \in \Sigma$.

Problem 2. Given a text x find all approximate inverted repeats with at most k-mismatches. Where an approximate inverted repeat is a factor of the text of the form $u\bar{u}^R$ or $ua\bar{u}^R$ s.t $u \equiv_k \bar{u}^R$ and $a \in \Sigma$.

Problem 3. Given a weighted text x find all weighted inverted repeats with probability at least μ. Where a weighted inverted repeat is a factor of the text of the form $u\bar{u}^R$ or $ua\bar{u}^R$ s.t $\pi(u\bar{u}^R) \geq \mu$ and $a \in \Sigma$.

Problem 4. Given a weighted text x find all approximate weighted inverted repeats with probability at least μ and at most k-mismatches. Where a weighted inverted repeat is a factor of the text of the form $u\bar{u}^R$ or $ua\bar{u}^R$ s.t $\pi(u\bar{u}^R) \geq \mu$ and $u \equiv_k \bar{u}^R$ and $a \in \Sigma$.

4   Algorithm

In this section we begin with a few definitions, then we give our approach for computing exact inverted repeats, the modifications to k-mismatches and finally we present the extensions of these 2 algorithms for the case of weighted strings.

We begin by defining the *centre* of an inverted repeat, an example of this can be seen in Example 1.

Definition 1. *An inverted repeat of length ℓ is centered around i iff $x[i - \ell ..i]\dot{=}x[i + 1 ..i + 1 + \ell]$ or $x[i - \ell ..i]\dot{=}x[i + 2 ..i + \ell + 2]$*

To compute all the inverted repeats of a string we initially compute all maximal inverted repeats from their centre. If we take a string x and its reverse complement $\bar{x}^R$ then the maximal inverted repeat centred at some position i is either the longest common prefix (LCP) of $x[i]$ and $\bar{x}^R[n - i]$ or the longest common prefix of $x[i]$ and $\bar{x}^R[n - i + 1]$. To report all of these in linear time we must be able to compute $lce(x[i], \bar{x}^R[n - i])$ and $lce(x[i], \bar{x}^R[n - i + 1])$ in constant time. For the rest of the article we focus on how to compute inverted repeats of the form $u\bar{u}^R$, clearly the other case can be handled similarly. To do that we make use of range minimum queries (RMQ). Give an array $A[0 .. n - 1]$ of integers and two indices, i, j, a RMQ asks for the index of the minimum value within the range $A[i .. j]$. To compute the exact inverted repeats we proceed as follows.

- Build the generalised suffix array [18], inverse suffix array and LCP array [6] of x and $\bar{x}^R$
- Preprocess the LCP array for RMQ queries
- Find the LCE for the suffixes starting at $(i_x, n - i_{\bar{x}^R})$, for $0 \leq i < n$.
- The result, for $0 \leq i < n$, will be the maximal inverted repeat centred at $x[i]$.

To find the longest common extension we use one of the simple algorithms presented in [10]. Following the preprocessing mentioned above we perform the following query in constant time to compute the LCE of any two suffixes.

$$\mathsf{LCE}(p, q) = \mathsf{LCP}[\mathsf{RMQ}_{\mathsf{LCP}}(\mathsf{iSA}[p] + 1, \mathsf{iSA}[q])]$$

Require: InvertedRepeat(x, n)
  Compute SA, iSA, LCP, and RMQ$_{\mathsf{LCP}}$ of $x \pounds \bar{x}^R \$$.
  for $i \leftarrow 0$ to $n - 1$ **do**
    $\delta_1 \leftarrow$ lce$(i, 2n - i)$;
    $\delta_2 \leftarrow$ lce$(i, 2n - i + 1)$;
    for $j \leftarrow 0$ to $\delta_1 - 1$ **do**
      Report$(i - j, 2 * j)$;
    end for
    for $j \leftarrow 0$ to $\delta_2 - 1$ **do**
      Report$(i - j, 2 * j + 1)$;
    end for
  end for

Following this scheme we can compute all maximal inverted repeats in time $\mathcal{O}(n)$. We then make use of the following fact, allowing us to report all inverted repeats in $\mathcal{O}(n + \alpha)$ where α is the number of inverted repeats.

Fact 2. *If the LCE of the suffixes* $i_x, n - i_{\bar{x}^R}$ *is of length* ℓ, *then every factor* $x[i - j .. i + j]$, *for* $1 \leq j \leq \ell$, *is an inverted repeat.*

Due to this, it is clear that we can report all occurrences directly in time proportional to their size as shown in Algorithm 4. So we get the following.

Theorem 3. *Algorithm aIR solves problem 1 in time* $\mathcal{O}(n + \alpha)$

Example 1. Given the string $GAGAGAACCGTACGGT$, then there exists a maximal inverted repeat, $ACCGTACGGT$ of length 10, centred at position 10. There also exists non-maximal inverted repeats of length $2, 4, 6$ and 8 also centred at position 10.

4.1   Inverted Repeats with k-Mismatches

To compute the maximal inverted repeats with up to k-mismatches we perform the same processing we did for exact inverted repeats but with a slight modification to the latter half of the algorithm. The approach we adopt is similar to that used in [3] and consists of performing $k + 1$ LCE queries instead of a single LCE query for each index. Assume we have performed one LCE query and a mismatch occurred at index p, q, we now simply take another LCE query from the suffixes starting at $p + 1, q + 1$. After performing this $k + 1$ times we find the maximal inverted repeats with at most k-mismatches.

With a slight modification to the technique we used to report all inverted repeats, see Fact 1, it can be applied to the approximate case as well. Applying Fact 1 allow us to report all inverted repeats under hamming distance in time $\mathcal{O}(kn + \alpha)$. A brief outline is as follows, after the i-th LCE query, with $1 \leq i \leq k$ we have the maximal inverted repeat with up to $i - 1$ mismatches. Assume we have already reported inverted repeats with up to $i - 2$ mismatches, we simply apply Fact 1 and report the new inverted repeats between the the last mismatch and the extension from the new LCE query. Doing this $k + 1$ times, we report all inverted repeats with at most k-mismatches. Note that this increases the processing done at each extension from constant to proportional to the number of new occurrences, but the total time remains $\mathcal{O}(kn + \alpha)$.

Theorem 4. *Algorithm akIR solves problem 2 in time* $\mathcal{O}(kn + \alpha)$

4.2   Extensions to Weighted Strings

For weighted strings our general approach is the same but some of the details are a little different. For instance we use space efficient suffix arrays for our previous algorithms, however, there is no such data structure for weighted sequences. So, we make use of the weighted suffix tree of [11]. First we give a brief introduction to the weighted suffix tree and then outline our algorithm.

The Weighted Suffix Tree. In [11], Iliopoulos et al. describe a method for constructing a weighted suffix tree (abbreviated WST) for storing a set of suffixes of a weighted sequence with probability of appearance greater than $\frac{1}{z}$ in linear time, where z is a given constant. The WST is distinct from the probabilistic suffix tree in that is does not model any stochastic process and is designed to work in the same way as a suffix tree, meaning that it maintains optimal search times; something not possible with the probabilistic suffix tree. As previously mentioned the WST is designed to behave exactly as a suffix tree and as such does not actually contain any information about probability distributions within the suffix tree itself, instead it only represents those suffixes which have probability of occurring $\frac{1}{z}$. As such the WST inherits all the interesting string manipulation properties of the standard suffix tree. We give an informal definition of the structure as follows: Let x be a weighted sequence, for every suffix starting at position i we define a list of possible weighted substrings so that the probability of appearance for each of them is greater than $\frac{1}{z}$. We denote each of them as $x_{i,j}$, where j is the sub-word rank in arbitrary numbering. We define $\mathsf{WST}(x)$ the weighted suffix tree of a weighted sequence x, as the compressed trie of a portion of all the weighted substrings starting with each suffix $x[i]$ of $x\$$, $\$ \notin \Sigma$,having a probability of appearance greater than $\frac{1}{z}$. Let $\mathsf{L}(v)$ denote the path-label of node v in $\mathsf{WST}(x)$, which results by concatenating the edge labels along the path from the root to v. Leaf v of $\mathsf{WST}(x)$ is labelled with index i if $\exists j > 0$ such that $\mathsf{L}(v) = x_{i,j}[i\,..\,n]$ and $\pi(x_{i,j}[i\,..\,n]) \geq \frac{1}{z}$, where $j > 0$ denote the j-th weighted sub-word starting at position i. We define the leaf-list $\mathsf{LL}(v)$ of v as a list of the leaf-labels in the subtree below v.

The resulting suffix tree has all the functionality a normal suffix tree, but it only consists of those factors which have probability greater than $\frac{1}{z}$. Due to this we can use existing algorithms to pre-process it for lowest common ancestor (LCA) queries and other classical algorithms applied to a suffix tree.

The outline of the algorithm is as follows.

- Build the generalised suffix tree of $x, \bar{x}^R$.
- Preprocess the suffix tree for LCA queries.
- Find the LCP for the suffixes starting at $(i_x, n - i_{\bar{x}^R})$, for $0 \leq i < n$.
- The result, for $0 \leq i < n$, is maximal inverted repeats centred at $x[i]$.

Although the general approach remains the same, we must take into consideration non-solid letters. If one of the start points is not a solid letter, then we can take the LCP of all the letters which could occur at both positions and take the longest one to be the maximal inverted repeat, assuming a constant sized alphabet this takes $\mathcal{O}(1)$ time. We now have inverted repeats where $\pi(u), \pi(\bar{u}^R) \geq \frac{1}{z}$. To ensure we only report inverted repeats of the form $\pi(u\bar{u}^R)$ we start by reporting inverted repeats from the center and keeping track of the current total probability of the inverted repeats as we go. As soon as the total probability of the inverted repeats drops below the threshold, we stop reporting more inverted repeats, this takes $\mathcal{O}(1)$ per inverted repeat.

Theorem 5. *Algorithm wIR solves problem 3 in time* $\mathcal{O}(n + \alpha)$

This is sufficient for the exact case, however, if we allow k-mismatches we now have some extra considerations to take into account. For example, applying the technique we used for regular strings we would actually find maximal inverted repeats of the form $u = u_1 m_1 \ldots u_k m_k$ where u_i are factors of probability greater than $\frac{1}{z}$ and m_i are mismatches. The problem being that the probability of u occurring might actually be less than $\frac{1}{z}$. To get over this problem we need to know the length of a factor associated with a leaf. If we know that the leaf we start with was only inserted up to some length ℓ, when performing LCA queries we can keep track of the length of the factor and if the length is greater than ℓ we simply cut it short at length ℓ, even if it has not reached k-mismatches. This can be done with a simple linear preprocessing scheme, a top down traversal of the suffix tree labelling each node with the corresponding length until we reach all leaves. Another problem, is that we may have non solid starting positions for each LCA query, and it may seem that this could up lead to σ^{k+1} queries. However, by [11] we know that the number of possible factors starting at a non-solid location is $\mathcal{O}(\sigma^q)$ where q is a constant derived from the value of z and thus $\mathcal{O}(1)$ possible factors starting at any position i. This fact combined with the approach of terminating any match with length $> \ell$ means we can never check more than the number of factors with probability $> \frac{1}{z}$ starting at any position. This leads us to the following.

Lemma 6. *For a position i, there will be at most $\mathcal{O}(k)$ LCA queries.*

Therefore.

Theorem 7. *Algorithm wkIR solves problem 4 in time $\mathcal{O}(kn + \alpha)$*

In both of the above cases we may apply Fact 1 to retrieve non maximal inverted repeats.

5   Conclusion and Future Work

In this paper we have given a number of linear time algorithms for finding inverted repeats in a number of different settings to aid in the identification of inverted repeats. The advantage of our approach over existing methods is twofold: we provide extensions which allow for the reporting of all inverted repeats, rather than simply the maximal ones and we present, to the best of our knowledge, the first algorithms for the weighted string problem. In this paper we show how to detect all exact and approximate hairpins, however, hairpin structures which are the longest are not necessarily the most biologically meaningful. It has been noted in various works that hairpin structures of length less than 3 are impossible and hairpins can contain 'bulges' caused by mismatches, which effect the stability of the loop. The stability is in fact affected by the length, the number of mismatches and the bases involved in the 'paired regions' ($G - C$ pairings are more stable than $A - U$ pairings). An interesting extension to the problem considered here would be the detection of the most stable hairpins at each position and the most stable hairpin with a bounded number of bulges. In addition to this these algorithms need to be implemented and tested to determine their practical efficiency.

References

1. Akutsu, T.: Dynamic programming algorithms for rna secondary structure prediction with pseudoknots. Discrete Applied Mathematics 104, 45–62 (2000)
2. Brown, M., Wilson, C.: Rna pseudoknot modeling using intersections of stochastic context free grammars with applications to database search. In: Pacific Symposium on Biocomputing, pp. 109–125 (1995)
3. Barton, S.P.P.C., Iliopoulos, C.S., Smyth, W.F.: Prefix tables & border arrays with k-mismatches & applications (2013) (submitted for publication)

4. Chen, J.-L., Greider, C.W.: Functional analysis of the pseudoknot structure in human telomerase rna. Proceedings of the National Academy of Sciences of the United States of America 102(23), 8080–8085 (2005)
5. Crochemore, M.: An optimal algorithm for computing the repetitions in a word. Information Processing Letters 12(5), 244–250 (1981)
6. Fischer, J.: Inducing the lcp-array. In: Dehne, F., Iacono, J., Sack, J.-R. (eds.) WADS 2011. LNCS, vol. 6844, pp. 374–385. Springer, Heidelberg (2011)
7. Gusfield, D.: Algorithms on strings, trees, and sequences: computer science and computational biology. Cambridge University Press, New York (1997)
8. Gusfield, D., Stoye, J.: Linear time algorithms for finding and representing all the tandem repeats in a string. Journal of Computer and System Sciences 69(4), 525–546 (2004)
9. Hsu, P.-H., Chen, K.-Y., Chao, K.-M.: Finding all approximate gapped palindromes. In: Dong, Y., Du, D.-Z., Ibarra, O. (eds.) ISAAC 2009. LNCS, vol. 5878, pp. 1084–1093. Springer, Heidelberg (2009)
10. Ilie, L., Navarro, G., Tinta, L.: The longest common extension problem revisited and applications to approximate string searching (2010)
11. Iliopoulos, C.S., Makris, C., Panagis, Y., Perdikuri, K.: Evangelos Theodoridis, and Athanasios Tsakalidis. The weighted suffix tree: An efficient data structure for handling molecular weighted sequences and its applications. Fundam. Inf. 71(2,3), 259–277 (2006)
12. Kandoth, C., Ercal, F., Frank, R.: A framework for automated enrichment of functionally significant inverted repeats in whole genomes. BMC Bioinformatics 11(suppl. 6), 1–10 (2010)
13. Kato, Y., Seki, H., Kasami, T.: Stochastic multiple context-free grammar for rna pseudoknot modeling. In: Proceedings of the Eighth International Workshop on Tree Adjoining Grammar and Related Formalisms, TAGRF 2006, Stroudsburg, PA, USA, pp. 57–64. Association for Computational Linguistics (2006)
14. Kolpakov, R., Kucherov, G.: Finding maximal repetitions in a word in linear time. In: Proceedings of the 1999 Symposium on Foundations of Computer Science, pp. 596–604. IEEE Computer Society (1999)
15. Lyngso, R.B., Pedersen, C.N.S.: RNA Pseudoknot Prediction in Energy-Based Models. Journal of Computational Biology 7(3-4), 409–427 (2000)
16. Main, M.G., Lorentz, R.J.: An o(n log n) algorithm for finding all repetitions in a string. Journal of Algorithms 5(3), 422–432 (1984)
17. Manacher, G.: A new linear-time "on-line" algorithm for finding the smallest initial palindrome of a string. J. ACM 22(3), 346351 (1975)
18. Nong, G., Zhang, S., Chan, W.H.: Linear Suffix Array Construction by Almost Pure Induced-Sorting. In: Data Compression Conference, pp. 193–202 (2009)
19. Pleij, C.W.A., Rietveld, K., Bosch, L.: A new principle of rna folding based on pseudoknotting. Nucleic. Acids Research 13(5), 1717–1731 (1985) C.W.A. Pleij, K. Rietveld, L. Bosch
20. Porto, A.H.L., Barbosa, V.C.: Finding approximate palindromes in strings (2002)
21. Stoye, J., Gusfield, D.: Simple and exible detection of contiguous repeats using a suffix tree. Theoretical Computer Science 270(12), 843–856 (2002)

Query Expansion with a Little Help from Twitter

Ioannis Anagnostopoulos[1], Gerasimos Razis[1], Phivos Mylonas[2],
and Christos-Nikolaos Anagnostopoulos[3]

[1] Computer Science and Biomedical Informatics Dpt., University of Thessaly
[2] Electrical & Computer Engineering School, National Technical University of Athens
[3] Cultural Technology and Communication Dpt., University of the Aegean
{janag,grazis}@ucg.gr, fmylonas@image.ntua.gr,
canag@ct.aegean.gr

Abstract. With the advent and rapid spread of microblogging services, web information management finds a new research topic. Although classical information retrieval methods and techniques help search engines and services to present an adequate precision in lower recall levels (top-k results), the constantly evolving information needs of microblogging users demand a different approach, which has to be adapted to the dynamic nature of On-line Social Networks (OSNs). In this work, we use Twitter as microblogging service, aiming to investigate the query expansion provision that can be extracted from large graphs, and compare it against classical query expansion methods that require mainly prior knowledge, such as browsing history records or access and management of search logs. We provide a direct comparison with mainstream media services, such as Google, Yahoo!, Bing, NBC and Reuters, while we also evaluate our approach by subjective comparisons in respect to the Google Hot Searches service.

Keywords: Query expansion, Microblogging services, Twitter, Social Data Mining.

1 Introduction

Microblogging is considered to be one of the most recent social raising issues of Web 2.0, being one of the key concepts that brought Social Web to the broad public. In other words, microblogging could be considered as a "light" version of blogging, where messages are restricted to less than a small amount of characters. Regarding their actual message content, this may be either textual data (e.g. short sentences), or even multimedia content (e.g. photos or hyperlinks to video sources). Yet, its simplicity and ubiquitous usage possibilities have made it one of the new standards in social communication; i.e., there is already a large number of social networks and sites that appear to have incorporated few or more microblogging functionalities; Twitter and Facebook being the most famous. The task of analyzing microblog posts and extract meaningful information from them in a (semi-)automated manner has been considered recently by some works in the literature, yet we believe their approaches are quite different to the one presented herein. Being part of a vast amount of information

L. Iliadis, H. Papadopoulos, and C. Jayne (Eds.): EANN 2013, Part II, CCIS 384, pp. 20–29, 2013.
© Springer-Verlag Berlin Heidelberg 2013

disseminated on the Web, it is very crucial for users to find relevant information in blogs or having recommendation in respect to their queries. Thus, modern information services provide a lot of mechanisms for suggestions in respect to users' information needs expressed by mostly syntactic queries. Research on query suggestion is highly related with query expansion [1], query substitution [2], query recommendation [3] or query refinement [4]. All are considered as similar procedures, aiming to adjust an initial user query into a revised one, which then returns more accurate results. In this work, we deviate from the traditional query suggestion proposal in a sense that users have their queries expanded directly from Twitter, and without having their queries or browsing history processed by search engines.

The remainder of this paper is organized as follows. In the next section we provide an overview over the related work on query analysis and expansion issues that need addressing in microblogging services. Section 3 provides the methodology we use, as well as the basic steps of our proposed algorithm. In Section 4, we describe a real case study in order to clearly show how our query expansion mechanism works. Finally, in Section 5, we evaluate our results against Google, Yahoo!, Bing, NBC and Reuters, while we also evaluate our approach by subjective comparisons in respect to the Google Hot Searches service. Section 6 concludes our work by summarizing the derived outcomes, providing in parallel some of our future directions.

2   Related Work

In general, microblogging posts [5] form a special category of user-generated data containing two major characteristics, that seriously affect linguistic analysis techniques [6], namely: a) they contain strong vernacular (acronyms, spelling changes, etc.) and, b) in principle they do not include any memorable repetition of words. Motivated by the observation that a microblog user retrieves information through queries formulation in order to acquire meaningful information, researchers focus on each post's characteristic features [7], whose quantitative evaluation could potentially affect the way in which the relevance between the user query and its returned results may be calculated. A first step towards this direction is discussed in [8], where authors identify two feature categories, i.e., features related to the user query and thus calculated as soon as the latter is formed and features that are not related to the specific query, but are inherent posts and thus calculated when the latter are modified, updated or added. In the context of social networking, query expansion techniques are of great inter-est using either previously constructed language models [9] or by taking into account personal user preferences, such as those resulting from user microblog posts and hashtags analysis [10]. The fact that microblog posts contain hashtags is also exploited in the literature towards query expansion methodologies in the direction of acquiring information that the user "is not aware of" and formulate queries that the user "does not know how to express" [11]. In [12], given a query, authors attempt to identify a number of hashtags relevant to the given query, that may be used to expand it and lead to better results; the proposed method is based solely on statistical techniques by building probabilistic language models for each available hashtag and by using a suit-able microblog posts corpus. Even in our own recent previous work

[13], we proposed the utilization of hashtags as the main source of information acquisition by searching the specific query terms within microblog posts under the condition that the former need to appear as hashtags. Then we calculated the most common hashtags that co-occur with the original query and thus expanded the query with the new hashtags. Finally, another broad related research category is the one formed by the observation that microblog posts are created during an actual event and contain comments or in-formation directly related to it, thus leading to event detection research efforts [14] based on posts and/or hashtags.

3   Proposed Query Expansion Mechanism

The microblogging service we use is Twitter and the Query Expansion (QE) mechanism is based on concatenated terms, known as hashtags (prefixed with "#"). Hashtags are actually a way Twitter users can semantically annotate the tweeted content, while there are no complicated syntactic rules, so Twitter users can annotate the information according to their will. This freedom of expression provide the best way to create a vast pool of crowd-sourced meta-data, leading to trends that best describe a social aware issue.

Prior to describing our QE method, we need to briefly introduce the context of capture-recapture experiments used in wildlife biological studies [15]. In these experiments, animals, birds, fishes or insects (subjects of investigation) are captured, marked and then released. If a marked individual is captured on a subsequent trapping occasion, then it is mentioned as "recaptured". Based on the number of marked individuals that are recaptured, and by using statistical models, we can estimate the total population size, as well as the birth, death and survival rate of each species under study. The sampling process is divided into k primary sampling periods, each of them consisting of l secondary sampling periods. Among primary sampling periods we assume that in the population we can have births, deaths and/or migration incidents. This population is called "Open". On the other hand, among secondary sampling periods the population is assumed closed, meaning that there are not gains of losses in the population [15]. During a secondary sampling period a set of different species is randomly selected, marked and keeping in parallel a history record of them, and then released back to nature. After a specific time interval, the second secondary sampling periods occurs and so forth, until the end of the last l secondary sampling period. Secondary periods are near and very short in time, while trapping occasions are considered instantaneous in order to assume that the populations under study are closed, meaning that no losses or gains occur during these time intervals. However, longer time intervals between primary sampling periods are desirable so evolution events can occur (e.g. survival, movement, and growth), as defined in the basic structure of Pollock's robust design model [16], which extends the Jolly-Seber model [17].

In wild-life experiments this model is applied to open populations, in which death, birth, and migration incidents possibly occur in the populations under study. In our case, birth means the appearance of a new Twitter hashtag, while death and migration incidents corresponds to evolution of hashtags. Moreover, a basic evolution metric we want to employ in our methodology is the survival rate of the subject under investigation (the expanded term in our paradigm). This actually highlights how durable in time hashtags are correlated to each other in the created real-time graph.

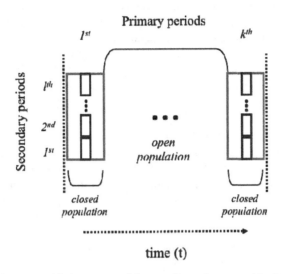

Fig. 1. The modified structure of the sampling scheme used in this work

However, since it is impossible to assume that due time the population of Twitter hashtags is closed, we modified the capture-recapture experimentations, in order to simultaneously conduct the secondary periods. This does not violate the assumptions of the Pollock's model, since in the real-life case, biologists set traps to different locations in the same space universe (e.g. a lake, a national park), while in the Twittersphere the space universe cannot be separated. Thus, Figure 1 depicts the modified structure of the sampling scheme we employ for the web paradigm. In our case, the individuals under study are Twitter hashtags. The trapping occasions are the keywords/seeds to be extended in our query. Each trapping occasion occurs in the primary sampling period, which we split in l secondary and simultaneously sampling occasions. In each of these l sampling instances we capture (and then mark) some hashtags with the same probability value p. By this we ensure that all secondary samplings are made in a "close" pool of instances under the basic principle of the Pollock's model. Then by investigating the recaptured instances in subsequent primary occasions, we calculate the survival probability of the examined hashtag according to Equation 1, where $(M_i - m_i)$ defines the marked hashtags not captured during the ith sampling period, while R_i are the hashtags captured at the ith period, marked, and then released for possible recapture in future samplings.

$$\tilde{\varphi}_i^e = \frac{\tilde{M}_{i+1}^e}{\tilde{M}_i^e - m_i^e + R_i^e} \tag{1}$$

4   Real-Life Experimentation: The Boston Marathon Bombing Case

As case study we use the "Boston Marathon Bombing" case, where during the Boston Marathon on April 15, 2013, two pressure cooker bombs exploded, killing 3 people and injuring 264 individuals[1]. This shocking event was among the breaking news globally for several weeks, since it was emotionally touched millions of people worldwide. Apart from mainstream media (e.g. TV/Radio), social media platforms covered all aspects of the incident disseminating an enormous amount of information, which was created from millions of users. Especially in Twitter, information contained not only shared information, but also personal opinions/thoughts, and constantly new links related to the crime, directly related to user-generated hashtags as semantic annotations. Having the experience for a previous work of ours [13], we selected the words "Boston" and "Marathon" as the basic terms someone enters in a search engine in order to find relevant information. It is worth noticing that even after one week after the event, most search engines in their main front-end environment did not suggest relevant terms after these terms. Our main contribution here is to provide query expansion to the user's submitted term(s) under the knowledge disseminated in Twitter, without having any other access or use of search engines' query logs.

In Figure 2, we can see the filtered graph that presents the entities with the 10% higher values in respect to eigenvector centrality (EiVC) values measured. This graph corresponds to a captured instance in one of the secondary sampling occasions conducted during our experiments. In this figure edges are classified as:

- Colour: black, name: *searchQuery*-from_user, explanation: This property is applied to edges between a Twitter user that used a queried term and the queried term,
- Colour: green, name: from_user-to_user, explanation: This property is applied in order to create an edge between a Twitter user that replied using a tweet to (an)other Twitter user(s),
- Colour: red, name: from_user-mentioned_user, explanation: This property is applied in order to create an edge between a Twitter user who mentioned at a tweet (an)other Twitter user(s) and the mentioned user(s),
- Colour: blue, name: from_user-tweeted_hashtag, explanation: This property is applied in order to create an edge between a Twitter user that included a hashtag in a tweet and the included hashtag,
- Colour: Yellow, name: from_user-tweeted_URL, explanation: This property is applied in order to create an edge between a Twitter user that included a Url in a tweet and the included Url,
- Colour: Purple, name: hashtag-URL, explanation: This property is applied in order to create an edge between hashtag and a Url in case that both of these hashtags are included in a tweet.

---

[1] Boston   Marathon   bombings,   http://en.wikipedia.org/wiki/Boston_Marathon_bombings, last accessed in May 13, 2013.

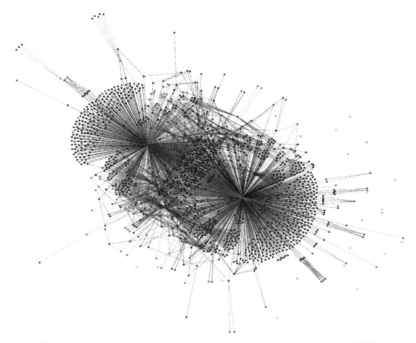

Fig. 2. A random selected secondary sampling occasion (April 18, 2013)

Table 1. Query expansion and suggested terms / case study evaluation

Query terms {Boston, Marathon}	Metrics		Also suggested by (in the same time period)			
Extended Term	Entity type	Score	Google	Ya-hoo!	Bing / NBC	Reu-ters
bostonstrong	#	0.2082	✗	✗	✗	✗
bostonbombing	#	0.1571	✔	✔	✔	✔
prayforboston	#	0.08359	✗	✗	✗	✗
news	#	0.04921	✔	✗	✗	✗
manhunt	#	0.04425	✗	✗	✗	✗
syria	#	0.04232	✗	✗	✗	✗

Similarly, nodes are classified as:

- Colour: Black, name: Trend/searchQuery, explanation: The queried term(s)
- from_user, colour: Black, explanation: A Twitter user that used the queried term(s)
- Colour: Green, name: to_userexplanation: A Twitter user that received one or more tweets as replies to a created tweet

- Colour: Red, name: mentioned_user, explanation: A Twitter user who is mentioned at other users' tweets
- Colour: Blue, name: tweeted_hashtag, explanation: A hashtag that is included in a tweet
- Colour: Yellow, name: tweeted_URL, explanation: A Url that is included in a tweet

5   Evaluation

In order to evaluate our QE mechanism we firstly compare the results derived from our case study in respect to the query suggestions of well-known search engines, like Google, Yahoo!, Bing, as well as mainstream Web media services, such as Reuter News and NBC. The second part of our evaluation, describes a generic evaluation, which involves subjective user ratings for results obtained from our QE mechanisms and Google.

5.1   Case Study Evaluation

Evaluation results for the "Boston Marathon" case are provided in Table I. Initial query terms ("Boston", "Marathon") are on the left side of the table, and then follow the expanded term(s). All suggested extra terms are according to a normalized survival probability of hashtag e during primary period i ($\tilde{\phi}_i^e$), as defined in Equation 1.

Finally, the last four columns of Table I indicate whether the specific expanded query has been suggested (even in different order of terms with respect to the seed term) by Google, Yahoo!, Bing, NBC and Reuters[2]. The date we performed this evaluation was April 18, 2013, just three days after the bombing incident. The trend analysis and the expanded terms in respect to the initial provided, performed between subsequent primary periods, where each of them consisted of two simultaneously secondary periods with capture probability equal to 0.3. We notice that apart a very good suggestion on related web sources for direct information (e.g. BBC World, 7news, etc.), our mechanism proposes two other really noteworthy acquisition terms like "manhunt" and "Syria", which are not suggested by the other four search services. It is worth noticing that during the next day (April 19, 2013) where one of the suspects was arrested, our algorithm "captured" two hashtags related with the suspect name (#dzhokhartsarnaev, #tsarnaev) having a quite high normalized scoring value (0.079 and 0.1302 respectively).

5.2   Evaluation against Google Hot Searches

In this sub-section we describe an evaluation of our QE mechanism in comparison with Google Hot Searches[3] service. For the purposes of this evaluation 17 individuals

---

[2] NBC search service is powered by Bing.
[3] http://www.google.com/trends/hottrends

were engaged. Their task was to subjectively rate the expanded queries against the respective service from Google. Each individual (postgraduate students from an MBA course class at National Technical University of Athens) was asked to select three different events from Google Hot Searches for a specific testing period (a specific week during March 2013). Each individual had to explicitly rate the suggested entities/hashtags derived by our QE mechanism, against their selected events as appear in Google Hot Searches (in U.S.). The rating performed upon a five-point Likert scale as indicated below:

1. Strongly disagree (totally irrelevant suggestion)
2. Disagree (not so good suggestion)
3. Neither agree nor disagree (nearly same suggestion)
4. Agree (potentially better)
5. Strongly agree (surely better)

After processing the one-week results we ended up with 87 unique related terms (as these were provided by Google Hot Searches) in 31 distinct events (20 out of the 51 events were identical). The average amount of suggested terms per tested event was 2.81, which practically means that nearly 3 terms in average expand the basic term that describe an event. At Figure 3, we can see some points that indicate the average evaluator rating for suggested hashtags, as derived from our QE mechanism in respect to scoring value φ. We noticed that the larger the φ is, the higher the mean subjective rate appears in the five-point Likert scale. More specifically, suggested hashtags that appeared as expanded terms having $\tilde{\varphi}_i^e \geq 0.7$ (even as suggested in concatenated format), were subjectively evaluated as more relevant presenting nearly

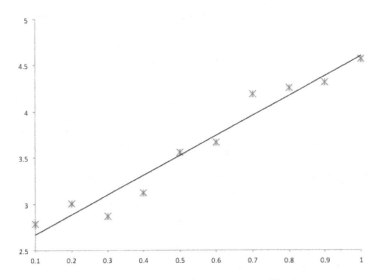

Fig. 3. Mean evaluator ratings vs. proposed hashtags

one-point higher level in the Likert scale. In other words, this means that through subsequent samplings in Twitter, several user-generated terms (hashtags) are more descriptive in comparison to a related query term in Google's log. This was somehow expected, since we performed short-term trend analysis rather than long-term log analysis, yet it is an indicative assumption that our QE method is in the right direction. We can also notice that the majority of subjective rates in average values (more than 60%) were close or slightly higher in the third level (point 3) in the Likert scale, thus indicating a "nearly same suggestion" in comparison to Google search service.

6   Conclusion – Future Directions

In this paper, we proposed a Query Expansion (QE) mechanism, which employs trend analysis issues from microblogging services (case study in Twitter). In order to efficiently analyze trends in terms of retrieving suggest terms for query expansion and reducing the sampling cost, we used the well-known capture-recapture methodology, which is mostly applied in biology experiments. For evaluating our proposal, we presented a recent real event, while we further evaluated it through subjective rates against Google Hot Search service, having as pool a class of 17 postgraduate students. Ongoing research is performed on how other Twitter entities like Twitter mentions (@s), URIs and other related information (e.g. images, geo-location, replies) can be part of suggested term expansion.

References

1. Xu, J., Croft, W.B.: Query expansion using local and global document analysis. In: 19th Annual International ACM SIGIR Conference on Research and Development in Information Retrieval (SIGIR 1996), pp. 4–11. ACM, New York (1996)
2. Jones, R., Rey, B., Madani, O., Greiner, W.: Generating query substitutions. In: 15th International Conference on World Wide Web (WWW 2006), pp. 387–396. ACM, New York (2006)
3. Baeza-Yates, R., Hurtado, C.A., Mendoza, M.: Query recommendation using query logs in search engines. In: Lindner, W., Fischer, F., Türker, C., Tzitzikas, Y., Vakali, A.I. (eds.) EDBT 2004. LNCS, vol. 3268, pp. 588–596. Springer, Heidelberg (2004)
4. Kraft, R., Zien, J.: Mining anchor text for query refinement. In: 13th International Conference on World Wide Web (WWW 2004), pp. 666–674. ACM, New York (2004)
5. Efron, M.: Information Search and Retrieval in Microblogs. Journal of the American Society for Information Science and Technology 62(6), 996–1008 (2011)
6. Massoudi, K., Tsagkias, M., de Rijke, M., Weerkamp, W.: Incorporating Query Expansion and Quality Indicators in Searching Microblog Posts. In: Clough, P., Foley, C., Gurrin, C., Jones, G.J.F., Kraaij, W., Lee, H., Mudoch, V. (eds.) ECIR 2011. LNCS, vol. 6611, pp. 362–367. Springer, Heidelberg (2011)
7. Huberman, B.A., Romero, D.M., Wu, F.: Social networks that matter: Twitter under the microscope. First Monday 14(1) (January 2009)

8. Tao, K., Abel, F., Hauff, C., Houben, G.-J.: Twinder: A Search Engine for Twitter Streams. In: Brambilla, M., Tokuda, T., Tolksdorf, R. (eds.) ICWE 2012. LNCS, vol. 7387, pp. 153–168. Springer, Heidelberg (2012)

9. Packer, H.S., Samangooei, S., Hare, J.S., Gibbins, N., Lewis, P.H.: Event Detection using Twitter and Structured Semantic Query Expansion. In: Proceedings of the 1st International Workshop on Multimodal Crowd Sensing (CrowdSens 2012), Sheraton, Maui Hawaii, pp. 7–14 (2012)

10. Zhou, D., Lawless, S., Wade, V.: Improving search via personalized query expansion using social media. Information Retrieval 15(3-4), 218–242 (2012)

11. Bouadjenek, M.R., Hacid, H., Bouzeghoub, M., Daigremont, J.: Personalized Social Query Expansion Using Social Bookmarking Systems. In: Proceedings of the 34th International ACM SIGIR Conference on Research and Development in Information Retrieval (SIGIR 2011), Beijing, China, pp. 1113–1114 (2011)

12. Efron, M.: Hashtag retrieval in a microblogging environment. In: Proceedings of the 34th International ACM SIGIR Conference on Research and Development in Information Retrieval, Geneva, pp. 787–788 (2010)

13. Anagnostopoulos, I., Kolias, V., Mylonas, P.: Socio-semantic query expansion using Twitter hashtags. In: Proceedings of the 2012 7th International Workshop on Semantic and Social Media Adaptation and Personalization (SMAP 2012), Luxembourg, pp. 29–34 (2012)

14. Packer, H.S., Samangooei, S., Hare, J.S., Gibbins, N., Lewis, P.H.: Event Detection using Twitter and Structured Semantic Query Expansion. In: Proceedings of the 1st International Workshop on Multimodal Crowd Sensing (CrowdSens 2012), Sheraton, Maui Hawaii, pp. 7–14 (2012)

15. Pollock, K.H., Nichols, J.D., Brownie, C., Hines, J.E.: Statistical inference for capture-recapture experiments. Wildlife Monographs 107 (1990)

16. Schwarz, C., Stobo, W.: Estimating temporary migration using the robust design. Biometrics 53, 178–194 (1997)

17. Jolly, G.: Explicit estimates from capture-recapture data with both death and immigration stochastic model. Biometrika 52, 225–247 (1965)

Recognizing Emotion Presence
in Natural Language Sentences

Isidoros Perikos and Ioannis Hatzilygeroudis

School of Engineering, Department of Computer Engineering & Informatics
University of Patras, 26500 Patras, Hellas, Greece
{perikos,ihatz}@ceid.upatras.gr

Abstract. Emotions constitute a key factor in human communication. Human emotion can be expressed through various mediums such as speech, facial expressions, gestures and textual data. A quite common way for people to communicate with each other and with computer systems is via written text. In this paper we present an emotion detection system used to automatically recognize emotions in text. The system takes as input natural language sentences, analyzes them and determines the underlying emotion being conveyed. It implements a keyword-based approach where the emotional state of a sentence is constituted by the emotional affinity of the sentence's emotional words. The system uses lexical resources to spot words known to have emotional content and analyses sentence structure to specify their strength. Experimental results indicate quite satisfactory performance.

Keywords: Sentiment Analysis, Emotion Recognition, Affective Computing, Human Computer Interaction, Natural Language Processing.

1   Introduction

Computer systems are increasingly involved in almost all aspects of everyday life. As the field of artificial intelligence matures and grows, it enhances the capabilities and the functionality of computer systems. A fundamental aspect of computer systems concerns the way that human interact and communicate with them. It becomes more and more important to be able to interact with them in a natural way, similar to the way we interact with other humans.

Emotions constitute a key factor of human nature, which colors the way of human communication. The role of emotions in human computer interaction was initially investigated by Picard, who introduced the concept of affective computing [12], indicating the importance of emotions in human computer interaction and drawing a direction for interdisciplinary research from areas such as computer science, cognitive science and psychology. The aim of affective computing is to enable computers to recognize and express emotions and bridge the gap between the emotional human and the computer by developing computational systems that recognize and adapt to the user's emotional states [3]. Automatically recognizing and responding to a user's affective states can enhance the quality of the interaction, thereby making a computer

L. Iliadis, H. Papadopoulos, and C. Jayne (Eds.): EANN 2013, Part II, CCIS 384, pp. 30–39, 2013.

interface more usable, enjoyable, and effective, which in learning systems may lead to a better adaptation to the learner [13].

Human emotion can be expressed through various mediums such as speech, facial expressions, gestures and textual data. The most common way for people to communicate with each other and with computer systems is via written text which is the main communication mean and the backbone of the web. Recognizing emotional presence in textual data is a domain with wide ranging applications. For example, conversational agents can greatly benefit from the identification of the emotional states of the participants, achieving more realistic interactions at an emotional level. Moreover, in intelligent tutoring systems, recognizing students' emotional states could significantly improve the learning procedure [16]. Furthermore, text-to-speech systems can adjust the voice based on the recognized emotional content of the text and thus achieving more smooth and vivid interactions.

In this paper, we address the problem of recognizing emotional presence in textual data. More specifically, we present a system developed to recognize the presence of the six basic emotions proposed by Ekman [6] in natural language sentences. The system analyses the sentence's structure using tools such as Stanford parser [4] and uses lexical resources such as Wordnet Affect [17] to spot words known to convey emotion. Then it specifies each emotional word strength and determines the sentence's emotional status based on the sentence's dependency graph.

The rest of this paper is structured as follows. Section 2 presents basic topics on sentiment analysis and emotion recognition and describes related work. Section 3 presents the developed system, describes its architecture and analyses its functionality. Section 4 presents the evaluation study conducted and the results gathered. Finally, section 5 concludes the paper and draws directions for future work.

2   Background and Related Work

2.1   Background

Emotion is considered to be a strong feeling deriving from one's circumstances, mood or relationships with others [5]. Studies have shown that analyzing and recognizing affect in text is considered to be a complex, 'NLP' complete problem and the interpretation varies depending on the context and the world knowledge [15].

Emotion representation is a basic aspect of an emotion recognition system. The most popular models for representing emotions are the categorical and the dimensional model. The *categorical model* assumes a finite number of basic, discrete emotions, each of which is serving a particular purpose. On the other hand, the *dimensional model* represents emotions on a dimensional approach, where an emotional space is created and each emotion lies in this space. A very popular categorical model is the Ekman emotion model [6], which specifies six basic human emotions: "anger, disgust, fear, happiness, sadness, surprise". It has been used in several studies/systems that recognize emotional text and facial expressions related to these emotional states. Another model is the OCC (Ortony/Clore/Collins) model [10] which specifies 22 emotion categories based on emotional reactions to situations and is mainly designed to model human emotions in general.

To enhance systems knowledge and efficiency in recognizing emotions and sentiments in natural language different lexical resources have been developed. A well-known lexical resource is WordNet Affect [17]. It is based on WordNet database and extends it by adding a subset of synsets suitable to represent affective concepts. More specifically, these synsets are annotated with one or more affective labels. Senti-WordNet [7] is based on WordNet and associates each synset with three numerical scores indicating the degree that the synset is objective, positive or negative.

In this work, we adopt the categorical model and more specifically the system developed implements the six basic emotion categories proposed by Ekman. In addition, the developed system utilizes WordNet Affect to spot words known to convey emotional content.

2.2 Related Work

The work presented in [1] explores automatic classification of sentences in children fairy tales. The authors developed a corpus consisting of fairy tales sentences, which were manually annotated with emotional information and explore sentence's classification according the Ekman's emotion categories. In [18], authors spot affective keywords and use a parser in order to identify the main objects of the sentence that are associated with the sentence's emotional words. The system generates emotional output only if emotional word(s) refer to the main object. The work presented in [9] recognizes Ekman's basic emotions in online blog posts. The authors analyze the posts using Machinese Syntax parser, spot emoticons and keywords that may appear in the posts and use rule-based approach to determine sentence emotional content. In [8] the OOC model is used to specify the students' emotions during they are playing educational games. The system analyses students' behavior and infers students' emotions in order to adapt interaction to each individual student's needs. In [11], a statistical approach is presented. The authors address the textual affect recognition task and classify emotions in text by using word counts.

Moreover, in [2] a system that classifies song lyrics into mood categories is presented. The song lyrics undergo pre-processing steps such as stemming, stop words and punctuation marks removal. Then the lyrics are classified into mood categories using a bag-of-words approach, where each word is accompanied by its frequency in the song and its tf-idf (term frequency-indirect document frequency) score.

3 Emotion Recognition System

In this section we present the developed system and analyze its functionality. In Fig.1, the system architecture is illustrated. The system performs emotion recognition on a sentence level. So, a given document is split into sentences by the *sentence splitter*. Then, the system analyses each sentence's structure. It uses a part-of-speech (POS) tagger to specify each word's grammatical role and base form (lemma) and creates the dependency tree (e.g. see Fig. 2), based on the words' relationships, using the Stanford parser. Words known to convey emotion are spotted using the lexical resources of the *knowledge base* (KB). Then, for each emotional word the system analyses its relations and the way it interacts with other words. Based on the words' relationships, identifies specific types of emotional word's interactions with quantification words, in order to specify its *emotional strength*. Finally, the overall sentence emotional status is specified by combining the emotion content of the sentence emotional parts. This process is implemented by the *sentence emotion extractor*.

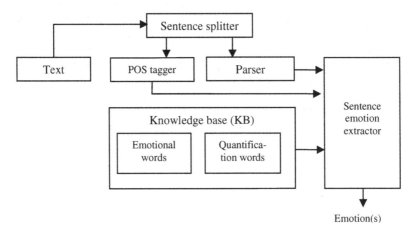

Fig. 1. The system architecture

So, the system for a given NL sentence proceeds as follows:

1. Use tree tagger to specify the words' lemmas and grammatical roles.
2. Use Stanford parser to analyze sentence structure and get the dependencies and the dependency tree
3. For each word use KB to determine whether it is emotional or not. If it is
   3.1 Analyze its relationships
   3.2 Check if a modification relationship with quantification words exists, analyze it and determine emotion strength.
   3.3 Analyze the dependency tree, recognize sentence pattern/structure and based on it determine the sentence's emotional content

The knowledge base (KB) stores information about emotional words, that is words known to convey emotion(s). The recorded emotional words are taken from the WordNet Affect lexical source, which we extended by manually adding 11 emotional words, and for each word its grammatical role (e.g. noun, verb, adjective etc.) and also its emotional category are stored. For example, an emotional word manually added is the verb "kiss", which is added to denote joy. Also KB stores quantification words that may modify the strength of emotional words. Examples of such words are: {all, none, very, quite, rather etc}. KB determines the emotional content of a sentence based on the analysis of the sentence as performed by the POS tagger and the Stanford parser and the linguistic knowledge it holds, as presented above.

3.1   Sentence Analysis

The system, given a NL sentence, firstly uses a POS tagger in order to determine the sentence's syntactic information. POS tagging is a fundamental process in a NLP system and gives a first level analysis of words' roles in the sentence. More specifically, Tree Tagger tool [14], a widely used statistical morphosyntactic tagger and lemmatizer, is used to specify for each word its base form (lemma) and its part of speech tag, indentifying its grammatical role in the sentence. Then, the system uses Stanford parser to get a deeper analysis of the sentence's structure. Stanford parser creates the parse tree and determines the type dependencies of the sentence providing helpful assistance in sentence analysis. Dependencies indicate the way that words are connected and interact with each other.

As an example, consider the sentence "She kissed her aunt with great happiness". In Figure 2 the sentence parse tree and the dependencies as specified by the Stanford parser are presented.

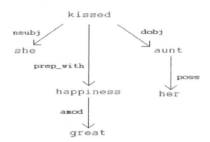

Fig. 2. Dependency tree of the sentence "She kissed her aunt with great happiness"

The dependency represents grammatical relations between the sentence's words. They are presented as triplets: name of the relation, governor and dependent. For example, nsubj (she,kissed) is an nominal subject relationship between the two words, defining that the word 'she' is the subject of the word 'kissed'.

3.2   Recognizing Emotional Words and Their Strength

After a given NL sentence is analyzed, the system proceeds in spotting emotional words. To do so, the system utilizes lexical resources in order to be able to spot words known to convey emotional context. KB, as mentioned above, stores information about (a) emotional words and (b) quantification words.

The system, for every word of the sentence, searches to see if it is stored as an emotional word in KB. If KB has recorded the word as emotional, it returns the emotional category the word belongs to. If the system does not find any same emotional word in the sentence, then it performs a deeper analysis of the sentence's words in terms of synonyms and antonyms. The assumption is that a word's synonym or an antonym may be recognized as an emotional word and thus an emotional content may underlies in the word.

KB also stores information about quantification words that quantify and modify the strength of emotional words through interacting with them. So, we developed a list of *quantification words*. Examples of such words are: {very, some, all, hardly, less etc}. A special category of the quantification words are words that denote negation. These *negation words*, when appear in a sentence, flip the polarity of the words that interact with. Examples of negation words are: {none, no, not, never, nobody}. For each one, its modification impact on words that interacts with is specified. So, words like 'very' and 'great' have a strong positive impact on words that are related with, increasing their emotional content, while negation words flip the emotional content. In Table 1, example quantification words and their impact on emotional words are presented.

Table 1. Example quantification words and their impact

Word	Modification impact
Very	High
Great	High
Hardly	Average
Quite	Average
Less	Low
Not	Flip

Low value is set to 20, average to 50 and high to 100. So, the quantification word 'extremely' when modifies an emotional word has an emotion strength set to 100 (max emotional strength), whereas the quantification word 'quite' sets the emotion strength to 20. These values have been specified based on empirical and experimental studies.

After emotional words are recognized, the system performs a deeper analysis regarding their role in the sentence and the type of their relationships/connections with other words. More specifically, it tries to analyze special types of relationships that may appear with quantification words. These relationships are recognized as 'mod' dependencies by the Stanford parser. So, these dependencies, connecting emotional words with quantification words, are analyzed and a quantification word's impact defines the strength of the connected emotional word. As an example, consider the sentence "She kissed her aunt with great happiness". KB recognizes the word 'kissed' as emotional word to denote 'joy' and also the word 'happiness' as emotional word to also denote 'joy'. The dependencies defined by the parser for the word 'kissed' are nsubj(kissed-2, She-1), root(ROOT-0, kissed-2), dobj(kissed-2, aunt-4), prep_with(kissed-2, happiness-7). There isn't any modification relationship so the emotional strength of the word is set to high (100).

The dependencies defined for the word 'happiness' are amod(happiness-7, great-6) and prep_with(kissed-2, happiness-7). The amod is recognized as a modification relationship and the word interacting with modification relationship on the emotional word 'happiness' is the word 'great'. This word is recorded by KB as having a positive high modification impact on the words it interacts with and thus adding to the word's happiness emotional strength, determining it to be high (100). So, the sentence is recognized to have two emotional parts that denote 'joy' with strength set to high (100) for the first and 'joy' with strength also set to high (100) for the second.

Finally, consider negation cases such as 'is not very furious', where the emotional word 'furious' is detected to denote 'anger', the word 'very' has a high quantification to the emotion and the negation detected flips/reverse the quantification to 'low' and thus this sentence part emotional content is determined to be low (20) 'anger'.

3.3   Determining Sentence Emotional Content

After the emotional words of the sentence are recognized and their strength is determined, the system specifies the sentence's overall emotional content. To do so, analyzes the sentence's structure. More specifically, it recognizes and analyses the basic pattern of the sentence consisting of the sentence's main verb, the object and the subject of it. So the pattern "Subject–Verb–Object" is extracted from the sentence based on the dependencies. This pattern is the backbone of the sentence structure and holds the core meaning of the sentence. Moreover, analyzing it can help the system in understanding the interactions of the sentence parts, that is the way the emotional parts are connected. So, the system processes the sentence structure as follows:

1. *Analyze the sentence dependencies and extract the subject-verb-object pattern.*
2. *For each grammatical role of the pattern (eg. object or verb or subject).*
   2.1 *Specify if it is an emotional part.*
   2.2 *Analyze its relationships with emotional parts (if any).*
   2.3 *Specify its emotional content.*
3. *Combine emotional contents of the parts to specify the sentence overall emotions.*

For example, consider once again the above sentence "She kissed her aunt with great happiness". The basic pattern extracted is 'she-kissed-aunt'. The word 'kissed', which is the verb of the pattern, is determined as emotional part denoting joy, and also is connected with the emotional part 'happiness' recognized to denote 'joy'. So, the verb of the pattern denotes 'joy' with strength set to high (100), the subject is neutral and the object is neutral. Thus, the sentence's emotional content is the verb's emotional content and thus the sentence's overall emotion is set to 'joy' with strength set to high (100). Actually, combination results are achieved via a rule-based approach.

4   Experimental Study

An experiment study was conducted to evaluate the system's performance. We created a corpus of 180 sentences and manually annotated them. More specifically, the corpus consisted of 60 neutral sentences and 120 emotional which were equally divided into the six emotional categories. The sentences where selected not to be lengthy and also not suffering from many ambiguities and anaphoric expressions and were selected from different sources such as, fairy tales, news headlines, web articles, scientific books and chatting conversations. During the annotation stage, for each sentence were determined (a) the existence and the (b) degree of each one of the six basic emotions. The emotion level ranges from 0 to 100, where 0 denotes the absence of a specific emotion and 100 denotes that the specific emotion is very strong. So, each sentence was associated with a vector representing the levels of the six emotions. Human annotation is used as a 'gold standard' for the system evaluation.

The evaluation consisted in two parts. In the first, the system was evaluated in characterizing the sentences as either emotional (in case that recognizes emotional content) or neutral (in case of emotional absence). Since the system classifies sentences either in emotional or in neutral class, we present the relevant metrics we used to evaluate it. Evaluation of a classification model is usually based on the following metrics: recall, precision, f_measure, which are defined as follows:

$$recall = \frac{TE}{TE + FN}, \qquad prec = \frac{TE}{TE + FE}, \qquad f\_measure = \frac{2 * prec * rec}{prec + rec}$$

where, TE (true-emotional) is the number of sentences classified correctly as emotional, FE (false-emotional) is the number of sentences that were incorrectly classified as emotional, TN (true-neutral) is the number of sentences correctly classified as non-emotional (i.e. neutral) and FN (false-neutral) is the number of sentences that are incorrectly classified as non-emotional (i.e. neutral). The evaluation results of the system's performance are presented in Table 2.

Table 2. Evaluation metrics

Metric	Value
Recall	0.850
Precision	0.935
F_measure	0.890

The results show a very good performance of the system. More specifically, from the corpus of 180 sentences that were tested, it correctly identified the presence or the absence of emotional content in 155 sentences. So, the general accuracy of the system is 0.861 (155/180), which also indicates a very good performance.

The second part of the study aims to evaluate the system's performance in specifying the strength of the emotion. So, the 155 sentences already characterized as emotional were selected as the study corpus. The system has determined for each sentence the existence of each one of the six basic emotions and their strength. The strength of an emotion is represented as an integer number ranging from 0 to100, where 0 denotes absence of a specific emotion and 100 denotes that the specific emotion, provoked by the sentence, is very strong. The same was done by a human annotator.

The system performance was analyzed and we performed a comparison with the human's annotations. For each sentence the emotion strength was cooperated with the strength specified by the human annotator. The difference between the system and the annotator on specifying the strength of an emotion was calculated and used as a metric for system's performance. More specifically, the difference is organized in three levels. A difference between 0-25 denotes that the system and the annotator were very close in determining emotion strength, a difference between 25-50 indicates adequate/average performance, while difference larger than 50 indicates that the system lacked in determining the emotion strength. The results are presented in Table 3.

Table 3. Analysis of the strength's differences

	[0-25]	[25-50]	[50<]
Anger	65	25	10
Disgust	30	50	20
Fear	55	30	15
Happiness	78	17	5
Sadness	71	21	8
Surprise	45	35	20
Average	57.33	29.66	13

The evaluation results indicate a good agreement between the system and the human annotator. More precisely, the system has specified the strength of the emotion in adequate agreement with the annotator in the 57,3% of the cases. A noticeable point is that the system had a very good agreement in identifying happiness and sadness emotion's strength. This is due to the fact that, in general and also in the corpus's sentences, these are very strong emotions and are almost always expressed explicitly with the use of emotional words. In contrast, surprise and disgust emotions may be implicitly denoted in the sentence. In addition, in our case this could be a result of the small number of sentences conveying these two emotions.

5   Conclusions and Future work

In this paper, we present a system developed to automatically recognize emotions in natural language sentences. It analyses a sentence's structure using Stanford parser and Tree Tagger and uses WordNet Affect lexical resource to spot emotional words. Then analyzes each emotional word's dependencies to specify the emotional word's strength and determines the overall sentence emotional status based on the sentence's dependency graph. Experimental study results indicate quite promising results.

However, there are points that the system could be improved. KB utilizes Affective WordNet and manually added words to identify sentence's emotional words. An extension of system's knowledge could be made by adding more lexical resources. Moreover, the patterns of KB could be extended to cover more complex sentences. Also, conducting a larger scale evaluation will give us a deeper insight of the system's performance. Finally, currently the system uses a rule-based approach to determine the sentence emotion presence, so another extension will be the addition of a neural network approach to assist in emotion recognition. Exploring this direction is a key aspect of our future work.

Acknowledgments. This work was supported by the Research Committee of the University of Patras, Greece, Program "Karatheodoris", project No C901.

References

1. Alm, C.O., Roth, D., Sproat, R.: Emotions from text: machine learning for text-based emotion prediction. In: Proceedings of the Conference on Human Language Technology and Empirical Methods in Natural Language Processing, pp. 579–586 (2005)
2. Brilis, S., Gkatzou, E., Koursoumis, A., Talvis, K., Kermanidis, K.L., Karydis, I.: Mood Classification Using Lyrics and Audio: A Case-Study in Greek Music. In: Iliadis, L., Maglogiannis, I., Papadopoulos, H., Karatzas, K., Sioutas, S. (eds.) AIAI 2012 Workshops, Part II. IFIP AICT, vol. 382, pp. 421–430. Springer, Heidelberg (2012)
3. Calvo, R.A., D'Mello, S.: Affect Detection: An Interdisciplinary Review of Models, Methods, and Their Applications. IEEE Transactions on Affective Computing 1(1), 18–37 (2010)
4. De Marneffe, M.C., MacCartney, B., Manning, C.D.: Generating typed dependency parses from phrase structure parses. In: Proceedings of LREC, vol. 6, pp. 449–454 (2006)
5. Dictionary, Oxford English. Oxford english dictionary (2008)
6. Ekman, P.: Basic emotions. In: Handbook of cognition and emotion, pp. 45–60 (1999)
7. Esuli, A., Sebastiani, F.: Sentiwordnet: A publicly available lexical resource for opinion mining. In: Proceedingsof LREC, vol. 6, pp. 417–422 (2006)
8. Katsionis, G., Virvou, M.: Adapting OCC theory for affect perception in educational software. In: The 11th International Conference on Human-Computer Interaction (2005)
9. Neviarouskaya, A., Prendinger, H., Ishizuka, M.: Textual affect sensing for sociable and expressive online communication. In: Paiva, A.C.R., Prada, R., Picard, R.W. (eds.) ACII 2007. LNCS, vol. 4738, pp. 218–229. Springer, Heidelberg (2007)
10. Ortony, A., Clore, G., Collins, A.: The Cognitive Structure of Emotions. Cambridge University Press, Cambridge (1988)
11. Osherenko, A., André, E.: Lexical affect sensing: Are affect dictionaries necessary to analyze affect? In: Paiva, A.C.R., Prada, R., Picard, R.W. (eds.) ACII 2007. LNCS, vol. 4738, pp. 230–241. Springer, Heidelberg (2007)
12. Picard, R.: Affective Computing. The MIT Press, Cambridge (1997)
13. Santos, O.C., Boticario, J.G., Arevalillo-Herraez, M., Saneiro, M., Cabestrero, R., del Campo, E., Salmeron-Majadas, S.: MAMIPEC-Affective Modeling in Inclusive Personalized Educational Scenarios. In: Bulletin of the IEEE Technical Committee on Learning Technology, vol. 14(4), p. 35 (2007)
14. Schmid, H.: Probabilistic Part-of-Speech Tagging Using Decision Trees. In: Proceedings of the International Conference on New Methods in Language Processing, pp. 44–49 (1994)
15. Shanahan, J.G., Qu, Y., Wiebe, J.: Computing attitude and affect in text: theory and applications, vol. 20. Springer (2006)
16. Shen, L., Wang, M., Shen, R.: Affective e-learning: Using "emotional" data to improve learning in pervasive learning environment. Educational Technology & Society 12(2), 176–189 (2009)
17. Strapparava, C., Valitutti, A.: WordNet-Affect: an affective extension of WordNet. In: Proceedings of LREC, vol. 4, pp. 1083–1086 (2004)
18. Zhe, X., Boucouvalas, A.C.: Text-to-emotion engine for real time internet communication. In: Proceedings of International Symposium on Communication Systems, Networks and DSPs, pp. 164–168 (2002)

Classification of Event Related Potentials of Error-Related Observations Using Support Vector Machines

Pantelis Asvestas[1], Errikos M. Ventouras[1], Irene Karanasiou[2],
and George K. Matsopoulos[4]

[1] Department of Medical Instruments Technology,
Technological Educational Institute of Athens, Greece
{pasv,ericvent}@teiath.gr
[2] School of Electrical and Computer Engineering,
National Technical University of Athens, Greece
{ikaran,gmatso}@esd.ece.ntua.gr

Abstract. The aim of this paper is to present a classification method that is capable to discriminate between Event Related Potentials (ERPs) that are the result of observation of correct and incorrect actions. ERP data from 47 electrodes were acquired from eight volunteers (observers), who observed correct or incorrect responses of subjects (actors) performing a special designed task. A number of histogram-related features were calculated from each ERP recording and the most significant ones were selected using a statistical ranking criterion. The Support Vector Machines algorithm combined with the leave-one-out technique was used for the classification task. The proposed approach discriminated the two classes (observation of correct and incorrect actions) with accuracy 100%. The proposed ERP-signal classification method provides a promising tool to study observational-learning mechanisms in joint-action research and may foster the future development of systems capable of automatically detecting erroneous actions in human-human and human-artificial agent interactions.

Keywords: Event-Related Potentials (ERPs), Support Vector Machines (SVM), observational-learning mechanisms.

1 Introduction

A significant part of the learning process in human development takes place through observation. The behavior of an observer might be influenced by the positive or negative consequences of a behavioral model. An observer will emulate the behavior of a model, if this includes characteristics which the observer deems attracting or desirable, such as talent, intelligence, power etc. Furthermore, the way through which the model is treated will influence the observer. If the model is rewarded, it is more probable that the observer will emulate the rewarded behavior, while the opposite is expected to happen when an observed behavior is reprimanded.

The results of studies in observational learning suggest that the mechanisms through which observation contributes to learning are similar to the mechanisms that

L. Iliadis, H. Papadopoulos, and C. Jayne (Eds.): EANN 2013, Part II, CCIS 384, pp. 40–49, 2013.
© Springer-Verlag Berlin Heidelberg 2013

contribute to learning through self-action [1]. It is known that when an incorrect action is committed by a person, or the person is reprimanded for an action, then a negative peak is present in that person's electroencephalographic Evoked Potentials (EPs, i.e., usually non-invasive scalp recordings of brain activity related to specific stimuli). The maximum of the peak takes place at around 80ms after the start of the wrong response/action. This peak is called Error-Related Negativity (ERN). Based on source localization studies, magnetoencephalographic studies and intracranial recordings, ERN has been related with activity in the anterior cingulate cortex (ACC). ERN has also been used as a quantitative marker of ACC activity in experiments which concerned the observation of errors [2]. It has been found that there is activity in the middle frontal lobe, which is dependent on the correctness of the activity of other persons who are observed. Van Schie et al. [2] found that ERNs are generated not only when errors are committed by the person whose EPs are recorded, but also when errors are observed by the person whose EPs are recorded, albeit with diminished amplitude and longer latency (time occurrence of the EP peak), than those recorded from the scalp of actors who behave incorrectly. Those findings strengthen the hypothesis that the same mechanisms are activated both when committing and when observing errors. Nevertheless, because it has been found that sometimes a negative ERN-like deflection is produced even for correct actions [3], something similar could happen when observation of the action of other persons takes place.

Whenever EP components present differentiation of their amplitude, latency, frequency etc., which is related to different aspects of cognitive processes, then automatic classification of the EP recordings in classes has been sought. Ventouras et al. investigated the classification of ERN Pe potentials by means of kNN and support vector machine classifiers [4]. In the work of Sveinsson et al. [5], EP data obtained from normal control subjects and chronic schizophrenic patients were classified, using a parallel principal component neural network. The proposed architecture provided overall classification accuracy up to 90%. In the work of Palaniappan and Paramesran, [6], genetic algorithms and a fuzzy ARTMAP classifier were combined to identify the discriminatory subset of the feature set for classification of alcoholics and non-alcoholics. The feature set consisted of spectral power ratios extracted from multi-channel visual evoked potential (VEP) recordings. Classification accuracy reached 95.9%. A classification system distinguishing patients with depression from normal controls using the P600 EP component was presented by Kalatzis et al. [7]. That system used a combination of Support Vector Machines (SVM) classifiers and a majority-vote engine.

The existence of differences in the ERPs of observers, when observing correct and incorrect actions, might foster the development of classification systems capable of detecting performance errors of a human - or an artificial agent – in need of being monitored in a joint-action situation. The primary aim of the present study was the development and implementation of a classification system for discriminating observations of correct and incorrect actions, based on scalp-recorded ERPs, using histogram-related features.

2   Material and Methods

2.1   Subjects and ERPs' Recording Procedure

The ERP data used in the present study were collected in previous research [8]. The data were acquired from eight (8) healthy volunteers (observers), who observed correct or incorrect responses of subjects (actors) performing a special designed task. In particular, the actors were faced in front of a table facing an experimenter, having in front of them, on the table, two joystick devices positioned to the left and right of a Led stimulus device. The actors were asked to respond to the direction of a center arrowhead surrounded by distracting flankers pointing either in the same direction as the center arrow, or in opposite direction. EEG activity of the observers was recorded from 47 electrodes, as well as vertical and horizontal electro-oculograms (Fig. 1) with sampling rate 250 Hz. Observations of correct and incorrect responses were averaged over a 800 ms epoch (baseline [-100 , 0] ms before response). Trials to be included in the averaging process had been selected according to an RT-matching procedure between correct and incorrect trials (described in [8]) to mitigate the differential contribution of stimulus-related activity in the ERP. A time window, starting at -6 msec and ending at 700 msec (corresponding to 176 samples) after the response, was selected for analysis. A total of 16×47 = 752 ERP recordings were available for analysis. From the available recordings, 8×47 = 376 recordings corresponded to observation of correct actions and the rest 8×47 = 376 recordings corresponded to observations of incorrect actions.

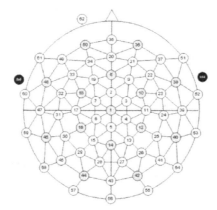

Fig. 1. Graphic representation of the electrode placement

2.2   Classification Methodology

The proposed methodology consists of three stages:

- Feature calculation
- Feature selection
- Classification

Each stage is described below.

Feature Calculation

Let $\mathbf{x} = [x_1, x_2, \ldots, x_{176}]$ be a vector with the samples of an ERP recording and $\{p_k\}$ ($k = 0,1,\ldots,M-1$) are estimates of the probability density function of the data vector $\mathbf{x}$ at points $\overline{c}_k$. Then, the following features of the probability density function can be calculated:

1. *Mean value*, which quantifies the central value of a distribution:

$$\mu = \sum_{k=0}^{M-1} p_k \overline{c}_k$$

2. *Standard deviation*, which is a measure of variability around the mean value:

$$\sigma = \sqrt{\sum_{k=0}^{M-1} (\overline{c}_k - \mu)^2 \, p_k}$$

3. *Skewness*, which characterizes the degree of asymmetry of a distribution around its mean:

$$skew = \frac{\sum_{k=0}^{M-1} (c_k - \mu)^3 \, p_k}{\sigma^3}$$

4. *Kurtosis*, which measures the relative peakedness or flatness of a distribution:

$$kurt = \frac{\sum_{k=0}^{M-1} (c_k - \mu)^4 \, p_k}{\sigma^4}$$

5. *Entropy*, which is a measure of uniformity of the histogram:

$$Entr = -\sum_{k=0}^{M-1} p_k \log_2 p_k$$

6. *Energy*: $Ener = \sum_{k=0}^{M-1} p_k^2$

7. *Median*, which is the number separating the higher half of a distribution, from the lower half :

$med = \overline{c}_{k_m}$, where k_m such that $\sum_{k=0}^{k_m} p_k \geq 0.5$ and $\sum_{k=k_m}^{M-1} p_k \geq 0.5$.

The estimates of the probability density function were calculated by means of the kernel density technique (Parzen window) [9], where the underlying distribution of the data is modeled by the mixture of Gaussian probability density functions.

Furthermore, the following features were calculated for each ERP recording:

8. Maximum value of samples: $\max_{n} \{y_n\}$

9. Minimum value of samples: $\min_{n} \{y_n\}$

10. Index of maximum value of samples: $\arg\max_{n} \{y_n\}$

11. Index of minimum value of samples: $\arg\min_{n}\{y_n\}$

In total, from each participant's ERPs, $47 \times 11 = 517$ features were calculated.

Feature Selection

Due to the high number of calculated features, it is necessary to eliminate features that are linearly correlated or carry no diagnostic information. Therefore, a process of feature selection is applied prior to classification, with the purpose of discovering the most significant features. In particular, the Wilcoxon test [10] was used as a criterion to assess the significance of every feature for separating the two classes (observation of correct actions, observation of incorrect actions). For every feature $j = 1, 2, \ldots, P$, ($P = 517$ in our case) the absolute value of the Wilcoxon test, Z_j, was calculated. Then, the most significant features were extracted using an iterative process as following:

Let $Y = \{1, 2, \ldots, P\}$ be the available features, D is the maximum number of features to be extracted ($D \leq P$) and X is the subset of Y with the selected features, then

- $X = \varnothing$ and $k = 0$
  - *Find the most significant feature:* $y = \arg\max_{i \in Y}\{Z_i\}$
  - *Add this feature in X and remove it from Y*
  - $k = k + 1$
- *while $k < D$*
  - *Calculate the mean value of cross-correlation, $\bar{\rho}$, of each feature in Y with all previously selected features in X*
  - *Multiply the Z score of each feature in Y by $1 - a \cdot \bar{\rho}$, to get a weighted Z score*
  - *Select the feature in Y with the highest weighted Z score*
  - *Add this feature in X and remove it from Y*
  - $k = k + 1$

The parameter a sets a weighting factor and assumes values between 0 and 1. When $a = 0$, potential features are not weighted. A large value of $\bar{\rho}$ (close to 1) outweighs the significance statistic; this means that features that are highly correlated with the features already picked are less likely to be included in the output list.

Classification

The classification task was performed by means of the Support Vector Machines (SVM) method, with radial basis function (RBF) kernel $k(\mathbf{x}, \mathbf{x}') = e^{-\gamma\|\mathbf{x}-\mathbf{x}'\|^2}$ ($\gamma > 0$) [11]. SVM is a category of kernel-based learning methods that have been successfully applied to various supervised classification problems, in various scientific fields.

The classifier is computed in a way that maximizes the margin between the training data and the decision boundary. The subsets of feature vectors that are closest to the decision boundary are called support vectors. Because training is based on the optimization of a convex cost function, there are no local minima where the training could be entrapped. Furthermore, the architecture of the learning machine that is constructed does not need to be found by experimentation and since the model constructed has an explicit dependence on the support vectors its interpretation is straightforward.

The classification accuracy was evaluated using the leave-one-out (LOO) cross-validation procedure [12]. The LOO procedure was adopted in order to evaluate the performance of the SVM classifier in a reliable manner, taking into account the limited number of cases available in the classes, and in the same time avoid over-training and achieving an acceptable generalization in the classification. According to this procedure, the SVM classifier was trained using feature vectors from observations of both types of actions (correct and incorrect), except from one observation (no matter whether it corresponded to a correct or incorrect actions), that was used for testing, afterwards. The generalization ability of the specific SVM classifier was then tested using the feature vector that was singled out. The above training-testing procedure was repeated, each time retaining a different feature vector for testing, until each feature vector was used once for testing. Using the LOO cross-validation procedure, the resulting SVM classifiers present slight differences between each other, by inference of the slight variation of the training and testing feature vectors sets used in each one. The classification accuracy was computed by the aggregate sums of correctly classified or misclassified observations of correct and incorrect actions.

3   Results

As was mentioned before, 517 features were calculated from each participant's ERPs. The probability density function for each ERP recording was estimated at $M = 100$ equally spaced points $\overline{c}_k$ ($k = 1, 2,...,M$). The feature selection algorithm was applied using the 16 available feature vectors. In order to determine the number of features to be selected (D) as well as the value of the weighting factor a, all the combinations (D,a) for $D = 1,2,...,10$ and $a = 0,0.1,0.2,...,1$ were investigated. For each combination, the SVM classifier with the LOO approach was applied.

The best classification accuracy was 100% and was obtained for $D = 4$ and $a = 0.2$. In Table I, the features and the corresponding electrodes that were finally selected are shown. It is also shown the mean value and the standard deviation in parenthesis of each feature for the two classes, namely observation of correct actions (Class 1) and observation of incorrect actions (Class 2).

Table 1. Extracted features and corresponding electrodes ranked in decreasing order with respect to their Z score

Feature	Electrode	Class 1	Class 2
Entropy	29	5.95 (0.11)	6.09 (0.09)
Standard Deviation	45	0.90 (0.20)	1.39 (0.49)
Entropy	39	5.97 (0.06)	6.08 (0.05)
Skewness	23	0.10 (0.30)	-0.51 (0.32)

The placement of the selected electrodes is shown in Fig. 2.

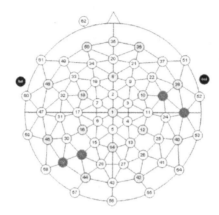

Fig. 2. Graphic representation of the electrodes that were finally selected

Considering the results that are listed in Table I, the following observations can be drawn:

- The entropies of electrodes 29 and 39 are in average slightly larger in Class 2 than in Class 1, which means that the corresponding histograms of Class 1 are slightly more uniform than these of Class 2.
- The standard deviation of electrode 45 is larger in Class 2 than in Class 1
- The skewness of electrode 23 in Class 1 (Class 2) is positive (negative), which in turn means that the corresponding histograms of ERPs a larger asymmetric tail towards positive (negative) values.

4   Discussion

As was mentioned in Section 3, the accuracy of the SVM classifier combined with the LOO technique was 100% and was obtained for $D = 4$ and $a = 0.2$, for the features shown in Table 1. However, there exist other combinations of a and D that provide equally satisfactory results. Fig. 3 shows a contour plot of the classification accuracy of the SVM classifier with respect to a and D. As can be seen, 100% classification rate can also be achieved for the following combinations:

- $D = 3$ and $a = 0.5, 0.6, 0.7, 0,8$ or 0.9
- $D = 4$ and $a = 0.2, 0.3, 0.4, 0,5$ or 0.5
- $D = 5$ and $a = 0.9$ or 1.0

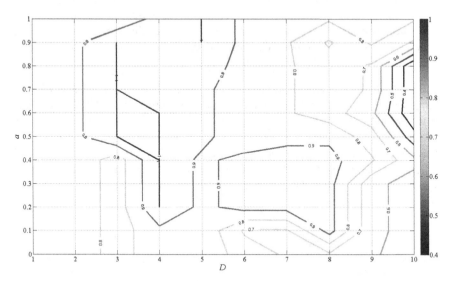

Fig. 3. Contour plot of classification accuracy for various combinations of (D,a)

An additional conclusion that can be drawn from Fig. 3 is that as the number of features increases, the classification accuracy deteriorates. This is probably due to the technique that was used for the feature selection. Specifically, the proposed technique is quite simple; it ranks each individual feature according to the Wilcoxon statistical test and selects the features with the highest scores that are not linearly correlated. It does not test combinations of features in order to find the optimum one with respect to some evaluation criterion. Thus, adding a new feature with a relatively high score does not necessarily mean improvement (or at least no change) of the classification accuracy.

It should be noted that although the results are quite promising, extensive trials on larger data sets are needed, in order to build a reliable classification system. Through such a system, indications could be provided in real time concerning whether or not an observer, who oversees the actions of another person, perceives the existence of errors in the actions of the supervised person. Apart from actions of humans, indications of critical parameters in display screens could also be the focus of observation, enabling the automatic evaluation of the performance of observers, which is very significant concerning the selection of suitable personnel, for example airplane pilots, or operators of critical installations. In applications in the more distant future, such classification systems might be able to detect errors in joint actions, either between two persons, or between a person and a robot.

5   Conclusions

In this paper, a methodology capable of discriminating between an subject's brain potentials that observe correct and incorrect actions was presented. The methodology consisted of two steps: the feature selection, which was based on statistical ranking, and the classification which was based on the SVM algorithm using a leave one out procedure. The proposed methodology reduced significantly the initial large number of features, providing satisfactory results.

Acknowledgment. The authors would like to thank Hein van Schie and Ellen de Bruijn from the Nijmegen Institute for Cognition and Information (NICI), The Netherlands, for kindly providing the data of their experiments and for their contribution to initial stages of the research.

This research has been co-funded by the European Union (European Social Fund) and Greek national resources under the framework of the "Archimedes III: Funding of Research Groups in TEI of Athens" project of the "Education & Lifelong Learning" Operational Programme.

References

1. Petrosini, L., Graziano, A., Mandolesi, L., Neri, P., Molinari, M., Leggio, M.G.: Watch how to do it! New advances in learning by observation. Brain Res. Rev. 42, 252–264 (2003)
2. van Schie, H., Mars, R.B., Coles, M.G.H., Bekkering, H.: Modulation of activity in medial frontal and motor cortices during error observation. Nature Neurosci. 7, 549–554 (2004)
3. Scheffers, M.K., Coles, M.G.: Performance monitoring in a confusing world: Error-related brain activity, judgments of response accuracy, and types of errors. J. Exp. Psychol. Human 26, 141–151 (2000)
4. Ventouras, E.M., Asvestas, P., Karanasiou, I., Matsopoulos, G.K.: Classification of Error-Related Negativity (ERN) and Positivity (Pe) potentials using kNN and Support Vector Machines. Comput. Biol. Med. 41, 98–109 (2011)
5. Sveinsson, J.R., Benediktsson, J.A., Stefansson, S.B., Davidsson, K.: Parallel principal component neural networks for classification of event-related potential waveforms. Med. Eng. Phys. 19, 15–20 (1997)
6. Palaniappan, R., Paramesran, R.: Using genetic algorithm to identify the discriminatory subset of multi-channel spectral bands for visual response. Appl. Soft. Comput. 2, 48–60 (2002)

7. Kalatzis, I., Piliouras, N., Ventouras, E., Papageorgiou, C.C., Rabavilas, A.D., Cavouras, D.: Design and Implementation of an SVM-based Computer Classification System for Discriminating Depressive Patients from Healthy Controls using the P600 Component of ERP Signals. Comput. Meth. Prog. Biol. 75, 11–22 (2004)
8. van Schie, H., Mars, R.B., Coles, M.G.H., Bekkering, H.: Modulation of activity in medial frontal and motor cortices during error observation. Nat. Neurosci. 7, 549–554 (2004)
9. Duda, R.O., Hart, P.E., Stork, D.G.: Pattern Classification. John Wiley & Sons, Inc., New York (2001)
10. Montgomery, D.C., Rumger, G.C.: Applied Statistics and Probability for Engineers. John Wiley & Sons, Inc., New Jersey (2003)
11. Steinwart, I., Christmann, A.: Support Vector Machines. Springer (2008)
12. Schenker, B., Agarwal, M.: Cross-validated structure selection for neural networks. Comput. Chem. Eng. 20, 175–186 (1996)

A Novel Hierarchical Approach
to Ranking-Based Collaborative Filtering

Athanasios N. Nikolakopoulos[1,2], Marianna Kouneli[1], and John Garofalakis[1,2]

[1] Computer Engineering and Informatics Department, University of Patras
[2] CTI and Press "Diophantus"
{nikolako,kounelim,garofala}@ceid.upatras.gr

Abstract. In this paper, we propose a novel recommendation method that exploits the intrinsic hierarchical structure of the item space to overcome known shortcomings of current collaborative filtering techniques. A number of experiments on the `MovieLens` dataset, suggest that our method alleviates the problems caused by the sparsity of the underlying space and the related limitations it imposes on the quality of recommendations. Our tests show that our approach outperforms other state-of-the-art recommending algorithms, having at the same time the advantage of being computationally attractive and easily implementable.

Keywords: Recommender Systems, Collaborative Filtering, Sparsity, Ranking Algorithms, Experiments.

1 Introduction

Collaborative Filtering (CF) is widely regarded as one of the most successful approaches to building Recommender Systems (RS). The great impact of CF on Web applications, and its wide deployment in important commercial environments, have led to the significant development of the theory over the past decade, with a wide variety of algorithms and methods being proposed [1]. In the majority of these algorithms the recommendation task, reduces to *predicting* the ratings for all the unseen user-item pairs, using the root mean square error (RMSE) between the predicted and actual ratings as the evaluation metric [1,2,3]. Recently, however, many leading researchers have pointed out that the use of RMSE criteria to evaluate RS is not an adequate performance index [4,5,6]; they showed that even sophisticated methods trained to perform well on MSE/RMSE, do not behave particularly well on the – much more common in practice – top-k recommendation task [5].

These observations have turned significant research attention to *ranking-based* recommendation methods which are believed to conform more naturally with how the recommender system will actually be used in practice [4,5]. Fouss et al. [7] follow a graph representation of the data and, using an approach based on random walks, they present a number of methods to compute node similarity measures, including the average commute time (normal CT and PCA-CT), and the pseudo-inverse of the Laplacian matrix ($\mathbf{L}^\dagger$), which they compare against

L. Iliadis, H. Papadopoulos, and C. Jayne (Eds.): EANN 2013, Part II, CCIS 384, pp. 50–59, 2013.
© Springer-Verlag Berlin Heidelberg 2013

standard approaches such as MaxF and Katz. Gori and Pucci [8] propose Item-Rank (IR); a PageRank-inspired scoring algorithm that produces a personalized ranking vector using an items' correlation graph. Zhang et al. [9] propose TR, an improvement of ItemRank based on topical PageRank algorithm that takes into account item genre and user interest profiles. Recently, Freno et al. [10] proposed a Hybrid Random Fields model (HRF) which they applied, together with a number of well-known probabilistic graphical models, including Dependency Networks (DN), Markov Random Fields (MRF) and Naive Bayes (NB), to predict top-N items for users.

Despite their success in many application settings, ranking-based CF techniques encounter a number of problems that remain to be resolved. The unprecedented growth of the number of users and listed items in modern e-commerce applications, make many current techniques suffer serious computational and scalability issues that restrain their applicability in realistic scenarios. Additionally, an even more important problem that limits the quality of recommendations arises when available data are insufficient for identifying similar elements and is commonly referred to as the *Sparsity* problem. Sparsity is intrinsic to recommender systems because users typically interact with only a small portion of the available items, and the problem is aggravated by the fact that new items with no ratings at all, are regularly added to the system.

In this work, based on the intuition behind a recently proposed Web ranking framework [11], we describe a new recommendation method that exploits the innately hierarchical nature of the underlying spaces to characterize inter-item relations in a macroscopic level. Central to our approach is the idea that blending together the *direct* as well as the *indirect* inter-item relations can help reduce the sensitivity to sparseness and improve the quality of recommendations. To this end, we develop *Hierarchical Itemspace Rank* (HIR); a novel recommender method that brings together the above components in a generic and mathematically attractive way. To verify the merits of our approach we run a number of experiments on the standard `MovieLens` dataset, which show that HIR outperforms other state-of-the-art recommending techniques in widely used metrics, proving at the same time to be less susceptible to sparsity related problems.

2 The HIR Framework

In this section, after describing formally the core components of our model (Section 2.1), we proceed to the rigorous mathematical definition of the involving matrices, and finally, we present the Hierarchical Itemspace Rank algorithm and discuss its computational and storage needs (Section 2.2).

2.1 Model Definition

Let $\mathcal{U} = \{u_1, u_2, \ldots, u_n\}$ be a set of *users* and $\mathcal{V} = \{v_1, v_2, \ldots, v_m\}$ a set of *items*. Let $\mathcal{R}$ be a set of tuples $t_{ij} = (u_i, v_j, r_{ij})$, where r_{ij} is a nonnegative number referred to as the *rating* given by user u_i to the item v_j. These ratings can either come from the explicit feedback of the user or inferred by the user's behavior and interaction with the system. We consider a partition $\{\mathcal{L}, \mathcal{T}\}$ of the ratings into a *training set* $\mathcal{L}$ and a *test set* $\mathcal{T}$. For each user u_i, we denote $\mathcal{L}_i$ the set of items rated by u_i in $\mathcal{L}$, and $\mathcal{T}_i$ the set of items rated by u_i in $\mathcal{T}$. Formally: $\mathcal{L}_i \triangleq \{v_k : t_{ik} \in \mathcal{L}\}$ and $\mathcal{T}_i \triangleq \{v_l : t_{il} \in \mathcal{T}\}$.

Each user u_i is associated with a vector $\boldsymbol{\omega}^i \triangleq [\omega_1^i, \omega_2^i, \ldots, \omega_m^i]$, whose nonzero elements contain the user's ratings that are included in the training set $\mathcal{L}$, normalized to sum to one. We refer to this as the *preference vector* of user u_i.

We consider a family of sets $\mathcal{D} = \{\mathcal{D}_1, \mathcal{D}_2, \ldots, \mathcal{D}_K\}$, defined over the underlying space, $\mathcal{V}$, according to a given criterion (e.g the categorization of movies into genres), such that $\mathcal{V} = \bigcup_{k=1}^{K} \mathcal{D}_k$, holds. We also define $\mathcal{G}_v$ to be the union of sets $\mathcal{D}_l$ that contain v and we use N_v to denote the number of different sets in $\mathcal{G}_v$. As we will see below, these sets will form the basis for the characterization of the indirect inter-item relations.

Having defined the parameters of our model, we are now ready to introduce matrices $\mathbf{C}$ and $\mathbf{D}$ that bring together the direct as well as the hierarchical structure of the underlying space, *in order to map each user's preference vector to a personalized distribution vector over the item space.*

Direct Association Matrix C: Matrix $\mathbf{C}$ is designed to capture the direct relations between the elements of $\mathcal{V}$. Generally, every such element will be associated with a discrete distribution $\mathbf{c}_v = [c_1, c_2, \cdots, c_m]$ over $\mathcal{V}$, that reflects the similarities between these elements. In our case (and for all the experiments presented in Section 3), we use the weighted mean of: (a) the correlation matrix [8] and (b) a row normalized version of the standard adjusted cosine similarity matrix [12]. These matrices are formally described below.

We first need to define a matrix $\mathbf{Q}$ whose ij^{th} element is $Q_{ij} \triangleq |\mathcal{U}_{ij}|$, where $\mathcal{U}_{ij} \subseteq \mathcal{U}$ denotes the set of users who rated both items v_i and v_j, i.e. $\mathcal{U}_{ij} \triangleq \{u_k : (v_i \in \mathcal{L}_k) \wedge (v_j \in \mathcal{L}_k)\}$ for $i \neq j$. Then, if we use $\hat{\mathbf{Q}}$ to denote the row normalized version of $\mathbf{Q}$, i.e. the matrix where every nonzero row sums up to 1, the resulting matrix will be defined as follows:

$$\mathbf{H} \triangleq \hat{\mathbf{Q}} + \frac{1}{m}\mathbf{a}\mathbf{e}^{\mathsf{T}} \tag{1}$$

where $\mathbf{a}$ is a vector that indicates the zero rows of matrix $\mathbf{Q}$ (i.e. its ith element is 1 if and only if $Q_{ij} = 0$ for every j) and $\mathbf{e}^{\mathsf{T}}$ is a properly sized unit vector. Thus, the final matrix $\mathbf{H}$ becomes a stochastic matrix. Similarly, the modified adjusted cosine similarity matrix is defined by:

$$\mathbf{S} \triangleq \hat{\mathbf{G}} + \frac{1}{m}\mathbf{a}'\mathbf{e}^{\mathsf{T}} \tag{2}$$

where matrix $\hat{\mathbf{G}}$ is the row normalized version of matrix $\mathbf{G}$, whose zero rows are indicated by $\mathbf{a}'$, and its ij^{th} element is formally defined by:

$$G_{ij} \triangleq 1 + \frac{\sum_{u_k \in \mathcal{U}}(r_{ki} - \bar{r}_{u_k})(r_{kj} - \bar{r}_{u_k})}{\sqrt{\sum_{u_k \in \mathcal{U}}(r_{ki} - \bar{r}_{u_k})^2}\sqrt{\sum_{u_k \in \mathcal{U}}(r_{kj} - \bar{r}_{u_k})^2}} \tag{3}$$

Here, $\bar{r}_{u_k}$ is the average of the ratings of user u_k, and r_{kj} is the rating assigned by user u_k to the item v_j (see [12]). The resulting direct proximity matrix is:

$$\mathbf{C} \triangleq \phi\mathbf{S} + (1 - \phi)\mathbf{H}$$

where $\phi < 1$ is a free parameter.

Hierarchical Proximity Matrix D: This matrix is created to depict the indirect connections between the elements of the item space that arise from its innate hierarchical structure. The existence of such connections is rooted in the idea that a user's rating, *except for expressing his direct opinion about a particular item, also gives a clue about his preferences regarding related elements of the item space.* For example, if Alice gives 5 stars to a specific comedy/romantic movie, except for testifying her opinion about that movie, also "hints" about her opinion regarding, firstly, comedy/romantic movies and, secondly, comedies and romantic movies in general. In the presence of sparsity the assistance of these indirect relations could be extremely helpful, as we sill see in Section 3.

Following this line of thought, we associate each row of matrix $\mathbf{D}$ with a probability vector $\mathbf{d}_v$, that distributes evenly its mass between the N_v different sets of $\mathcal{D}$, comprising $\mathcal{G}_v$, and then, uniformly to the included items of every such set. Formally, the ij^{th} element of matrix $\mathbf{D}$, that relates item v_i with item v_j, is defined as:

$$D_{ij} \triangleq \sum_{\mathcal{D}_k \in \mathcal{G}_{v_i}, v_j \in \mathcal{D}_k} \frac{1}{N_{v_i}|\mathcal{D}_k|} \tag{4}$$

Example 1. To clarify the definition of matrix $\mathbf{D}$, we give the following example:

	$\mathcal{D}_1$	$\mathcal{D}_2$	$\mathcal{D}_3$	N_v	$\mathcal{G}_v$
v_1	✓	–	–	1	$\{v_1, v_2, v_4, v_6\}$
v_2	✓	✓	–	2	$\{v_1, v_2, v_4, v_5, v_6\}$
v_3	–	–	✓	1	$\{v_3, v_6\}$
v_4	✓	✓	–	2	$\{v_1, v_2, v_4, v_5, v_6\}$
v_5	–	✓	–	1	$\{v_2, v_4, v_5\}$
v_6	✓	–	✓	2	$\{v_1, v_2, v_3, v_4, v_6\}$

The item space $\mathcal{V}$ consists of 6 movies, that belong to 3 Genres. In the last column we have computed the proximal sets. The corresponding matrix $\mathbf{D}$ is presented in Figure below:

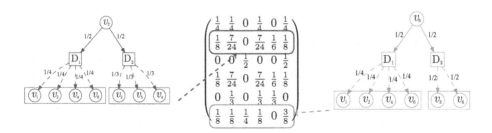

Fig. 1. We see the matrix $\mathbf{D}$ that corresponds to example 1, as well as a detailed computation of $\mathbf{d}_{v_2}$ and $\mathbf{d}_{v_6}$

2.2   The Hierarchical Itemspace Rank Algorithm

We are now ready to define the personalized ranking vector of HIR as the probability distribution over the itemspace produced by the algorithm presented below:

Algorithm 1. Hierarchical Itemspace Rank (HIR)

Input: Matrices $\mathbf{C}, \mathbf{D} \in \Re^{m \times m}$, parameters $\alpha, \beta > 0$ such that $\alpha + \beta < 1$, and the personalized preference vector $\boldsymbol{\omega} \in \Re^m$
Output: The ranking vector for the user, $\boldsymbol{\pi} \in \Re^m$

1: $\boldsymbol{\pi}^{\mathsf{T}} \leftarrow (1 - \alpha - \beta)\boldsymbol{\omega}^{\mathsf{T}}$
2: **for all** $\omega_j \neq 0$ **do**
3:     $\boldsymbol{\pi}^{\mathsf{T}} \leftarrow \boldsymbol{\pi}^{\mathsf{T}} + \omega_j(\alpha \mathbf{c}_j^{\mathsf{T}} + \beta \mathbf{d}_j^{\mathsf{T}})$ ▷ where $\mathbf{c}_j^{\mathsf{T}}, \mathbf{d}_j^{\mathsf{T}}$ the jth row of matrices $\mathbf{C}, \mathbf{D}$
4: **end for**
5: **return** $\boldsymbol{\pi}$

Theorem 1. *For every preference vector $\boldsymbol{\omega}$, the personalized vector $\boldsymbol{\pi}$ produced by the HIR algorithm denotes a distribution vector.*

Proof. Clearly, $\boldsymbol{\pi}$ is a non-negative vector. Thus, it suffices to show that $\boldsymbol{\pi}^{\mathsf{T}}\mathbf{e} = 1$.

$$\boldsymbol{\pi}^{\mathsf{T}}\mathbf{e} = \left((1 - \alpha - \beta)\boldsymbol{\omega}^{\mathsf{T}} + \alpha \sum_{j:\omega_j \neq 0} \omega_j \mathbf{c}_j^{\mathsf{T}} + \beta \sum_{j:\omega_j \neq 0} \omega_j \mathbf{d}_j^{\mathsf{T}} \right) \mathbf{e}$$

$$= (1 - \alpha - \beta)\boldsymbol{\omega}^{\mathsf{T}}\mathbf{e} + \alpha \sum_{j:\omega_j \neq 0} \omega_j \mathbf{c}_j^{\mathsf{T}}\mathbf{e} + \beta \sum_{j:\omega_j \neq 0} \omega_j \mathbf{d}_j^{\mathsf{T}}\mathbf{e}$$

But since the elements of the preference vector $\boldsymbol{\omega}^{\mathsf{T}}$ are by definition normalized to sum to 1, and matrices $\mathbf{C}$ and $\mathbf{D}$ are row stochastic, we get:

$$\boldsymbol{\pi}^{\mathsf{T}}\mathbf{e} = (1 - \alpha - \beta) + \alpha \sum_{j:\omega_j \neq 0} \omega_j + \beta \sum_{j:\omega_j \neq 0} \omega_j = (1 - \alpha - \beta) + \alpha + \beta = 1$$

and the proof is complete. □

Computational and Storage Needs. HIR needs to store the direct and the hierarchical proximity matrices. Matrix $\mathbf{C}$ is innately sparse and scales very well with the increase of the number of users; the addition of a new user to the system could result only in an increase of the number of nonzero elements of $\mathbf{C}$, since the dimension of the matrix depends solely on the cardinality of the item space which in most real applications increases slowly. In case of matrix $\mathbf{D}$, from the definition of the family of sets $\mathcal{D}$, it becomes intuitively obvious that whenever $K < m$, matrix $\mathbf{D}$ is a low-rank matrix. Furthermore, a closer look at the definition 4 above, suggests a very useful factorization of matrix $\mathbf{D}$, that can be exploited in order to achieve efficient storage. In particular, if we define an "aggregation" matrix $\mathbf{A} \in \mathfrak{R}^{m \times K}$, whose ik^{th} element is 1, when $v_i \in \mathcal{D}_k$ and zero otherwise, and letting $\mathbf{X}$ and $\mathbf{Y}$ denote the row-normalized versions of $\mathbf{A}$ and $\mathbf{A}^\mathsf{T}$ respectively, we observe that matrix $\mathbf{D}$ can be expressed as $\mathbf{D} = \mathbf{XY}$. Now, if we take into account the fact that for any reasonable decomposition of the item space, $K \ll m$ holds, the storage needs for the hierarchical proximity matrix become very small; we just have to store 2 sparse and thin matrices instead of the dense square matrix $\mathbf{D}$. This allows in realistic scenarios the introduction of more than one decompositions, according to different criteria, that could lead to better recommendations. From a computational point of view, we see that step 3 of our algorithm involves $O(|\mathcal{V}|)$ operations and it is executed $|\mathcal{L}_i|$ times. Typically, $|\mathcal{L}_i| \ll m$ since users interact with only a very small fraction of the available items, so the resulting burden is small.

3   Experimental Evaluation

To evaluate HIR, we apply it to the classic movie recommendation problem, employing a number of widely used performance indices. Our experiments were done using the publicly available `MovieLens` dataset.

MovieLens Dataset. The `MovieLens` dataset is a standard benchmark for recommender system techniques, containing 100,000 ratings from 943 users for 1,682 movies. Every user has rated 20 or more movies, in order to achieve a greater reliability for user profiling. Rating scores are integers between 1 and 5. The dataset comes with 5 predefined splittings, each with 80% of ratings as training set and 20% as test set (as described in [12]). `MovieLens` dataset also comes with information that relates the included movies to genres. For the experimental evaluation of our method, we use this as the simple criterion of decomposition behind the definition of matrix $\mathbf{D}$. Of course, in realistic situations, when more information about the data is available, there can be more than one decomposition of the underlying space according to different criteria.

To simplify the definition of the metrics presented below, we define $\overline{\mathcal{W}_i} \triangleq \overline{(\mathcal{L}_i \cup \mathcal{T}_i)}$, which denotes a set of movies that are neither in the training set nor in the test set of user u_i (i.e. these movies have not been watched by the user). Unless stated otherwise, the values of the free parameters used for the following experiments are: $\alpha = 0.8, \beta = 0.1, \phi = 0.7$.

3.1   Experiments

Degree of Agreement. The first performance index we used is the *degree of agreement* (DOA); a measure commonly used in the literature to evaluate the quality of ranking-based recommendation methods [7,10,8,9]. DOA is a variant of the Somers'D statistic, defined as follows:

$$\text{DOA}_i \triangleq \frac{\sum_{v_j \in \mathcal{T}_i \wedge v_k \in \overline{\mathcal{W}}_i} [\pi^i(v_j) > \pi^i(v_k)]}{|\mathcal{T}_i| * |\overline{\mathcal{W}}_i|}$$

where $\pi^i(v_j)$ is the ranking score of the movie v_j in user's u_i ranking list, and $[S]$ equals 1, if statement S is true and zero otherwise. Then, macro-averaged DOA (macro-DOA) is the average of all DOA_i and micro-averaged DOA (micro-DOA) is the ratio between the aggregate number of movie pairs in the correct order and the total number of movie pairs checked (for details, see [7,10]).

Table 1. Average performance and standard deviation between HIR and other state-of-the-art recommendation algorithms ([13,10,7,8,9]), computed over the same standard predefined folds. The metrics used for the comparison are the micro-DOA and macro-DOA.

micro-DOA		macro-DOA		
DN	81.33 ± 0.43	DN	80.51 ± 1.23	MRF	89.47 ± 0.44
HRF	88.07 ± 0.59	HRF	89.83 ± 0.52	NB	88.87 ± 0.22
IR	87.06 ± 0.10	IR	87.76 ± 0.27	**HIR**	**89.99 ± 0.20**
HIR	**88.85 ± 0.29**	Katz	85.83 ± 0.24	CT	84.09 ± 0.01
MRF	88.09 ± 0.50	$L^\dagger$	87.23 ± 0.84	PCA CT	84.04 ± 0.76
NB	86.66 ± 0.30	MaxF	84.07 ± 0.00	TR	89.08 ± 0.11

In Table 1, we present the micro-DOA and macro-DOA values measured by 5-fold cross-validation. The test employs the publicly available predefined partitioning of the dataset into five pairs of training and test sets, which allows easy comparisons with the different results to be found in the literature. All the DOA scores included in this table refer to the same predefined splitting. We see that HIR outperforms all other state-of-the-art techniques considered, by obtaining a macro-DOA value of **89.99** which is about 6.8% greater than the baseline (MaxF). The same is true for the micro-DOA measure, where HIR achieves an **88.85** value opposite to 88.09 of MRF and 88.07 of HRF having at the same time better standard deviation.

Sparsity. In order to demonstrate the merits of our method in dealing with the problems caused by the low density of the item space, we conduct the following experiment. For each of the 5 predefined splittings, we simulate the phenomenon of sparseness by randomly selecting to include 80%, 60%, and 40% of the ratings on a new artificially sparsified version of the dataset, which we then use to test the quality decay caused by sparseness. To isolate the positive effect of the exploitation of hierarchical proximity and the related matrix **D**, we run HIR against two algorithms related to its basic sub-components (namely the ItemRank and the SimRank[1] algorithms) and we evaluate their performance running the standard degree of agreement tests.

In Figure 2, we clearly see that HIR performs better than the other algorithms in both micro- and macro-DOA metrics, exhibiting very good results even when

---

[1] SimRank is a simple variant of ItemRank that correlates the items using the adjusted cosine-based similarity matrix defined by relation 2, instead of the correlation matrix defined by relation 1.

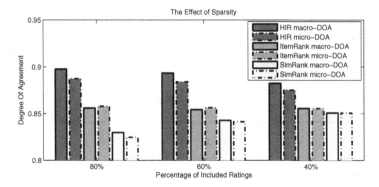

Fig. 2. The effect of sparsity on the quality of recommendations for artificially sparsified versions of the MovieLens dataset

only 40% of the ratings are available. These results verify the intuition behind HIR; even though the direct item-item relations of the dataset collapse with the exclusion of such many ratings, the indirect relations captured by matrix **D** decay harder and thus, preserve longer the coarser structure of the data. This results in a recommendation ranking vector that proves to be less sensitive to the sparsity of the underlying space.

Localized Sparsity. One very common and interesting manifestation of the sparsity problem is through the introduction of new items to the system. Naturally, because these items are new, they have been rated fewer times and thus, their relation with the rest of the elements of the item space is not yet clear. This, in many cases, could result in unfair treatment.

To evaluate the performance of our algorithm in coping with the newly added items bias problem, we run the following experiment. We randomly select a number of movies having 30 ratings or more, and we randomly delete 90% of their ratings. The idea is that the modified data represent an "earlier version" of the dataset, when these movies were new to the system, and as such, had fewer ratings.

We run several instances of HIR for different values of β, varying from $\beta = 0$ (where matrix **D** is not included and the ranking is induced by the weighted combination of the direct sub-components) to $\beta = 0.4$, keeping the sum $\alpha + \beta = 0.9$. Then, we compare the rankings induced on the modified data with their corresponding original rankings. The measure used for this comparison is Kendall's τ correlation coefficient. High value of this metric suggests that the two ranking lists are "close", which means that the newly added movies are more likely to receive treatment similar to their original one. Finally, to have a measure of the quality of the original list for each HIR instance, we also run the DOA tests and we present the results in Table 2.

Table 2. HIR performance in dealing with the localized sparsity problem

# New Movies	$\alpha = 0.9$ $\beta = 0$	$\alpha = 0.85$ $\beta = 0.05$	$\alpha = 0.8$ $\beta = 0.1$	$\alpha = 0.7$ $\beta = 0.2$	$\alpha = 0.6$ $\beta = 0.3$	$\alpha = 0.5$ $\beta = 0.4$
100	0.8736	0.8776	0.8812	0.8878	0.8945	0.9020
200	0.7814	0.7886	0.7949	0.8066	0.8186	0.8317
300	0.6843	0.6940	0.7025	0.7190	0.7369	0.7572
macro-DOA	89.63 ± 0.18	89.91 ± 0.20	$\mathbf{89.99 \pm 0.20}$	89.58 ± 0.22	88.45 ± 0.23	86.62 ± 0.25
micro-DOA	88.47 ± 0.28	88.75 ± 0.29	$\mathbf{88.85 \pm 0.29}$	88.50 ± 0.29	87.42 ± 0.29	85.63 ± 0.29

The experiment shows that the introduction of even a very small β induces a positive effect against localized sparsity. This is in accordance with the way HIR views the problem. In our method, the ranking score of the items is not exclusively determined by their ratings alone; their proximal sets also matter, because they "sketch" the relations of these newly added items to the other elements of the item space. Thus, even though they lack sufficient number of ratings, they are treated more fairly.

As expected, the value of Kendall's τ increases with β for all the parametric range tested. However, when the value of β becomes 0.2 or larger, the quality of the original recommendations begins to drop, as the direct inter-item relations get increasingly ignored. Intuitively, the proper selection of parameters is expected to vary with the sparsity of the underlying space. Our experiments with the MovieLens dataset suggest that a choice of β between 0.1 and 0.15 and a choice of α between 0.8 and 0.75, give very good results in both quality of recommendations and sparseness insensitivity.

4 Conclusions and Future Work

In this paper we proposed Hierarchical Itemspace Rank; a novel recommender method that exploits the innate hierarchical structure of the itemspace to provide an elegant and computationally efficient method for generating ranking based recommendations.

One very interesting research direction we are currently pursuing involves the effect of the granularity of the decompositions: intuitively, there seems to be a trade-off between the sparseness insensitivity, which is generally assisted by coarse grained decompositions, and the quality of recommendations, which seems to be supported by more detailed categorizations. Another interesting path that remains to be explored involves the introduction of more than one decompositions based on different criteria, and the effect it has to the various performance metrics. In this work we considered the single decomposition case. Our experiments suggest that HIR with the exploitation of the hierarchical proximity properties of the itemspace, produces recommendations of higher quality than the other state-of-the-art methods considered, and at the same time, helps alleviating commonly occurring sparsity related problems.

References

1. Desrosiers, C., Karypis, G.: A comprehensive survey of neighborhood-based recommendation methods. In: Recommender Systems Handbook, pp. 107–144. Springer (2011)
2. Koren, Y., Bell, R.: Advances in collaborative filtering. In: Recommender Systems Handbook, pp. 145–186. Springer (2011)
3. Koren, Y., Bell, R., Volinsky, C.: Matrix factorization techniques for recommender systems. Computer 42(8), 30–37 (2009)
4. Balakrishnan, S., Chopra, S.: Collaborative ranking. In: Proceedings of the Fifth ACM International Conference on Web Search and Data Mining, WSDM 2012, pp. 143–152. ACM, New York (2012)
5. Cremonesi, P., Koren, Y., Turrin, R.: Performance of recommender algorithms on top-n recommendation tasks. In: Proceedings of the Fourth ACM Conference on Recommender Systems, pp. 39–46. ACM (2010)
6. Marlin, B.M., Zemel, R.S.: Collaborative prediction and ranking with non-random missing data. In: Proceedings of the Third ACM Conference on Recommender Systems, pp. 5–12. ACM (2009)
7. Fouss, F., Pirotte, A., Renders, J., Saerens, M.: Random-walk computation of similarities between nodes of a graph with application to collaborative recommendation. IEEE Transactions on Knowledge and Data Engineering 19(3), 355–369 (2007)
8. Gori, M., Pucci, A.: Itemrank: a random-walk based scoring algorithm for recommender engines. In: Proceedings of the 20th International Joint Conference on Artifical Intelligence, IJCAI 2007, pp. 2766–2771. Morgan Kaufmann Publishers Inc., San Francisco (2007)
9. Zhang, L., Zhang, K., Li, C.: A topical pagerank based algorithm for recommender systems. In: Proceedings of the 31st Annual International ACM SIGIR Conference on Research and Development in Information Retrieval, SIGIR 2008. ACM, New York (2008)
10. Freno, A., Trentin, E., Gori, M.: Scalable pseudo-likelihood estimation in hybrid random fields. In: Proceedings of the 15th ACM SIGKDD International Conference on Knowledge Discovery and Data Mining, KDD 2009, pp. 319–328. ACM, New York (2009)
11. Nikolakopoulos, A.N., Garofalakis, J.D.: NCDawareRank: a novel ranking method that exploits the decomposable structure of the web. In: Proceedings of the Sixth ACM International Conference on Web Search and Data Mining, WSDM 2013, pp. 143–152. ACM, New York (2013)
12. Sarwar, B., Karypis, G., Konstan, J., Riedl, J.: Item-based collaborative filtering recommendation algorithms. In: Proceedings of the 10th International Conference on World Wide Web, WWW 2001, pp. 285–295. ACM, New York (2001)
13. Domingos, P., Pazzani, M.: On the optimality of the simple Bayesian classifier under zero-one loss. Mach. Learn. 29(2-3), 103–130 (1997)

Mimicking Real Users' Interactions on Web Videos through a Controlled Experiment

Antonia Spiridonidou, Ioannis Karydis, and Markos Avlonitis

Dept. of Informatics, Ionian University, Kerkyra 49100, Greece
{p10spyr,karydis,avlon}@ionio.gr

Abstract. The huge volume of available video content calls for methods that offer insight to the content without necessitating burdensome users' extra effort or being applicable to specific types or conditions. Preliminarily experimentation on collective users' interactions with the web video interface has been shown to offer such information to a great extend. This work, reports on the design, execution and results of a controlled user-experiment wherein participants are requested to view a video and identify their opinion on the importance of the scenes viewed in a realistic web-based video content viewing scenario, based on the interface of prominent video web-streaming provider. Initial results on the data collected show increased interaction on areas that were expected to attract attention whether related to the content or not.

Keywords: User-experiment, users' interactions, collective intelligence, stochastic patterns.

1 Introduction

Nowadays, video content consumption and creation is easier than ever. On one side, widespread penetration of fast and highly interactive internet allows to an ever increasing number of users enjoying video content while on the other hand affordable storage as well as high-quality capturing devices have made creating such content an ubiquitous process with unprecedentedly high demand. The most popular web streaming video content service, YouTube [8], serves more than 1 billion unique users per month, while storing 72 hours of video every minute [9]. Accordingly, being able to make sense of the available content in a computerised manner, that is, being able to extract new and interesting information that is otherwise very difficult to be done due to the sheer volume of data, is of paramount importance.

Traditional content-based methodologies for the aforementioned data mining processes examine the actual content of each video in order to extract information. Nevertheless, their performance and capabilities fall short in certain occasions and thus research has recently focused on contextual or user-based semantics[1]. Such semantics rely on a broad spectrum of interactive behavior

---

[1] For a detailed discussion about the complementary character of the two approaches see [1].

L. Iliadis, H. Papadopoulos, and C. Jayne (Eds.): EANN 2013, Part II, CCIS 384, pp. 60–69, 2013.

and "social activities" users exhibit and perform in relation to video content consumption such as sharing with others, assigning comments/tags, producing replies by means of other videos or even just expressing their preference/rating on the content. Rich as these "social metadata" may be, they have also been critiqued [1] as offering extra burden in the usual consumption process, that mainly includes viewing and browsing, by necessitating extra user effort, leading to the long-tail effect as to their existence. Thus, "social metadata" aside, research [3,5] has examined the interaction information during the core processes of video content consumption, i.e. during viewing and browsing.

The previously mentioned increased interactivity that Web 2.0 offered for the consumption of video content additionally assisted, through web-oriented architectures, in exposing content providers' functionality that other applications leverage and integrate in order to provide a set of much richer applications. In order to enhance the effectiveness of these applications, there is need for extensive studies of large users' interaction data. To this end, the design and implementation of controlled users' experiments has gained increasing attention. Indeed, controlled experiments provide sets of data of (almost) any desirable size under controlled conditions giving thus the possibility to study specific users' interactions properties for specific Web content. Accordingly, Gkonela and Chorianopoulos [3] utilising the SocialSkip platform [2,6] collected a pioneering user-based interaction dataset by conducting a controlled experiment during video content consumption providing a clean set of data that was easier to analyse. The platform integrated custom interface videos from YouTube with querying form functionalities of the Google Docs API in order to create an environment that would allow for video content consumption as well as user querying in order to accumulate data.

In this way, the collective behavior of Web users watching the video content emerged by means of characteristic patterns in their activity leading to *collective intelligence* as to the importance of video content solely from users' interactions with the video player.

1.1   Motivation and Contribution

The dataset introduced at [3] was based on the following assumptions:

- all viewing interface buttons were made similar to a typical VCR device in order to take advantage of the existing cognitive model most users have related to controls of VCR devices,
- in order to fill-in a questionnaire, users browsed the available video content searching for answers within a specific time limit,
- the significance of scenes was predefined based on the questions users' were asked.

To begin with, the typical video content consumption interface supported by key players in the area such as YouTube and Vimeo [7] is highly differentiated to the interface found in VCRs. Web streaming interfaces usually begin immediately without necessitating to press the play button, only include pause and play buttons while the main feature for browsing is the slider that allows arbitrary moving of the content's time as per the user's will backwards and forward. In addition, the querying process of the experimentation should be free of time limits in order to simulate the realistic usage scenario. Finally, in order to fully take advantage of the collective intelligence of users that may lead to previously unexpected results, experimentation should allow users to freely designate segments of the content thought of as important.

To address the requirements posed, the experiment proposed herein adheres to the following principles:

- the viewing interface includes the controls found in YouTube in order to simulate realistic user video content consumption,
- content viewing does not have any time limit, again in order to simulate realistic user video content consumption,
- the questionnaire requests free-text replies in order to ensure that users' declare the segments they thought of as most important without interference,
- the significance of segments is not predefined, allowing for true collective intelligence.

The rest of the paper is organised as follows: Section 2 presents the participants (Section 2.1), materials (Section 2.2) and the procedure (Section 2.3) of the methodology utilised in order to conduct the experiment. Next, Section 3 details the results received from the experiment conducted and the paper is concluded in Section 4.

2   Methodology

This Section details the experiment conducted under the principles discussed in Section 1.1.

2.1   Participants

Being delivered through web application, the experiment did not require the physical presence of the subjects and thus allowed users to undertake it at their own time and location of preference. The dissemination phase included informing the users of the experiment's URL link through emails as well as facebook messages.

Following the received link, users were requested to voluntarily participate in the experiment the duration of which would not exceed six and a half minutes to watch the video and then a few minutes answer the questions. Users were also given simple and concise instructions for both parts of the experiment through the web interface.

The participants that undertook the experiment were 103 in number, 36 of who were male and 67 female. The age range varied from 17 to 35 years old and all of them were interested in cookery and The Greek Guide Association. All participants had experience in using the internet and video streaming services such as YouTube.

2.2   Materials

To design such an experiment, the open-source SocialSkip platform was modified according to the principles discussed in Section 1.1. Natively, the platform intentionally supports YouTube videos in order to expand the content available for experimentation.

The chosen video for the experiment presented herein was required to be interesting to a lot of people so as to motivate to be viewed by a satisfying number of participants. Additionally, it was required to be interesting to the participants, again to motivate users' interactions on the areas they deemed important.

Thus a set of criteria were identified for the video content selection process. To begin with, an important criterion on the selection of the content was the duration of the video. According to the creators' initial programming of the Socialskip the video should not be more than 10 minutes as otherwise it would be boring and users will not watch it until the end. Another key criterion was that the video should not be structured as the less structured it is, the more important the postdata for the future viewers would be [2]. Accordingly, the video required should be without montage, that is, without replay, slow motion or pause from the director. Moreover, the video should not have areas of increased importance too close to one another. Accordingly, a video about "Hot Wine" [2] was selected that lasts 6 minutes and 14 seconds was captured by The Greek Guide Association in Komotini, Greece.

2.3   Procedure

Before viewing a video, participants were offered a set of instructions as to the process of the experiment, stating that:

1. You may view this video as many times required but after selecting to reply to the questions associated, video viewing will not be possible. Thus, you are advised to memorise the parts of the video that you found interesting in order to describe these later,
2. You can use the slider while watching the video in any way you want (back, forward, pause),
3. When sure you have memorised the parts you found interesting press the link to move to the questions phase.

After the participants had selected to move to the questions phase they were given the following instructions: *"Describe the scenes that you consider important in the video (indicative number of scenes A, B, C, D, E). Write your description after you have given the letter of the scene. (ex. A the scene where the cooking instructions are given)"*. Intuitively, the instructions were made in this way in order to gather from each user the most important scenes for them, while receiving unguided postdata.

---

[2] http://www.youtube.com/watch?v=XyZcS5GCF2k

Users' postdata were collected in web-documents supported by the Google docs [4] service. Every time a user initiated the experiment, a set of new records were created automatically in the web-documents, one per each interaction with the interface's buttons. This registration, as shown in Figure 1, has five fields: a unique id, the time-stamp the user began to view the content, the id of button interacted with, the time of the video in seconds the interaction lead to and the user's id as a number. In order to ensure anonymity of the participants, in both Figure 1 and the data user id's have been removed or replaced with random unique equivalent ids.

1-20 Next 20 ›

ID/Name	date	interaction	time	user
id=1	2012-02-29 09:40:11.268000	5	83	<null>
id=2	2012-02-29 09:40:11.258000	5	40	<null>
id=3	2012-02-29 09:40:11.258000	5	246	<null>
id=4	2012-02-29 09:40:11.258000	5	230	<null>
id=5	2012-02-29 09:40:11.268000	5	366	<null>
id=1001	2012-03-30 12:42:50.365000	1	333	<null>
id=1002	2012-03-30 12:42:50.365000	2	333	<null>
id=2001	2012-03-31 08:56:15.883000	5	0	<null>
id=2002	2012-03-31 08:56:15.883000	5	16	<null>
id=2003	2012-03-31 08:56:15.883000	5	60	<null>
id=2004	2012-03-31 08:56:15.883000	5	55	<null>
id=2005	2012-03-31 08:56:15.883000	5	61	<null>
id=2006	2012-03-31 08:56:15.883000	5	49	<null>
id=2007	2012-03-31 08:56:15.883000	5	45	<null>
id=2008	2012-03-31 08:56:15.883000	5	76	<null>
id=2009	2012-03-31 08:56:15.883000	6	147	<null>
id=2010	2012-03-31 08:56:15.883000	5	138	<null>
id=2011	2012-03-31 08:56:15.883000	5	131	<null>
id=2012	2012-03-31 08:56:15.883000	5	131	<null>
id=2013	2012-03-31 08:56:15.883000	5	159	<null>

Delete Flush Memcache 1-20 Next 20 ›

Fig. 1. Fields of the registrations of the users interactions

During the video content's presentation, the slider was available just under the viewing area allowing the user to randomly navigate through the content at will. A "pause" button was also available for the participants' convenience (Figure 2) that freezed the video on the current scene it was pressed. The existence of both these buttons was dictated by the requirement to ensure realistic and familiar usage to the users. Furthermore, under the viewing area, the participants could see the total duration of the video as well as the relative current viewing video time.

Finally, to conclude the procedure participants had to press the "Submit" and "Exit" buttons as otherwise their answers and interactions would not register (Figure 2).

3 Results

Preliminarily results revealed that the "pause" button was not used a lot since users had the option to browse the content using the slider, an interface very close to real scenarios. Similarly, the "play" button is was not used a lot as it was only required to restart the video after pausing.

After having studied the interactions, a common viewing pattern emerged: at the beginning every user wanted to move the video forward to see the end.

ΠΑΙΧΝΙΔΙ ΜΝΗΜΗΣ

1. Θα παρακολουθήσετε το επερχόμενο βίντεο μία και μοναδική φορά. Θα σας παρακαλούσαμε να απομνημονεύσετε σημεία ή περιοχές του βίντεο που κρίνετε ενδιαφέροντα έτσι ώστε να είστε σε θέση να τα περιγράψετε στη συνέχεια.
2. Για την απομνήμόνευση, κατά τη διάρκεια του βίντεο μπορείτε να χρησιμοποιήσετε την μπάρα με όποιον τρόπο θέλετε (μπροστά, πίσω, παύση).
3. Όταν θα είστε σίγουροι ότι έχετε απομνημονεύσει τα σημεία που κρίνετε ενδιαφέροντα πατήστε τον σύνδεσμο για να εμφανιστούν οι ερωτήσεις.

ΠΡΟΣΟΧΗ
Στην φόρμα των ερωτήσεων πατήστε SUBMIT / ΥΠΟΒΟΛΗ ώστε να καταχωρηθούν οι απαντήσεις σας ΚΑΙ πριν βγείτε απο το σύστημα SYBMIT AND EXIT !!!! Είναι 2 διαφορετικές επιλογές!

Fig. 2. Snapshot of socialskip

Then, the participant would use the slider to go back to the parts thought of as interesting so that memorising potential answers to the questions given later would be possible. The common viewing pattern was additionally supported by the fact that most users exhibited the same interactions. Moreover, most users's browsing approach included moving the video time back and forth as they were trying to identify a specific segment of the content. In any case most interactions were at the points of intuitive interest in the content. Figure 3 shows the cumulative users' interaction for the conducted experiment: the y-axis indicates the cumulative number of interactions while the x-axis shows the equivalent relative time of the content the interaction occurred. Three peaks (areas of interest) are easily distinguishable and while these are not even, it is obvious that some parts of the video prevail.

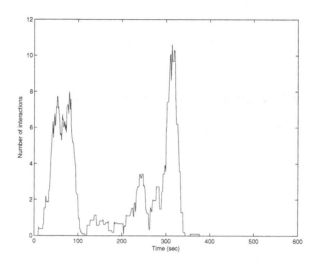

Fig. 3. The collected interaction signal

The first part, where several interactions occurred, was the part where the ingredients are given. In this scene the first secret for the recipe is revealed. In the following picture we can see a part of that scene. (Figure 4)

In this scene there is a detailed presentation of the ingredients which will be used to make the "hot wine". It is one of the most important scenes because when a video is about a recipe the part where the ingredients are presented is important. Moreover, the sequence the ingredients will is presented, also a important for the success of the recipe. Thus, according to the users' interactions it was apparent that this scene was one of the most important.

Continuing further, based on users' interactions as shown in Figure 3, the second part found important was that of a description of how to make an ingredient used in the recipe, Figure 5. In this scene the actors divulge how to make the spice which should be added in the recipe in a specific way to ensure successful blending with the rest ingredients and aroma of the recipe. Thus, it is one of the key parts of the content as well as of the recipe as it is crucial to its success. Furthermore, the actors prefaced the scene by announcing that "the second secret ..." thus pointing out the importance of the scene.

Fig. 4. The part where the ingredients are given

Fig. 5. The scene of how to make an ingredient used in the recipe

In order to test further the notion of what may be deemed as interesting in video content by collective intelligence, although the video is clearly about a recipe for making "hot wine", at the end of it, a scene not related to the recipe was intentionally included. In this scene, one of the two actors re-appears dressed as Santa Claus, as shown in Figure 6. The scene registered as the third more significant part in the content based on users' collective intelligence. Although a scene that is unrelated to the content of the video is expected not to receive special attention, the experimental results showed that most interactions had appeared in this part of the video, thus indicating that users thought of it as equally important. This result challenges common notions as to what should be thought of as important.

Fig. 6. The entrance of Santa Claus

4   Conclusion

In this work, a controlled user-experiment was conducted in order to identify the segments of video content participants thought of as most important by means of registering their interactions with the video interface. The experiment was designed and implemented in order to achieve high degree of realism to the typical contemporary scenario of web video-streaming services.

Initial results on the data collected showed high correlation of the areas (video time segments) that received more interactions from the users with their free-text submitted replies on what they thought of as important. In other words, the most important scenes according to the participants of the experiment, that form the so called "ground truth", highly coincide with the patterns emerged in users' interaction time signal.

The existence of a signal (here the signal counts how many times the slider was located at a specific second) that can carry the information about the most important video scenes could be a valuable tool for Wed applications. More-over, the existence of the aforementioned users' signal can provide the basis for new series of metrics in order to study new characteristics of users' interactions. Indeed, the identification of most important scenes is just a first order video characteristic. On a higher level, one could search for second order characteris-tics: what is the duration of each important scene, how popular each scene is (there could be more than one popular scenes in each video but what is their gradation popularity) and how one can predict the most important scenes from low quantity early data of users' interactions. The above issues can be addressed in a very rigorous manner since these features can be mapped to well known functions in standard signal processing theories. These aspects will be addressed in a future work.

References

1. Avlonitis, M., Karydis, I., Chorianopoulos, K., Sioutas, S.: Semantic Multimedia Analysis and Processing. In: Treating Collective Intelligence in Online Media. CRC Press / Taylor & Francis (2013)
2. Chorianopoulos, K., Leftheriotis, I., Gkonela, C.: Socialskip: pragmatic understanding within web video. In: Proceddings of the 9th International Interactive Conference on Interactive Television, EuroITV 2011, pp. 25–28 (2011)
3. Gkonela, C., Chorianopoulos, K.: VideoSkip: event detection in social web videos with an implicit user heuristic. In: Multimedia Tools and Applications, pp. 1–14 (2012)
4. Google: Google docs - online documents, spreadsheets, presentations (2013), https://docs.google.com/
5. Karydis, I., Avlonitis, M., Sioutas, S.: Collective intelligence in video user's activity. In: Iliadis, L., Maglogiannis, I., Papadopoulos, H., Karatzas, K., Sioutas, S. (eds.) AIAI 2012 Workshops, Part II. IFIP AICT, vol. 382, pp. 490–499. Springer, Heidelberg (2012)
6. SocialSkip: User-based video analytics (2013), https://code.google.com/p/socialskip/
7. Vimeo: Your videos belong here (2013), https://vimeo.com/
8. YouTube: Share your videos with friends, family, and the world (2013), http://www.youtube.com/
9. YouTube: Statistics (2013), http://www.youtube.com/t/press_statistics

Mining Student Learning Behavior and Self-assessment for Adaptive Learning Management System

Konstantina Moutafi[1], Paraskevi Vergeti[1], Christos Alexakos[1],
Christos Dimitrakopoulos[1], Konstantinos Giotopoulos[1],
Hera Antonopoulou[2], and Spiros Likothanassis[1]

[1] Pattern Recognition Lab., Dept. of Computer Engineering and Informatics,
University of Patras, Greece
{moutafi,vergetip,alexakos,dimitrakop,
kgiotop,likothan}@ceid.upatras.gr
[2] Technological Educational Institute of Patras, Greece
santonop@teipat.gr

Abstract. The specific contribution aims to provide a web-based adaptive Learning Management System (LMS), named EVMATHEIA, which integrates specific innovative fundamental aspects of Student Learning Style and Intelligent Self-Assessment Mechanisms. More specifically the proposed adaptive system encapsulates an integrated student model that facilitates the decision about the learning style of the student monitoring his/her behavior. Furthermore, the platform utilizes semantic modeling techniques for the representation of the knowledge, semantically annotated educational material and an intelligent mechanism for the self-assessment and recommendation process.

Keywords: learning management systems, adaptive e-learning, user modeling, personalized learning, learning styles, assessment.

1 Introduction

It is commonly known that e-learning environments are widely spread in all levels of human education. The specific aspect has imposed scientific research to enhance the efforts in the field of adaptive and intelligent learning platforms aiming to contribute significantly in the provision of high quality services towards the end users of e- learning systems. Over the last years scientific research aim to provide integrated systems that are intelligent and adaptive, and special attention has been given to specific key features of the learning style of the student and the self-assessment mechanisms.

The proposed EVMATHEIA platform aims to the provision of personalized learning adapted to student's receptivity. Its main target is to deliver knowledge, through a web platform, to individuals based on their interactions with the system, reducing the interference of the tutor and the collaboration with other students. System's key characteristic is the identification of student's learning style providing educational content aiming to a faster study and easier learning. Furthermore, it provides a mechanism detecting student knowledge's weakness through intelligent evaluation of self-assessment questionnaires and stimulates him/her to study specific additional material.

L. Iliadis, H. Papadopoulos, and C. Jayne (Eds.): EANN 2013, Part II, CCIS 384, pp. 70–79, 2013.
© Springer-Verlag Berlin Heidelberg 2013

In order to realize these aspects, EVMATHEIA encapsulates a *student model* storing information about student's preference, knowledge and learning style. The last denotes the way a student better perceives the provided knowledge. The student model is accompanied by a monitoring mechanism of the student behavior and decides his/her learning style. System's *knowledge* (subject of education) is represented on a semantic model setting as basic knowledge unit named the *concept*. The result is an ontological network of concepts that depicts the various relations between them (i.e. if a concept prerequisites another one).   Another key element is the semantically annotated *educational material* which contains the appropriate information for the personalized and recommendation mechanisms. Finally, system affords an *intelligent mechanism* which interacts with the student during the *self-assessment process*, deciding the level of the accumulated knowledge for each concept. The results are processed by a recommendation mechanism that takes into account the learning style and the education material providing suggestions to the student for further reading.

Section 2 presents a brief reference to the work done the past years on the relative fields of learning style and learner assessment in personalized e-learning systems. An overview of the proposed EVMATHEIA architecture is depicted in Section 3. Sections 4, 5 and 6 present the realization of student modeling, educational material personalization and student assessment in the system. Finally, some conclusion is given in Section 7. Work presented in this paper has been partially developed in the framework of the project LOC PRO II-Support and Promotion of Local Products and SMEs through ICT, Operational Programme Greece – Italy 2007-2013, s.c.: I1.12.01.

2   Learning Style and Student Assessment in LMS

The integration of learning styles in the adaptation process of Learning Management Systems (LMS) has become a major field of study the last years. Many models for learning styles are proposed and many techniques have been used to infer the learning style from the behavior of the student. One of the most widely used is the Felder-Silverman Learning Style Model (FSLSM) [1] which is proposed for engineering students. According to this model (which Felder revised in 2002) a student is classified according to his/her preference for one of the categories in each of the four learning style dimensions [2]: a) *Sensing* (concrete, practical, oriented toward facts and procedures) or *Intuitive* (conceptual, innovative, and oriented towards theories and meanings) depending on the type of information he/she prefers to perceive. b) *Visual* (prefer visual representations of presented material: pictures, diagrams, flow charts) *or Verbal* (prefer written and spoken explanations) depending on the way he/she prefers to receive information. c) *Active* (learn by trying things out, working with others) or *Reflective* (learn by thinking things through, working alone) depending on the way he/she prefers to process information. d) *Sequential* (linear, orderly, learn in small incremental steps) or *Global* (holistic, systems thinkers, learn in large leaps) learners (learn by thinking things through, working alone) depending on the way he/she prefers to process information. For the assessment of the preferences on the four dimensions of FSLSM, Felder and Soloman developed the Index of Learning

Styles (ILS), a questionnaire consisting of 44 questions, 11 for each learning style dimension [3]. According to the answers, student receives a score for each dimension: a) *balance* b) *moderate* or c) *strong* according to its preference for one of the two categories.

A crucial issue in automatic student modeling is to determine which student's behavior is indicative about his/her learning style. Graf et al [4] and García et al [5] are utilizing the FSLSM as their basic learning style model. Popescu at [6] proposes a combination of learning styles models. These approaches describe a great number of navigational, temporal and performance indicators identifying the learning style preferences according to FSLSM and propose thresholds that are necessary to classify the behavior of students. Furthermore, experimental studies were conducted investigating the behavior of students with different learning styles in online courses [7, 8].Useful conclusions derived from these studies contributed to the selection of patterns from the literature for online learning.

A key aspect of adaptive LMS is related with the intelligent self-assessment mechanism. Several efforts have been made towards this direction utilizing Computational Intelligence Techniques to support self-assessment in LMS [9]. The main idea behind this approach is to use Bayesian Networks and Genetic Algorithms simplifying the assessment by predicting student's answers [10, 11]. Several issues from the aforementioned efforts resulted in the present contribution. It is important also to note the efforts in personalized e-learning system with self-regulated learning assisted mechanisms to help learners promote their self-regulated learning abilities [12].

3   System Overview

Figure 1 depicts the main architectural components used for the realization of the main concepts in the EVMATHEIA approach. The core platform is a full-functional web-based application providing the main e-learning services to the users (students and tutors) such as registration, course structure, presentation and management.

Student model is a database segment that stores all the information needed from the system regarding student's characteristics and preferences. *Assessment questions* and *Educational Material* contain the questions related to each knowledge concept and the digital real educational material respectively. Two ontologies are used in order to define the knowledge and the educational material. *Knowledge ontology* contains the provided concepts, defining the student's knowledge and the relationship between educational material and the knowledge's concepts. *Educational Material Annotation Ontology* provides the annotation layer to the stored material.

Three supported modules execute the additional functionalities of the proposed system. *Student modeling module* collects information about student's behavior and updates student model. *Student assessment module* assists the student to the self-assessment process by providing the appropriate questions and evaluates the results. It also evaluates the answers and updates the student's accumulated knowledge. In addition, it forwards the results to Education Material Selection Module which by utilizing the information of *Educational Material Annotation Ontology* recommends additional material to the student. The third module is also responsible to provide student with the appropriate material according to his/her learning style.

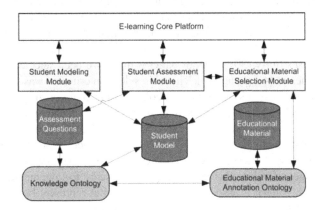

Fig. 1. Proposed system architecture

4   Student Model

Student model contains the information about the student that is needed by EVMATHEIA modules in order to deliver the desired level of personalization. The basic groups of information stored in the student model are: a) Student's *Personal Information* saves data such as name, surname, age, communication details, etc. b) *Language Knowledge* depicts student's level of knowledge (average, good, very good) of various languages. *c) Student's knowledge* is an overlay of system's knowledge which depicts the level of student's understanding on each concept. *d) Accessibility* contains special information regarding student's capability to access and study digital educational material. This group stores information such as visual, hearing, physical and cognitive disabilities. *e) Learning Style* includes student's score in every dimension of FSLSM.

4.1   Student Learning Style Modeling

The identification of student's learning style plays a key role in the EVMATHEIA system and as a result a modeling mechanism is developed combining the basic educational and psychological concepts around this issue. The modeling procedure is conducted in two phases: a) an initial approximation from student's answers in the classic ILS questionnaire and b) the continuous monitoring of student's behavior in the system and the re-calculation of student's learning style.

The indicative behavior patterns of the student's learning style preferences are based on the literature regarding the FSLSM and the features of our system. An online course in the system consists of sections and each section presents learning objects

(LOs) for a set of concepts. For each section educational material is provided that contains content in different types (text, video, sound, image, etc.), exercises and examples. At the end of each section the knowledge can be assessed through self-evaluation tests. Students navigate in the course through the course outline, the navigation menu and Next-Previous buttons. Thus the set of monitored behavior patterns consists of navigational, temporal and performance indicators correlated to the above features.

The system uses thirty (30) behavior patterns that have been selected after analyzing the approaches of [4], [5] and [6]. The main selection criterion was the monitoring feasibility of a pattern in a system without collaborating functionalities. Some examples of the behavior patterns are: percentage of time spent on examples, relative time spent by the student on content type text versus the relative study time for content type text and relative number of visits of content type video versus the total relative number of content type video available in the course. For each pattern, two thresholds define three ranges of values that conclude pattern's score to low, medium and high. Furthermore each pattern is associated with weights that represent how indicative is the respective behavior in the online course on identifying student's learning preference. These weights mainly derived from [7] and [8].

A set of N relevant patterns (P_{ij}) has been assigned in each dimension (D_j) of FSLSM. Since the two categories related to each dimension of FSLSM are opposed, if a high value of a pattern is associated with one category of a specific dimension, the low value of the same pattern is associated with the other category of the same dimension. Therefore calculations can be done only for one category in each dimension. Equation 1 defines the calculated score S_j that corresponds to the dimension D_j of FSLSM.

$$S_j = \frac{\sum_{1 \le i \le N} w_{ij} * p_{ij}}{\sum_{1 \le i \le N} w_{ij}} \tag{1}$$

The p_{ij} is the numerical value of i-th pattern P_{ij} of a particular category of the dimension D_j. The numerical values are (-) 1 for low, (-) 2 for medium and (-) 3 for high values. The positive values are if the pattern corresponds to the particular category and negative values for the opposite one. Pattern's weights w_{ij} are enumerated with 0.2, 0.5 and 1, indicating low, medium and high importance respectively. The calculated S_j is a number ranging from -3 to 3. According to absolute value of S_j, the student is classified for the pointed category as balanced ($0 \le S_j \le 1$), moderate ($1 < S_j \le 2$) or strong ($2 < S_j \le 3$) preference.

5   Knowledge and Educational Material

The proposed system utilizes semantic annotation defining knowledge and educational material, permitting their combination with the student model in order to infer the personalized presentation of the educational material to the student.

5.1   Knowledge Representation

An ontological approach is used for the representation of the knowledge structure, which is simply based on concepts with given relations between them. The ontology used derives from the education domain ontology proposed in [13], in which three kinds of relations are given: a) HasPart (an inclusion relation), b) IsRequiredBy (an order relation) and c) SuggestedOrder (a 'weak' order relation). These relations form a graph, where the nodes are the concepts and the edges are the relations. Figure 2 depicts an example of the knowledge semantic representation graph.

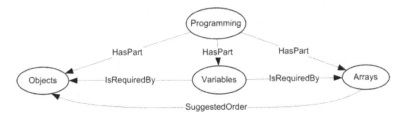

Fig. 2. Example portion of knowledge representation

5.2   Educational Material Annotation

EVMATHEIA educational material is defined/annotated using the OWL LOM ontology proposed by Hartonas C. [14] utilizing the IEEE Learning Object Metadata (IEEE LOM) [15] standard. LOM is the most common standard used for the description of learning material. The structure of LOM consistes of nine categories of the educational material: general, lifecycle, meta-metadata, technical, educational, rights, relation, annotation and classification. The schema used in the system is a subset of LOM, consisting of metadata mainly from the educational category. The metadata entities that support the adaptation mechanism are the following: language, format, typical age range, semantic density, interactivity type, typical learning time and associated concept. *Language* is a general characteristic referring to the language or languages that are used in the educational material. *Format* typically is the technical data type of the learning material, but here we are borrowing the set of values defined in [16]. These values are text, image, streaming media and application. *Typical age range* of the intended users is an educational characteristic to match the age of the learner. *Semantic density* of the educational material is the degree of conciseness and its value space is: very low, low, medium, high and very high. *Interactivity type* according to LOM takes values: active, expositive and mixed. The *typical learning time* of the material denotes the average time a student needs to study it. This characteristic has link to the active/reflective characteristic of the learning style of the student. Therefore, for a student with active learning style will be more accurate to propose an 'active' material, e.g. a questionnaire. *Associated concept* defines the relationship of the particular material to the relative concept in the knowledge ontology graph. Each material describes at least one concept.

5.3   Student Personalized Educational Material

The annotation of educational material in conjunction with student model permits the system to decide the best material for each student. The decision mechanism rates the available material based on a set of rules from the definition of FSLSM dimensions and the student's preferences. First the system chooses the relative, to a particular concept, educational material taking into account student's previous knowledge and his/her language. Then, for each retrieved educational material k the Sc_k score is calculated by the equation 2.

$$Sc_k = \frac{\sum_{1 \leq i \leq max\_associations} LS_i * ws_i}{\sum_{\substack{1 \leq i \leq max\_associations \\ LS_i \neq 0}} ws_i} + Ln_s + A_s \tag{2}$$

The LS_i values define the association of educational material characteristics to student's FSLSM dimensions score. LS_1 depicts the Active/Reflective dimension, the LS_2 the Sensing/Intuitive and the LS_3 the Visual/Verbal. The respective weights ws_i denote student's trend for a particular axis, i.e. strong verbal. Weights vales are 0, 0.5 and 1 for "well balanced", "moderate" and "very strong" score respectively.

For LS_1, value equal to 1 is assigned in case student's score to Active/Reflective dimension is moderate or above in the corresponding axis of material's *Interactivity type* value (active or expositive). In other case 0 is assigned. In case of the educational material's definition as the value 0.5 is assigned. For LS_2, value equal to 1 is assigned to this parameter when following combinations are true: a) student tends to sensing, material's interactivity type is active and its semantic density is low or very low and b) student tends to intuitive, material's interactivity type is expositive and it's semantic density is high or very high. Otherwise value equals to 0. The conditions are derived from the hypothesis that intuitive students prefer non active and of high semantic density educational material, whereas sensing learners prefer active and of low semantic density material. Finally, LS_3 is graded with 1 when the format of the educational material is corresponding to the student's learning style in the Visual/Verbal dimension. When a student is verbal then prefers text format. In the opposite case, when a student is visual prefers image, streaming media or application formatted material.

Ln_s parameter takes values of 0.5, 1 and 2 regarding student's level of knowledge (average, good, very good) in the material's language. The value of A_s is set to 0.5 in case student's age is contained to material's *typical age range* and to 0 otherwise. Finally, the educational material with the higher Sc_k is proposed to the student.

6   Student Assessment and Recommendation

The self-assessment mechanism aims to precisely identify student's acquired knowledge and to find the concepts that he/she has weakly learned. The proposed assessment procedure consists of a set of questions, related with one of the section's knowledge concepts, at the end of each section. The assessment questionnaire is created on the fly from a pool of questions for each concept. The questions are selected randomly, trying to avoid the repetition when a student conducts the questionnaire several times. The algorithm identifies the level of student's understanding in each concept. The idea is to provide him/her questions and predict the answer; if the answers are similar with the prediction then the algorithm concludes regarding how well the student understands a concept. In this concept the algorithm tries to be indifferent to student's random answers that are correct and to give a precise result, taking into account the correlations between concepts and the fact that the more "sincere" the student is, the faster he/she will complete the self-assessment.

6.1   Question Answer Prediction - Simple Majority Voting

For a particular student k, the algorithm sets the state of a question to +1 if the question has been answered correctly and to -1 if the question has been answered incorrectly. Next, the algorithm selects an unanswered question i by the student k repeatedly, and sets its state s_i to -1 or +1 according to the following rule:

$$S_i = sgn\left(\sum_{1 \leq j \leq n_i} S_j - \theta\right) \tag{3}$$

Where n_i is the number of questions answered from the student k either correctly or incorrectly and belong to the same concept that the unanswered question belongs, s_j is the state of brother j and θ is an activation threshold. The questions that belong to the same concept with question i and have been answered from the student k previously are considered as "brothers" of question i. The right hand side of this equation computes the sum of the states of the brothers of question *i* and sets its state to +1 if the sum is $> \theta$, and to -1 otherwise. In this implementation the activation threshold was chosen for each student and concept independently and was set at an integer value that yielded the best F-measurement score in cross-validation tests. Finally, the algorithm predicts the answer to a question to be "Correct" by the student k if the state computed was set to +1.

Furthermore, the algorithm incorporates the information of linked concepts in order to improve the prediction performance of the algorithm. For each unanswered question q_i, the state of a particular question q_j is set to +1, if the question is answered correctly by the student k and belongs to the same concept that q_i belongs **or** if the question is answered correctly by the student k and belongs to a concept that is linked with the concept that q_i belongs. Otherwise the state of the unanswered question q_i is set to -1 (in case the question is answered incorrectly and belongs to either the concept of q_i or to a concept that is linked to that of q_i). Next, the unanswered question q_i is assigned a state of +1 or -1 using the same rule as before. Now the brothers (n_i) of the unanswered question q_i for student k are chosen as the questions that belong to the same concept with q_i or to a concept linked to that of q_i, and have been answered from student k previously either correctly or incorrectly. Finally, the unanswered question q_i is predicted to be answered correctly by the student k if its state was set to +1. The activation threshold was optimized for every student/concept separately by using cross-validation. The activation rule in the case of the linked concepts is modified as follows:

$$S_i = sgn\left(w_n\left(\sum_{1 \leq j \leq n_i} S_j - \theta_n\right) + w_m\left(\sum_{1 \leq j \leq m_i} S_j - \theta_k\right)\right) \tag{4}$$

where n_i are the questions that belong to the same concept with q_i and m_i the questions that are assigned to a concept linked to that of q_i. We suppose that the concept of the question is more significant from the concepts linked to the question. As a result we define two weights $w_n=0.8$ and $w_m=0.2$ that represent the significance of the question's concept and of the linked concepts.

6.2   Self-assessment Result and Recommendation

The aforementioned algorithm is used in the EVMATHEIA system in order to ensure a valid result of student's assessment procedure. If the student has answered a considerable amount of questions then his/her answers are used to predict his/her answers to the remaining unanswered questions. If the remaining questions have been predicted to be answered correctly by the student k in a high degree then the assessment for the particular concept stops and the concept is considered as adapted by the student in a high (80%) or fundamental (100%) extend. If the remaining questions' prediction cannot lead to a certain result (same or near to same proportion of correctly and incorrectly answers predicted) or the remaining questions have been predicted to be answered incorrectly in a high degree, the system continues to assess the student with questions until a valid concept adaptation prediction is detected or the available questions are finished. If the questions finish without assimilation halt, then the student is assigned a below 80% knowledge extend.

Following this approach, the system predicts the level of knowledge for each assessed concept. After this, the system updates the student model using the procedure depicted in section 4.1 and the student's score for each knowledge concept. For the concepts ranked below 80% score, the system recommends material using the rules defined in section 5.3 and rejecting educational material that he/she has studied.

7   Conclusion

The proposed EVMATHEIA platform provides a new perspective in the provision of personalized learning adapted to student's receptivity. The main focus of the current work is to monitor the interaction of each individual student and reduce the interference of the tutor and the collaboration of the students. A self-managed learning process has been presented aiming to deliver knowledge in a personalized approach.

The key features that were exploited are the identification of the learning style of the student aiming to minimize the time needed for learning process. Furthermore, new and emerging concepts were presented in the specific field aiming to enhance the research in the specific filed and to open the path for future related work.

Last, but not least, it is important to emphasize in the intelligent assessment and recommendation mechanism that utilizes a new algorithm for the identification the level of student's understanding in each concept.

References

1. Felder, R.M., Silverman, L.K.: Learning and Teaching Styles in Engineering Education. Engineering Education 78(7), 674–681 (1988)
2. Felder, R.M., Spurlin, J.E.: Applications, Reliability, and Validity of the Index of Learning Styles. Intl. Journal of Engineering Education 21(1), 103–112 (2005)
3. Felder, R.M., Soloman, B.A.: Index of Learning Styles Questionnaire, North Carolina State University (2001), `http://www4.ncsu.edu/unity/lockers/users/f/felder/public/ILSpage.html` (accessed at April 2013)
4. Graf, S., Kinshuk, L.T.-C.: Supporting Teachers in Identifying Students' Learning Styles in Learning Management Systems: An Automatic Student Modelling Approach. Educational Technology & Society 12(4), 3–14 (2009)
5. García, P., Amandi, A., Schiaffino, S., Campo, M.: Evaluating Bayesian networks' precision for detecting students' learning styles. Computers & Education 49(3), 794–808 (2007)
6. Popescu, E.: Diagnosing Students' Learning Style in an Educational Hypermedia System. In: Mourlas, C., Tsianos, N., Germanakos, P. (eds.) Cognitive and Emotional Processes in Web-based Education: Integrating Human Factors and Personalization. Advances in Web-Based Learning Book Series, pp. 187–208. IGI Global (2009)
7. Graf, S., Kinshuk: Analysing the Behaviour of Students in Learning Management Systems. In: Wallace, M., Angelides, M.C., Mylonas, P. (eds.) Advances in Semantic Media Adaptation and Personalization. SCI, vol. 93, pp. 53–73. Springer, Heidelberg (2008)
8. Popescu, E.: Learning Styles and Behavioral Differences in Web-based Learning Settings. In: 9th IEEE International Conference on Advanced Learning Technologies, pp. 446–450. IEEE Press, New york (2009)
9. Giotopoulos, K., Alexakos, C., Beligiannis, G., Stefani, A.: Bringing AI to E-learning: the case of a modular, highly adaptive system. International Journal of Information and Communication Technology Education (IJICTE) 6(2), 24–35 (2010)
10. Mangalwede, S.R., Rao, D.H.: Application of Bayesian Networks for Learner Assessment. E-Learning Systems International Journal of Computer Applications 4(4), 23–28 (2010)
11. Dascălu, M.-I.: Application of Particle Swarm Optimization to Formative E-Assessment in Project Management. Informatica Economica 15(1/2011), 48–61 (2011)
12. Chen, C.-M., Huang, T.-C., Li, T.-H., Huang, C.M.: Personalized E-Learning System with Self-Regulated Learning Assisted Mechanisms for Promoting Learning Performance. In: 7th IEEE International Conference on Advanced Learning Technologies, pp. 637–638. IEEE Press, New York (2007)
13. Gaeta, M., Orciuoli, F., Ritrovato, P.: Advanced ontology management system for personalized e-Learning. Knowledge-Based Systems 22, 292–301 (2009)
14. Hartonas, C.: A Learning Object Metadata Ontology (2013), `http://wdok.cs.uni-magdeburg.de/Members/miotto/diplomarbeit/ontologien-und-beispiel/lom.owl/view` (accessed at April 2013)
15. Learning Technology Standards Comittee of the IEEE, Draft Standard for Learning Objects Metadata, IEEE Draft P1484.12.1/D6.412 (2002)
16. Nikolopoulos, G., Kalou, A., Pierrakeas, C., Kameas, A.: Creating a LO Metadata Profile for Distance Learning: An Ontological Approach. In: Dodero, J.M., Palomo-Duarte, M., Karampiperis, P. (eds.) MTSR 2012. CCIS, vol. 343, pp. 37–48. Springer, Heidelberg (2012)

Exploiting Fuzzy Expert Systems in Cardiology

Efrosini Sourla[1], Vasileios Syrimpeis[2], Konstantina-Maria Stamatopoulou[1],
Georgios Merekoulias[4], Athanasios Tsakalidis[1], and Giannis Tzimas[3]

[1] Computer Engineering & Informatics Department, University of Patras,
26500 Patras, Greece
[2] General Hospital of Patras "Agios Andreas", 26335 Patras, Greece
[3] Department of Applied Informatics in Management & Economy,
Faculty of Management and Economics, Technological Educational Institute of
Messolonghi, 30200 Messolonghi, Greece
[4] Public Health Department, Medical School, University of Patras,
26500 Patras, Greece
{sourla,stamatok,tsak}@ceid.upatras.gr,
{siva,gmerek}@upatras.gr, tzimas@teimes.gr

Abstract. This paper presents a Sugeno-type Fuzzy Expert System
(FES) designed to help mostly Cardiologists and General Practitioners
in taking decisions on the most common cardiological clinical dilemmas.
FES is separated in five sub-systems; Coronary Disease, Hypertension,
Atrial Fibrillation, Heart Failure and Diabetes, covering a wide range of
Cardiology. The Fuzzy Rules of the sub-systems start counting from 30
till 300. FES is verified and validated from three different Medical Doc-
tors User Groups (A, B & C) for three basic criteria, which are Medical
Reliability, Assistance in Work and Usability. In addition, FES proved
to be a valuable educative tool for Cardiology Medical Residents and
Medical Students.

Keywords: fuzzy logic, cardiology, cardiovascular disease, artificial
intelligence.

1 Introduction

Artificial intelligence (AI) in bioinformatics is considered to be a great step towards disease classification or even disease treatment, since it provides a variety
of tools available to be exploited, from rule-based expert systems and fuzzy
logic to neural networks and genetic algorithms. [4], [13], [18]. Artificial intelligence gives the opportunity through artificial neural networks (ANN) to process
medical information and classify patterns, something that can lead to disease
diagnosis or even treatment, since it is widely believed to have greater predictive
power than signal analysis techniques [7].

However, since there is always the factor of uncertainty in decision making,
especially as far as decision making in medical applications is concerned, fuzzy
logic is considered to be one of the most suitable approximations, since it deals
with reasoning that is approximate rather than fixed and exact, thus closer to

L. Iliadis, H. Papadopoulos, and C. Jayne (Eds.): EANN 2013, Part II, CCIS 384, pp. 80–89, 2013.
© Springer-Verlag Berlin Heidelberg 2013

human reasoning. Therefore, based on human expert knowledge, they are capable of modelling complex phenomena [8], [19].

This paper presents the fuzzy approach of modelling five diseases, using material acquired from expert knowledge: coronary disease, hypertension, atrial fibrillation, heart failure, diabetes. We implemented a Sugeno-type Fuzzy Expert System (FES) which is part of an ongoing project [16], accrued from the collaboration between the Computer Engineering and Informatics Department and School of Medicine of the University of Patras and the General Hospital of Patras "Agios Andreas".

The rest of the paper is organized as follows: Related Work is presented in Section 2. Section 3 presents a thorough description of the implemented FES, followed by System Evaluation in Section 4. Finally, Section 5 hosts Conclusions and Future Work.

2   Related Work

There is a great deal of research projects [9], [11], [12], [14] that reflect the integration of artificial intelligence in medicine. More specifically in [11], the models used in Decision Support Systems (DSS) are examined in combination with Genome Information and Biomarkers to produce personalized result for each individual. In [14], advanced biocomputing tools for cancer biomarker discovery and multiplexed nanoparticle probes for cancer biomarker profiling, in addition to the prospects for and challenges involved in correlating biomolecular signatures with clinical out- come are discussed. This bio-nano-info convergence holds great promise for molecular diagnosis and individualized therapy of cancer and other human diseases. In [9], a useful tool to translate the gene expression signatures into clinical practice for personalized medicine is proposed. More specifically, a new learning method that implements: (a) feature selection using the k-TSP algorithm and (b) classifier construction by local minimax kernel learning is devised in order to categorize patients as cancer or healthy and receive the appropriate treatment in each scenario. In [12], recent advances in the development of classification algorithms using micro array technology for prediction of anticancer sensitivity are reviewed, the availability of ensemble methods for prediction models is discussed and there is a presentation of data regarding the identification of potential responders to a certain medication therapy using random forests algorithms. In [7], an ANFIS model combining the neural network adaptive capabilities and the fuzzy logic qualitative approach is proposed in order to classify electroencephalogram signals through feature extraction, using the wavelet transform (WT) and the ANFIS trained with the backpropagation gradient descent method in combination with the least squares method in order to evaluate the patient's health condition. All of the above projects are aiming in either disease diagnosis or even prediction of the most suitable treatment in each case. Most of those applications are addressed to medical stuff mainly for educational reasons or to medical doctors as simply advisory.

As far as cardiology is concerned, AI has become a promising method in the diagnosis of heart diseases. AI, out of invasive and noninvasive diagnostic tools, becomes the promising method in the diagnosis of heart diseases. In [1], a comparison is presented of multilayered perceptron neural network (MLPNN) and support vector machine (SVM) on determination of coronary artery disease (CAD) existence upon exercise stress testing (EST) data. In [15], neural networks are used as the most suitable solution to outcome prediction tasks in postoperative cardiac patients. An AI-based Computer Aided Diagnosis system is designed in [6] to assist the clinical decision of nonspecialist staff in the analysis of heart failure patients. The system computes the patient's pathological condition and highlights possible aggravations, using four AI-based techniques: a Neural Network, a support vector machine, a decision tree and a fuzzy expert system whose rules are produced by a Genetic Algorithm. Neural networks achieved the best performance with an accuracy of 86%. Another application domain for AI is nuclear cardiology imaging, since the automatic interpretation of nuclear cardiology studies is a complex and difficult task, and a variety of expert systems, neural networks, and case-based reasoning approaches have been attempted in this area [17]. In [3], [5], personal health care systems are proposed which allow any patient to self record ECGs with some smart portable device that analyses the signal inputs and through a set of rules and patient's risk factors can estimate the severity of the condition of life threatening threat episodes. In the aforementioned projects extra hardware is necessary in order to produce the ECGs and export the final results. In [8], a Sugeno-type fuzzy inference system (FIS) that predicts the effect of regular aerobic exercise on blood pressure (BP), based on the exercise dose variables, exercise frequency and intensity, as well as demographics (age, gender, ethnicity) and the baseline BP of a person, is described. Since BP response to exercise varies largely between individuals, the system takes an initial step towards personalized prediction.

3   Description of the Fuzzy Expert System

3.1   Introduction to Fuzzy Logic

Fuzzy logic idea is similar to the human being's feeling and inference process. Unlike classical control strategy, which is a point-to-point control, fuzzy logic control is a range-to-point or range-to-range control. The output of a fuzzy controller is derived from fuzzifications of both inputs and outputs using the associated membership functions. A crisp input will be converted to the different members of the associated membership functions based on its value. From this point of view, the output of a fuzzy logic controller is based on its memberships of the different membership functions, which can be considered as a range of inputs [2].

To implement fuzzy logic technique to a real application requires the following three steps: (a) *Fuzzification*: convert classical data or crisp data into fuzzy data or Membership Functions (MFs), (b) *Fuzzy Inference Process*: combine membership functions with the control rules to derive the fuzzy output, and (c)

Defuzzification: convert the fuzzy output back to the crisp or classical output to the control objective [2].

All machines can process crisp or classical data such as either '0' or '1'. In order to enable machines to handle vague language input, the crisp input and output must be converted to linguistic variables with fuzzy components. Generally, *Fuzzification* involves two processes: derive the membership functions for input and output variables and represent them with linguistic variables. In practice, membership functions can have multiple different types, such as the triangular waveform, trapezoidal waveform, Gaussian waveform, etc [2]. In our Fuzzy Expert System we use triangular membership functions, as we need significant dynamic variation in a short period of time. In the second step, to begin the *Fuzzy Inference Process*, one needs to combine the Membership Functions with the control rules to derive the control output. The control rule is the core of the fuzzy inference process, and those rules are directly related to a human being's intuition or expertise [2]. In our system, it is up to the clinical doctor to decide the severity of every risk factor, according to his/her experience and knowledge, thus making the system adjustable. The conclusion or control output derived from the combination of input and output membership functions and fuzzy rules is still a fuzzy element, and this process in called fuzzy inference. To make that fuzzy output available to real applications, a defuzzification process is needed. The fuzzy conclusion or output is still a linguistic variable, and this linguistic variable needs to be converted to the crisp variable via the *Defuzzification* process [2]. In all our sub-systems we use the weighted average defuzzification method. In Fig. 1 we can see an example of a membership function from our system.

Our Fuzzy Expert System consists of five sub-systems: (a) Coronary Disease, (b) Hypertension, (c) Atrial Fibrillation, (d) Heart Failure, (e) Diabetes. The first four sub-systems concern exclusively heart diseases, while the fifth sub-system concerns a systematic disease, very common in the population that strongly affects the functionality of the heart and the circulatory system (vessels and arteries).

3.2   The Coronary Disease Sub-system

The Coronary Disease sub-system (CDss) is designed to help MDs in answering one of the most common clinical questions in cardiology, that is the medical approach and guidance of a patient with a possible coronary artery disease. The clinical decision (*Outcome*) is separated in five discrete values: (1) No Further Evaluation, (2) Re-Evaluation After 3 Months, (3) Magnetic Tomography (MRI), (4) Computerized Tomography (CT) and (5) PCI.

The medical criteria that need to be taken under consideration in such a dilemma are too many with different importance indexes and present fuzziness. A knowledge acquisition procedure resulted in the grouping of the most severe criteria in larger medical entities that are reflected in the input variables of the CDss: (a) Family History, (b) Risk Factors, (c) Laboratory Findings, (d) Myocarditis Propability and (e) Other Reasons.

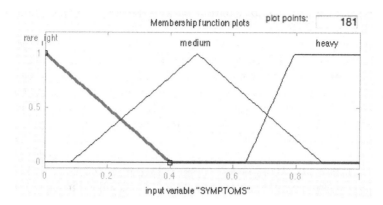

Fig. 1. An example of a triangular membership function used in our system

The severity of the above factors is fuzzified into three membership functions **min, med, max**. The rules constructed for CDss are close to 300 and present the below form:

IF *Family History* IS **min** AND *Risk Factors* IS **med** AND *Laboratory Findings* IS **med** AND *Myocarditis Propability* IS **min** AND *Other Reasons* IS **min** THEN *Outcome* IS **MRI**.

3.3   The Hypertension Sub-system

The Hypertension sub-system (HPTss) is designed to help MDs in answering the dilemma of when a patient should start or not taking medication for treating hypertension. The clinical decision (*Outcome*) is separated in five discrete values: (1) No Further Evaluation, (2) Life Style Changes & Re-Evaluation After 3 Months, (3) Life Style Changes & Re-Evaluation After 12 Months, (4) Evaluation For 2 Consecutive Weeks and (5) Drug Therapy.

The knowledge acquisition procedure resulted in the grouping of the most severe criteria in larger medical entities that are reflected in the input variables of the HPTss: (a) Risk Factors, (b) Laboratory Findings and (c) Echo.

The severity of the above factors is fuzzified into three membership functions **min, med, max**. The rules constructed for HPTss are close to 30 and present the below form:

IF *Risk Factors* IS **med** AND *Laboratory Findings* IS **med** AND *Echo* IS **min** THEN *Outcome* IS **Evaluation For 2 Consecutive Weeks**.

3.4   The Atrial Fibrillation Sub-system

The Atrial Fibrillation sub-system (AFss) is designed to help MDs in answering one of the most common clinical dilemmas that is how to approximate a patient with a possible atrial fibrillation. The clinical decision (*Outcome*) is separated in nine discrete values: (1) Regular Assessment ECG, (2) Holter, (3) Echo, (4)

Cardioversion-Ablation, (5) Treatment Underlying Disease, (6) Rhythm Control, (7) Rate Control, (8) Aspirin and (9) Oral Anticoagulant.

The knowledge acquisition procedure resulted in the grouping of the most severe criteria in larger medical entities that are reflected in the input variables of the AFss: (a) CHA2DS2VASC Score, (b) Family History, (c) Symptoms, (d) Risk Factors and (e) Documentation of Arrhythmia.

The severity of the above factors is fuzzified into three membership functions **min, med, max**. The rules constructed for AFss are close to 200 and present the below form:

IF *CHA2DS2VASC Score* IS **med** AND *Family History* IS **med** AND *Symptoms* IS **min** AND *Risk Factors* IS **min** AND *Documentation of Arrythmia* IS **med** THEN *Outcome* IS **Rhythm Control.**

3.5   The Heart Failure Sub-system

The Heart Failure sub-system (HFss) is designed to help MDs in diagnosing and classifying heart failure on a patient. The clinical decision (*Outcome*) is separated in five discrete values: (1) No Heart Failure, (2) Heart Failure Class I (NYHA), (3) Heart Failure Class II (NYHA), (4) Heart Failure Class III (NYHA) and (5) Heart Failure Class IV (NYHA).

The knowledge acquisition procedure resulted in the grouping of the most severe criteria in larger medical entities that are reflected in the input variables of the HFss: (a) Family History, (b) Risk Factors, (c) Signs & Symptoms and (d) Echo.

The severity of the above factors is fuzzified into three membership functions **min, med, max**. The rules constructed for HFss are close to 80 and present the below form:

IF *Family History* IS **min** AND *Risk Factors* IS **med** AND *Signs & Symptoms* IS **med** AND *Echo* IS **min** THEN *Outcome* IS **Heart Failure Class III (NYHA)**.

3.6   The Diabetes Sub-system

The Diabetes sub-system (DBss) is designed to help MDs in approaching patients with possible Diabetes. The clinical decision (*Outcome*) is separated in four discrete values: (1) No Further Evaluation, (2) Periodical Fasting Glucose, (3) Fasting Glucose And 2 Hours Postprandial Glucose Every 2 Years and (4) Fasting Glucose And 2 Hours Postprandial Glucose Every Year And HgbA1C.

The knowledge acquisition procedure resulted in the grouping of the most severe criteria in larger medical entities that are reflected in the input variables of the DBss: (a) Family History, (b) Risk Factors, (c) Laboratory Findings and (d) Cardiovascular Problem.

The severity of the above factors is fuzzified into three membership functions **min, med, max**. The rules constructed for HFss are close to 80 and present the below form:

IF *Family History* IS **min** AND *Risk Factors* IS **med** AND *Laboratory Findings* IS **med** AND *Cardiovascular Problem* IS **min** THEN *Outcome* IS **Fasting Glucose And 2 Hours Postprandial Glucose Every 2 Years**.

3.7   Choosing the Proper Fuzzy Method

There are two types of fuzzy methods: Mamdani and Sugeno. Mamdani method is widely accepted for capturing expert knowledge. It allows us to describe the expertise in more intuitive, more human-like manner. However, Mamdani-type Fuzzy Interference System (FIS) entails a substantial computational burden. On the other hand, Sugeno method is computationally efficient and works well with optimization and adaptive techniques, which makes it very attractive in control problems, particularly for dynamic non linear systems. These adaptive techniques can be used to customize the membership functions so that fuzzy system best models the data [20].

The most fundamental difference between Mamdani-type FIS and Sugeno-type FIS is the way the crisp output is generated from the fuzzy inputs. While Mamdani-type FIS uses the technique of defuzzification of a fuzzy output, Sugeno-type FIS uses weighted average to compute the crisp output. The expressive power and interpretability of Mamdani output is lost in the Sugeno FIS, since the consequents of the rules are not fuzzy. But Sugeno has better processing time since the weighted average replace the time consuming defuzzification process. Due to the interpretable and intuitive nature of the rule base, Mamdani-type FIS is widely used in particular for decision support application [10].

In this work, we implemented a Sugeno-type fuzzy system as it is a more compact and computationally efficient representation than a Mamdani and lends itself to the use of adaptive techniques for constructing fuzzy models.

4   System Evaluation

The five expert systems were tested, verified and evaluated by three different groups of MDs. The first group, group A, was consisted of Cardiologists that did not participate at all in the design of the FES and present the more strict group. Group B was made up from General Practitioners (GPs) and group C was composed from medical students. All three groups were called to grade in a scale from 1 to 10 the FES for the criteria of Medical Reliability, Assistance in Work & Usability, see Table 1.

Medical Reliability criterion was graded with 7/10 from Group A and Group B which is an acceptable threshold for a first version of an expert system. Although the FES presented an 80%-90% of correct proposals in the testing phase, Cardiologists and GPs did not acknowledge it. Group C graded Medical Reliability with 8/10, higher than the previous groups, maybe because it was made up from medical students that were less experienced on the field.

Assistance in Work criterion was graded with a 5/10 from Group A, which is a low grade that is justified because it is given from already experts on the field.

Table 1. Results from Groups A, B & C for Medical Reliability, Assistance in Work and Usability

	Medical Reliability	Assistance in Work	Usability
Group A	7	5	5
Group B	7	7	6
Group C	8	8	8

On the other hand, less experts on the field such as GPs, Group B, and medical students, Group C, find the FES much more helpful given higher grades such as 7/10 and 8/10 respectively.

Usability criterion was graded with a 5/10 and 6/10 from Group A and Group B respectively while a higher grade, 8/10, was given by Group C. The lack of a user- friendly interface in combination with the higher average age of Group A and Group B users are the main reasons for the low grades. On the other hand, Group C made up of younger users more patient in adopting new technologies than the elderly, did not seem to find major difficulties in using the Fuzzy Logic Toolbox.

In addition, Cardiologists that participated in the design of the FES installed Matlab and Fuzzy Logic Toolbox in personal PCs that use in their every day clinical practice in the General Hospital of Patras (GHP). Their comments were contradictory. *Most of them were not satisfied with the interface since they were not familiarized with the forms and the way of "running" the *.fis files.* Apart from that, some did not seem to want to use them, because they thought that since their experience was imported into the FES, there were no added values to use. On the other hand, an unexpected positive result from this effort was that there were cases where the physicians although they had taken their decisions upon a clinical problem, they tried to check what was going to be the FESs decision. In an 83% of the cases, the FES proposed the same decisions with the physicians. In a 5% there were cases where the physicians disagreed at first with the FES's decision, but given a second thought they seemed to re-think their decision and finally agreed with the decision of the FES. This phenomenon is mostly observed were situations of stress or other emotional situations are interfered in a physicians decision.

5   Conclusions and Future Work

Clinical decisions in medicine are always based on the experience and capabilities of the specific attending physician. Although medical guidelines are continuously updated and wide known, most of the times they allow to the MDs a variety of different clinical choices. Several times guidelines themselves expressing the clinical experience of the Medical Institution that are produced from, are characterized as "offensive" or "conservative" highlighting the different pathways physicians can follow when they have to take decisions upon a specific clinical problem. The five sub-systems implemented in this paper, is an effort to import the most commonly used guidelines for five very common clinical problems, in expert systems that try to follow the way physicians think and work.

Further customization of the fuzzy system is strongly suggested adopting more AI technologies. Neuro-fuzzy systems are fuzzy systems, which use ANNs theory in order to determine their properties (fuzzy sets and fuzzy rules) by processing data samples. Neuro-fuzzy systems harness the power of the two paradigms: fuzzy logic and ANNs, by utilizing the mathematical properties of ANNs in tuning rule-based fuzzy systems that approximate the way humans process information. A specific approach in neuro-fuzzy development is the adaptive neuro-fuzzy inference system (ANFIS), which has shown significant results in modeling nonlinear functions. In ANFIS, the membership function parameters are extracted from a data set that describes the system behavior. The ANFIS learns features in the data set and adjusts the system parameters according to a given error criterion [7].

The future work concerns the incorporation of neural networks to our FES for automated optimization instead of the so far used "trial" and "error" method. Different data sets reflecting more "offensive" or "conservative" approaches are to be collected for the tuning of the FES presented in this paper. Moreover, a "doctor- friendly" interface will be implemented for augmenting the user-friendliness and usability of the system.

Acknowledgements. This research has been co-financed by the European Union (European Social Fund - (ESF)) and Greek national funds through the Operational Program "Education and Lifelong Learning" of the National Strategic Reference Framework (NSRF) - Research Funding Program: Heracleitus II. Investing in knowledge society through the European Social Fund.

References

1. Babaoğlu, İ., Baykan, Ö.K., Aygül, N., Özdemir, K., Bayrak, M.: A comparison of Artificial Intelligence Methods on Determining Coronary Artery Disease. In: Papasratorn, B., Lavangnananda, K., Chutimaskul, W., Vanijja, V. (eds.) IAIT 2010. CCIS, vol. 114, pp. 18–26. Springer, Heidelberg (2010)
2. Bai, Y., Wang, D.: Fundamentals of Fuzzy Logic Control - Fuzzy Sets, Fuzzy Rules and Defuzzifications. In: Advanced Fuzzy Logic Technologies in Industrial Applications, pp. 17–36. Springer (2006)
3. Bellos, C., Papadopoulos, A., Fotiadis, D.I., Rosso, R.: An Intelligent System for Classification of Patients Suffering from Chronic Diseases. In: 32nd Annual International Conference of the IEEE EMBS, Buenos Aires, Argentina (2010)
4. Chu, A., Ahn, H., Halwan, B., et al.: A Decision Support System to Facilitate Management of Patients with Acute Gastrointestinal Bleeding. Artificial Intelligence in Medicine 42(3), 247–259 (2008)
5. Fayn, J., Rubel, P.: Toward a Personal Health Society in Cardiology. IEEE Transactions on Information Technology in Biomedicine 14(2) (2010)

6. Guidi, G., Iadanza, E., Pettenati, M.C., Milli, M., Pavone, F., Biffi Gentili, G.: Heart Failure Artificial Intelligence-based Computer Aided Diagnosis Telecare System. In: Donnelly, M., Paggetti, C., Nugent, C., Mokhtari, M. (eds.) ICOST 2012. LNCS, vol. 7251, pp. 278–281. Springer, Heidelberg (2012)
7. Gulera, I., Ubeyli, E.D.: Adaptive Neuro-Fuzzy Inference System for Classification of EEG Signals using Wavelet Coefficients. Journal of Neuroscience Methods (2005)
8. Honka, A.M., van Gils, M.J., Parkka, J.: A Personalized Approach for Predicting the Effect of Aerobic Exercise on Blood Pressure Using a Fuzzy Inference System. In: 33rd Annual International Conference of the IEEE EMBS, Boston, Massachusetts, USA (2011)
9. Jones, L.K., Zou, F., Kheifets, A., Rybnikov, K., Berry, D., Choon Tan, A.: Confident Predictability: Identifying Reliable Gene Expression Patterns for Individualized Tumor Classification using a Local Minimax Kernel Algorithm. BMC Medical Genomics 4(10) (2011)
10. Kaur, A., Kaur, A.: Comparison of Mamdani-Type and Sugeno-Type Fuzzy Inference Systems for Air Conditioning System. International Journal of Soft Computing and Engineering (IJSCE) 2(2) (2012) ISSN: 2231-2307
11. Kouris, I., Tsirmpas, C., Mougiakakou, S.G., Iliopoulou, D., Koutsouris, D.: E-Health towards Ecumenical Framework for Personalized Medicine via Decision Support System. In: 32nd Annual International Conference of the IEEE EMBS, Buenos Aires, Argentina (2010)
12. Midorikawa, Y., Tsuji, S., Takayama, T., Aburatani, H.: Genomic Approach Towards Personalized Anticancer Drug Therapy. Pharmacogenomics 13(2), 191–199 (2012), doi:10.2217/pgs.11.157
13. Patel, V.L., Shortliffe, E.H., Stefanelli, M., et al.: The Coming of Age of Artificial Intelligence in Medicine. Artificial Intelligence in Medicine 46(1), 5–17 (2009)
14. Phan, J.H., Moffitt, R.A., Stokes, T.H., Liu, J., Young, A.N., Nie, S., Wang, M.D.: Convergence of Biomarkers, Bioinformatics and Nanotechnology for Individualized Cancer Treatment. Trends in Biotechnology 27(6) (2009)
15. Rowan, M., Ryan, T., Hegarty, F., OHare, N.: The Use of Artificial Neural Networks to Stratify the Length of Stay of Cardiac Patients based on Preoperative and Initial Postoperative Factors. Artificial Intelligence in Medicine 40(3), 211–221 (2007)
16. Sourla, E., Sioutas, S., Syrimpeis, V., Tsakalidis, A., Tzimas, G.: CardioSmart365: Artificial Intelligence in the Service of Cardiologic Patients. In: Advances in Artificial Intelligence, Article ID 585072, vol. 2012, 12 pages. Hindawi Publishing Corporation (2012)
17. Wallis, J.W.: Invited commentary: Use of Artificial Intelligence in Cardiac Imaging. Journal of Nuclear Medicine 42(8), 1192–1194 (2001)
18. Zhou, X., Chen, S., Liu, B., et al.: Development of Traditional Chinese Medicine Clinical Data Warehouse for Medical Knowledge Discovery and Decision Support. Artificial Intelligence in Medicine 48(2-3), 139–152 (2010)
19. Fuzzy Logic, http://en.wikipedia.org/wiki/Fuzzy_logic
20. Mathworks Documentation Center, Comparison of Sugeno and Mamdani Systems, http://www.mathworks.com/help/fuzzy/
comparison-of-sugeno-and-mamdani-systems.html

The Strength of Negative Opinions

Thanos Papaoikonomou, Mania Kardara, Konstantinos Tserpes,
and Theodora Varvarigou

National Technical University of Athens
{tpap,nkardara,tserpes,dora}@mail.ntua.gr

Abstract. We investigate signed social networks, in which users are connected via directional signed links indicating their opinions on each other. Predicting the sign of such links is a crucial task for many real world applications like recommendation systems. Based on the premise that like-minded users tend to influence each other more than others, we present a logistic regression classifier built on evidence drawn from the users' ego-networks. The main focus of this work is to examine and compare the relative strength of positive and negative opinions investigating to what extent each type of link affects the overall prediction accuracy. We evaluate our approach through a thorough experimental study that comprises three large-scale real-world datasets.

Keywords: signed social networks, edge sign prediction.

1 Introduction

During the recent years online social networks (OSNs) have seen a dramatic rise in popularity, with most users spending a significant amount of their time on them. Facebook[1] is a social network of 800 million active users, more than half of which will log on at least once per day. Inside OSNs, users create relationships with each other, express opinions on certain subjects or even other users and interact, in general.

It is critical for the success of an OSN platform to assure a certain level of trust for its users. A person is considered *trusted* if we regard her actions as not malicious. In most popular online social networks trust is expressed implicitly. For example, Facebook users confirm friend requests from other users, and thus allow them to view personal data (e.g. photos, status updates) that would otherwise be inaccessible. Similarly in Twitter, where connections are asymmetric, a directed link from user A (*follower*) to user B (*followee*), implies that A values B's opinions on certain topics. It is equivalent to say that A believes that B will not behave maliciously e.g. act as a *spammer*. In the context of such applications, all links between users are positive as they indicate a certain level of trust between them. There are, however, web sites, which allow their users to explicitly annotate their disposition towards others as either positive or negative. For example in Slashdot, a technology news web site, after the introduction of the Zoo feature users are able to indicate their preferences, by tagging each other as *friend* or *foe*. Another example is that of Epinions, a product review web site whose members are able to express trust or distrust towards other community members.

---

[1] http://www.facebook.com/press/info.php?statistics

L. Iliadis, H. Papadopoulos, and C. Jayne (Eds.): EANN 2013, Part II, CCIS 384, pp. 90–99, 2013.
© Springer-Verlag Berlin Heidelberg 2013

In sociological research, aspects related to trust are described by the general term *trust metrics*, which can be divided into two large categories, namely global and local. Global trust metrics assign a unique trust score to every user in the network. For example, in eBay, after a transaction has occurred the participants (buyer and seller) can leave positive, neutral or negative feedback for each other. The trust score of a user is then calculated as the ratio of positive to total feedback and can be viewed by every other user in the network. On the other hand, in local trust metrics, the trust score of a target user is tailored to the preferences of the user viewing the trust score. In general, local trust metrics tend to be more precise, since they are specific to each user, but they are more computationally intensive, because they require more executions.

Although, due to its applicability in a variety of commercial applications, trust in social networks has been studied extensively in the recent years, until recently researchers have largely ignored the significance of distrust. A first attempt to study the connectivity patterns of *signed* graphs (*i.e.* graphs with positive and negative edges) was the *structural balance theory* that was articulated by Heider [8] in the 1940s, and was later reformed in graph-theoretic language by Cartwright and Harary [3]. According to this theory, triads of individuals with an odd number of positive ties are more probable than others. Recently, Lescovec et al. in [11] introduced a new theory of *status*, which examines pairs of adjacent nodes considering of higher status the recipient of a positive link and of lower status the recipient of a negative link.

The main challenge is whether we can exploit the patterns of the connections in a social graph so that we can gradually derive trust scores for the social network users. In [19] the authors show that there is a positive correlation between interpersonal trust and interest similarity. In other words, it seems that we trust more the individuals that we are similar. We adopt this outcome as our main assumption that will lead us to the construction of our algorithm in the following sections. The rest of the paper is organized as follows. In Section 2 we discuss related work. Section 3 states the problem definition and introduces the basic notation. In Section 4, we present our classifier and in Section 5, we evaluate through a thorough experimental study on three large-scale datasets. Finally, Section 6 concludes.

2 Related Work

In [7] Guha et al. present one of the earliest attempts to address the propagation of both trust and distrust in the signed network of Epinions. They develop a framework of trust propagation schemes, based on the exponentiation of the adjacency matrix. In [9] Kunegis et al. study the friend/foe network of the Slashdot Zoo, introducing the signed variants of global, node-level and link-level network characteristics. In [13], Massa et al. show the usefulness of local trust metrics in the identification of controversial users. In [4] DuBois et al, present a probabilistic interpretation of trust based on random graphs with a modified spring-embedding algorithm. In [17] Victor et al. derive trust and distrust metrics by employing bilattice-based aggregation approaches and investigate how they can be improved by using ordered weighted averaging techniques. In [14] the authors propose an algorithm to compute the bias and prestige of nodes in networks where the edge weight denotes the trust score.

The most recent work on trust/distrust propagation is that of Lescovec et al. in [10], which we consider as a direct predecessor. The authors propose a logistic regression model that maps each edge to a high-dimensional feature space, categorized into two classes. The first set examines the correlation between the out-degree of the node originating the link and the in-degree of the receiver node. The second set of features deals with the number and type of triads that are defined by the endpoints of the edge and their common neighbors. We consider the results of this work as our baseline.

3   Problem Definition and Notation

This work focuses on the *edge sign prediction problem*. As input we are given a signed social graph $G(V,E,L)$, where V is the set of vertices, E is the set of edges and L is a set of labels that can be assigned to the edges. Two users are neighbors in the social graph, if they have been engaged in some social interaction, i.e. expressed an opinion for each other, with the label $l \in L$ of the edge indicating the type of interaction. We refer to the user making the assessment as the *source* user and the receiver of this action as the *target* user. For the social networks that we investigate it is sufficient to set L to $\{-1, +1\}$ which classifies the edges of the graph into negative and positive respectively, with *sign(source, target)* denoting the sign of the label. The semantics are pretty straightforward: A positive link from a node u to another node v, indicates that user u has expressed trust towards user v. Likewise, a negative edge is an expression of distrust.For the rest of this section we will introduce definitions and metrics that will help us formulate our approach in Section 4. The first definition comes from the social network analysis domain and it is used to describe the local social structure around a focal node:

Definition 1. *The ego-network of a node contains this node, its neighbors and all edges among the selected nodes*

For reasons that will become apparent in the next sections, we extend the notion of the ego-network by defining the extended-k ego network which includes the nodes that are maximally k hops away from the focal node.

Definition 2. *The extended k-ego-network of a node u contains this node, the nodes that are k or less hops away from this node and all edges among the selected nodes. We denote it as $ego_k(u)$*

The cost of extracting an extended-k ego-network from a social graph is similar to running a Breadth First Algorithm (BFS), starting from the focal node until depth k, and hence its time complexity is $O(b^k)$, where b is the average degree of the nodes. The next definition introduces four graph operators that categorize the neighbors of a node according to the direction and the type of the link:

Definition 3. *Let A be a node in the social network. We denote as $\Gamma_{in}^+(A)$ the set of nodes which direct a positive link to node A, and $\Gamma_{out}^+(A)$ the set of nodes which receive a positive link from node A. In a similar manner, we define $\Gamma_{in}^-(A)$ and $\Gamma_{out}^-(A)$ in the case of negative links.*

Next, we present the similarity metrics that will be used throughout this paper. We considered three metrics: common neighbors [12], Jaccards coefficient [16] and Adamic/Adar [1]. Due to the fact that these measures were originally defined for unsigned, undirected networks, we wil describe how they can be adapted to our case. We start with the common neighbors metric, which simply counts the number of common assessments of two users.

Definition 4. *Let u, v be nodes in a signed social graph. Their common neighbors metric, denoted as commons(u,v) is defined as follows:*

$$common(u, v) = \sum_{\Gamma} |\Gamma(u) \cap \Gamma(v)|$$

where $\Gamma = \{\Gamma_{in}^+, \Gamma_{out}^+, \Gamma_{in}^-, \Gamma_{out}^-\}$, and $|.|$ is the cardinality of the set

The common neighbors metric is equivalent to the edge embeddedness metric that is also used in the literature [10].

Jaccard's coefficient is a commonly used metric in information retrieval, that measures the probability that two datapoints x, y have a feature f, for a randomly selected feature f that either x or y has [12]. For our purposes, we define it as follows:

Definition 5. *Let u, v be nodes in a signed social graph. Their Jaccard 's coefficient metric , denoted as jaccard(u,v) is defined as follows:*

$$jaccard(u, v) = \frac{\sum_{\Gamma} |\Gamma(u) \cap \Gamma(v)|}{\sum_{\Gamma} |\Gamma(u) \cup \Gamma(v)|}$$

where $\Gamma = \{\Gamma_{in}^+, \Gamma_{out}^+, \Gamma_{in}^-, \Gamma_{out}^-\}$, and $|.|$ is the cardinality of the set

Finally, the Adamic/Adar[1] metric counts the common features by putting more weight on the "rarer" ones. We associate the "rareness" of a node with its total degree and we interpret the Adamic/Adar metric as follows

Definition 6. *Let u, v be nodes in a signed social graph. Their Adamic/Adar score denoted by AdamicAdar(u,v) is given by:*

$$AdamicAdar(u, v) = \sum_{t \in \Gamma(u) \cap \Gamma(v)} \frac{1}{log(degree(t))}$$

where $degree(t) = |\Gamma_{in}^+(t)| + |\Gamma_{out}^+(t)| + |\Gamma_{in}^-(t)| + |\Gamma_{out}^-(t)|$ is the total degree of node t

Worth noting is that all three measures lookup for the common neighbors of the examined nodes, which means that the nodes should be connected by a path of length two or less in the social graph. In other words, a node u may have non-zero similarity score with nodes that exist only in its extended-2 ego-network. Second, these measures are symmetric, which makes them more adequate for large-scale recommendation systems, since they are computed once but can be used for both endpoints. The selection of these measures is further justified by their good balance between simplicity and performance.

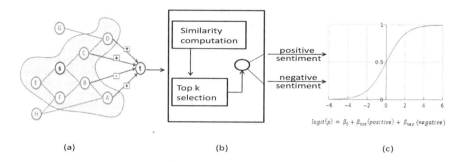

Fig. 1. Summarization of the algorithm.a)The algorithm extracts the ego_2 subgraph of the source node s and keeps all the nodes that point an edge to the target node t(C_i set). b)Similarity scores and sentiments computation c) The logistic classifier is trained using the positive,negative sentiment as predictors.

4   Approach

In this section we present our algorithm using the notation introduced in the previous section. In order to predict the disposition of some user A towards another user B, we will look for users similar to A that have expressed their opinion for user B and then we will exploit their opinions to fit a predictive statistical model. Our approach is depicted in Figure 1. At first, we retrieve the extended-2 ego-network of the source user, and we keep only those nodes that direct a link (positive or negative) to the specified target node. Then, we apply the proposed similarity measures to the resulting user list and we rank it in descending order. We extract the top k members of the list forming the *candidate influencers* set that we will denote it as C_i. We investigated a large range of possible values for k and we concluded that a value of $k = 15$ achieves the optimal trade-off between prediction accuracy and dataset coverage.

In the second step, we aggregate the contributions of the candidate influencers, which are weighted by their similarity score to the source user. We end up with two distinct quantities that capture the overall positive and negative sentiment of the candidate influencers towards the target node, and are computed as follows:

$$pos = \sum_{u \in C_i} score(u, source) \cdot I\{sign(u, target) = +1\}$$
$$neg = - \sum_{u \in C_i} score(u, source) \cdot I\{sign(u, target) = -1\}$$

where I is the indicator function ($I\{x\} = 1$, if x is true and 0 otherwise) The final step involves fitting a logistic regression classifier using the positive and negative sentiment of the second step as features. The fitted model has the form $logit(p) = \beta_0 + \beta_{pos} \cdot pos + \beta_{neg} \cdot neg$ where p is the probability of a positive edge, and pos, neg are the positive and negative sentiments computed in the second step. We train our classifier on the 60% of the available data and we evaluate on the rest 40%.

Table 1. Generalization properties of the classifier.We train on the row dataset and we measure the AUC on the column dataset.

Dataset	Epinions	Slashdot	Wikipedia
Epinions	94.50%	91,68%	87.01%
Slashdot	94.12%	91,61%	86.79%
Wikipedia	93,70%	91.56%	86,69%

Table 2. Dataset statistics. The number of nodes, the number of edges and the percentage of the positive edges in the graph are reported.

Dataset	#nodes	#edges	+edges%
Epinions	131,828	841,200	85.0%
Slashdot	81,867	549,202	77.4%
Wikipedia	7,194	103,747	78.7%

Table 3. Logistic model coefficients, 0/1 loss and AUC for the full and balanced datasets, and for all the similarity metrics

Metric	Common neighbours					Jaccard					Adamic/Adar				
Dataset	β_0	β_{pos}	β_{neg}	AUC	0-1 loss	β_0	β_{pos}	β_{neg}	AUC	0-1 loss	β_0	β_{pos}	β_{neg}	AUC	0-1 loss
Epinions full	2,158	0,002	-0.007	91,43%	93,96%	1.560	3.014	-10.240	94,50%	95,52%	2,155	0,015	-0.043	91,98%	94,08%
Epinions balanced	0,525	0.001	-0.009	92,72%	85,00%	0.072	2.216	-13.761	94,89%	88,38%	0,525	0.011	-0.061	92,96%	85,56%
Slashdot full	1,412	0.013	-0.052	89,72%	86,88%	0,922	5.653	-14.644	91,61%	87,99%	1,372	0.087	-0.291	90,04%	87,13%
Slashdot balanced	0.525	0.010	-0.063	89,82%	82,38%	0.067	4.372	-17.321	91,90%	84,14%	0.500	0.063	-0.350	90,31%	82,85%
Wikipedia full	1.049	0.002	-0.005	82,76%	83,64%	0.785	1.472	-2.776	86,69%	85,82%	1.064	0.011	-0.022	82,91%	83,55%
Wikipedia balanced	-0.197	0.002	-0.006	84,52%	77,70%	-0.429	1.393	-3.538	86,52%	78,90%	-0.160	0.010	-0.031	83,94%	77.16%

5   Evaluation

5.1   Dataset Description

We have conducted our experiments on three large-scale datasets from real-world social networks. All three datasets were retrieved from the SNAP [10] website. The first dataset comes from Epinions which is a product review website. Epinions members may express trust or distrust to each other forming a signed social graph with positive and negative edges. The data span a period from 1999 to 2003, comprising 130k distinct users and about 840k relationships among them. The second dataset comes from the technology blog, Slashdot. Slashdot's users can tag each other as *friend* or *foe* indicating trust and distrust relationships respectively. The dataset is a snapshot taken in February 2009 and contains around 82k users and 549k relationships.The third dataset contains vote history data from Wikipedia's administrator elections. In order for a user to become administrator, a *Request for adminship* is issued, and the Wikipedia community is called upon to decide whether to accept or not the candidacy. The dataset comprises data from 7k users and about 100k total votes.Table 2 presents summary statistics for all three datasets.

5.2   Predictive Performance

In this section, we evaluate the predictive performance of our classifier. The major problem that we faced during this phase is related to the high *skewness* of social network data. As it is shown in Table 2, about 80% of the edges are positively signed allowing

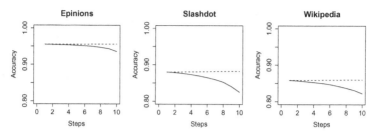

Fig. 2. Predictive accuracy of the transformed datasets. The dashed line represents the dataset where the positive links were removed. The solid line refers to the dataset where the negative links were removed.

even a random classifier (i.e. assigning signs at random with a $p = 0.8$ probability of giving a positive sign) to achieve remarkable accuracy. We tried to address this issue in two ways. First, we evaluated our algorithm on a metric that is insensitive to the unequal class distribution [5], namely the AUC (Area Under Curve) of the ROC (Receiver Operator Characteristic) curve. Second, we estimated the robustness of our method by applying it on balanced datasets i.e. datasets with an equal number of positive and negative edges. This is more generally known as *undersampling* and it is a standard approach when dealing with highly imbalanced data [15] [18]. To accomplish this step, we followed the testing framework proposed by Guha et al. in [7]: For each social network, we kept all the negative edges and then randomly selected an equal number of positive ones. We repeated this procedure 10 times to minimize the effect of randomness. Table 3 summarizes the results from the evaluation of our classifier to the datasets and for each similarity metric. The first three columns present the coefficients of the fitted logistic regression model while the last two columns provide the measurements for the AUC and 0/1 loss(number of correct guesses).

Some conclusions can be easily drawn by analyzing the results of Table 3. First, it seems that the Jaccard's coefficient metric achieves the maximum performance in all three cases and from now on, we will consider it as the default metric. Second, the AUC measurements for the full and balanced datasets give approximately the same results, as expected. On the other hand, the bias introduced by the skewness of the datasets, gives a greater 0/1 loss performance when the full dataset is considered. Third, it is apparent that the performance of our classifier is quite good. We reach an AUC maximum at 95,52% for the Epinions dataset and a minimum of 86,69% for the Wikipedia dataset employing the Jaccard's coefficient metric. Compared to the work of [10] which we consider as a baseline, we demonstrate a slight improvement ($\sim 2\%$) in the case of Epinions and a larger improvement in the case of Wikipedia ($\sim 6\%$). On the other hand, we perform slightly worse in Slashdot ($\sim -2\%$)

Finally, it is worth mentioning that the coefficients for the positive and negative sentiment differ barely between the full and balanced cases, when applied to the same dataset and on the same similarity metric. The transition from the full to the balanced case is mostly "absorbed" by the intercept coefficient β_0, which gives the overall probability of a positive edge. This is why β_0 has a larger value in the full dataset, and approaches zero in the balanced one. The results of this section have confirmed our hypothesis, that an individual is largely affected by her "close companions" in her judgments towards other social network users.

5.3   Generalization Properties

In this subsection we evaluate the generalization properties of our classifier following the method of [10]. In short, this involves training our classifier on one dataset and evaluating it on the other. For brevity, we have performed these experiments using only the Jaccard's coefficient metric and Table 1 summarizes the results. The off-diagonal elements show the actual generalization measurements, while the diagonal elements are just a repetition of Table 3. By analyzing the performance of the classifier in all cases, we arrive at the following conclusions. First, the worst performance is achieved when we test on the Wikipedia dataset, without regard to which dataset we use for training. Second, it is remarkable the fact that we achieve success rates 93,70% and 91.56% on Epinions and Slashdot respectively, when we train in the Wikipedia dataset, even though itself attains only 86,69% accuracy. Third, we observe that the AUC measurements are almost identical for each dataset, independently of the training dataset. For example, we get approximately 94% for the Epinions dataset, around 91,6% for Slashdot, and about 87% for Wikipedia.

5.4   The Strength of Negative Opinions

In this subsection we investigate the relative strength of negative links compared to the positive ones and how this affects the overall predictive accuracy. In other words, we aim to identify whether one link type bears more information than the other. Our findings confirm that the negative edges in a signed social graph, can be considered more "important" for the edge sign prediction problem.

We start by considering the coefficients of the logistic regression presented in Table 3. Logistic regression models the relationship between the logit transformation of the outcome variable and the linear combination of the predictor variables. In our case, the independent variables are the positive and negative sentiment generated by the "close companions" of the source node, which are mixed according to the coefficients β_{pos} and β_{neg} respectively. It is easily discernible that for all datasets and for each similarity metric, the coefficient for the negative sentiment β_{neg} is always larger (in absolute values) than that of the positive sentiment β_{pos}. Of course, this fact alone cannot guarantee statistical significance but it constitues an early indication.

In order to validate our claim, we resort to *postestimation* testing and more specifically to a *Likelihood ratio* test. Our goal is to prove that the coefficients for the positive and negative sentiment differ significantly. Therefore, we build two separate models:The first is the model we already have, which we call the *full* model, because it has distinct coefficients β_{pos} and β_{neg} for the two features. The second model, which we will call the *nested* model results from the full model plus the constraint that the two coefficients β_{pos} and β_{neg} are equal. Thus, the nested model is of the form

$$logit(p) = \gamma_0 + \gamma_{common} \cdot (pos + neg),$$

where the γ_{common} is the coefficient of the combination of the two sentiments. The purpose of the likelihood ratio test is to compare the fit of the two models and reject the nested model in the case of a significant difference. Our findings support the rejection of the nested model at a significance level of $p < 0.001$ for all the three datasets.

So far, we have strong evidence that the presence of a negative link affects significantly more than a positive link. In this last part, we will try to quantify the superiority of the negative links and how they affect the overall predictive accuracy. We set up the following experiment: We specify a constant number m and given a dataset, we construct two new sets. For the first set, we remove m negative links while we keep all the positive links, and in the second set we keep all the negative links but we remove m positive links. We will use these two sets to re-train our classifier and finally we re-evaluate on the full dataset, measuring the 0/1 loss. We will call S^- the set from which the negative links were removed and S^+ the other set, with the superscript denoting the link type of the removed edges. S^+ and S^- sets have equal sizes but different proportions.

We iterated for $k = 10$ steps and each time we removed $k \cdot 10\% \cdot |E^-|$ edges of the same type from the inital dataset, where $|E^-|$ is the total number of negative edges in the graph. Again, we repeated 10 times to diminish the effect of randomness. The results are depicted in Figure 2. The dashed line refers to the S^+ set and the solid line to the S^- set. In Epinions, both lines start from a common 95,52% predictive rate. The performance of the S^+ set remains steady throughout the test, while that of the S^- gradually decreases and finally attains predictive rate of 93,54%. Similar patterns arise in the case of Wikipedia, where the initial 85,83% success rate becomes 85,98% for the S^+ set and 82,22% for the S^- set. Larger variance is noted for Slashddot, where the graphs for the S^+ , S^- sets start from a common 88,01% rate and end at 88,27% and 82,54% respectively. The results are in agreement with our hypothesis. The removal of even a small portion of the negative links, resulted in a significant loss in the predictive accuracy, which may be critical for real-world applications. Taking into account the negative opinions of the users, we can model more accurately the relationships among them and provide more effective recommendations. Our findings are also in agreement with the work of Garcia et al. in [6]. There, the authors analyzed three established lexica of affective word usage in English, German, and Spanish. They found that words with a positive emotional content are more frequently used, which is in accordance with the Pollyanna hypothesis [2] that there is a positive bias in human expression. The authors conclude that negative words contain more information than positive words.

6   Conclusions

This work focused on the edge sign prediction problem, where given an edge in a signed social graph, we attempted to predict its label. A positive edge was assumed to be an indication of trust, whereas a negative edge an indication of distrust. We built a logistic regression classifier that drew evidence from the local neighbourhood of the user originating the link, which achieved high accuracy. The second part of our contribution focused on the relative strength between the positive and negative links. We set up an experiment to measure their importance in the overall prediction accuracy and we concluded that the negative links bear more information that the positive ones.

References

1. Adamic, L.A., Adar, E.: Friends and neighbors on the web. Social Networks 25(3), 211–230 (2003)
2. Boucher, J., Osgood, C.E.: The pollyanna hypothesis. Journal of Verbal Learning and Verbal Behavior 8(1), 1–8 (1969)
3. Cartwright, D., Harary, F.: Structural balance: a generalization of Heider's theory. Psychological Review 63(5), 277–293 (1956)
4. DuBois, T., Golbeck, J., Srinivasan, A.: Predicting trust and distrust in social networks. In: SocialCom/PASSAT, pp. 418–424. IEEE (2011)
5. Fawcett, T.: An introduction to roc analysis. Pattern Recogn. Lett. 27(8), 861–874 (2006)
6. Garcia, D., Garas, A., Schweitzer, F.: Positive words carry less information than negative words. CoRR, abs/1110.4123 (2011)
7. Guha, R., Kumar, R., Raghavan, P., Tomkins, A.: Propagation of trust and distrust. In: Proceedings of the 13th International Conference on World Wide Web, WWW 2004, pp. 403–412. ACM, New York (2004)
8. Heider, F.: Attitudes and Cognitive Organization. Journal of Psychology 21, 107–112 (1946)
9. Kunegis, J., Lommatzsch, A., Bauckhage, C.: The slashdot zoo: mining a social network with negative edges. In: Proceedings of the 18th International Conference on World Wide Web, WWW 2009, pp. 741–750. ACM, New York (2009)
10. Leskovec, J., Huttenlocher, D., Kleinberg, J.: Predicting positive and negative links in online social networks. In: Proceedings of the 19th International Conference on World Wide Web, WWW 2010, pp. 641–650. ACM, New York (2010)
11. Leskovec, J., Huttenlocher, D., Kleinberg, J.: Signed networks in social media. In: Proceedings of the SIGCHI Conference on Human Factors in Computing Systems, CHI 2010, pp. 1361–1370. ACM, New York (2010)
12. Liben-Nowell, D., Kleinberg, J.: The link prediction problem for social networks. In: Proceedings of the Twelfth International Conference on Information and Knowledge Management, CIKM 2003, pp. 556–559. ACM, New York (2003)
13. Massa, P., Avesani, P.: Controversial users demand local trust metrics: an experimental study on epinions.com community. In: Proceedings of the 20th National Conference on Artificial Intelligence, vol. 1, pp. 121–126. AAAI Press (2005)
14. Mishra, A., Bhattacharya, A.: Finding the bias and prestige of nodes in networks based on trust scores. In: Proceedings of the 20th International Conference on World Wide Web, WWW 2011, pp. 567–576. ACM, New York (2011)
15. Provost, F.: Machine learning from imbalanced data sets 101 (extended abstract)
16. Salton, G., McGill, M.J.: Introduction to Modern Information Retrieval. McGraw-Hill, Inc., New York (1986)
17. Victor, P., Cornelis, C., De Cock, M., Herrera-Viedma, E.: Practical aggregation operators for gradual trust and distrust. Fuzzy Sets Syst. 184(1), 126–147 (2011)
18. Zhang, J., Mani, I.: KNN Approach to Unbalanced Data Distributions: A Case Study Involving Information Extraction. In: Proceedings of the ICML 2003 Workshop on Learning from Imbalanced Datasets (2003)
19. Ziegler, C.-N., Golbeck, J.: Investigating interactions of trust and interest similarity. Decis. Support Syst. 43(2), 460–475 (2007)

Extracting Knowledge from Web Search Engine Using Wikipedia

Andreas Kanavos, Christos Makris, Yannis Plegas, and Evangelos Theodoridis

Computer Engineering and Informatics Department,
University of Patras, Greece, 26500
{kanavos,makri,plegas,theodori}@ceid.upatras.gr

Abstract. Nowadays, search engines are definitely a dominating web tool for finding information on the web. However, web search engines usually return web page references in a global ranking making it difficult to the users to browse different topics captured in the result set. Recently, there are meta-search engine systems that discover knowledge in these web search results providing the user with the possibility to browse different topics contained in the result set. In this paper, we focus on the problem of determining different thematic groups on web search engine results that existing web search engines provide. We propose a novel system that exploits semantic entities of Wikipedia for grouping the result set in different topic groups, according to the various meanings of the provided query. The proposed method utilizes a number of semantic annotation techniques using Knowledge Bases, like WordNet and Wikipedia, in order to perceive the different senses of each query term. Finally, the method annotates the extracted topics using information derived from clusters which in following are presented to the end user.

1 Introduction

Search engines are an inestimable tool for retrieving information from the Web. However, they lack in presenting ambiguous queries that usually result in web page references mapped to different meanings mixed together in the answer list. The knowledge discovery in the results, that common web search engines give, is a plausible solution to this problem. Extracting knowledge and grouping the results returned by a search engine into groups or a hierarchy of labeled clusters, is a very important task that modern search engines have recently started taking into consideration[1]. By providing category clustered results, the user may focus on a general topic by entering a generic query and then selecting these themes that match his interest.

Clustering web search results is usually performed in two steps: at the first step the retrieval is achieved based on a query from a public web search engine and following, the clustering is performed. The aforementioned problem can be seen

---

[1] Google:http://www.google.com/insidesearch/features/search/knowledge.html

L. Iliadis, H. Papadopoulos, and C. Jayne (Eds.): EANN 2013, Part II, CCIS 384, pp. 100–109, 2013.
© Springer-Verlag Berlin Heidelberg 2013

as a particular subfield of clustering concerned with the identification of thematic groups of items in web search results. The input of the clustering algorithms is a set of web search results S obtained in response to a user query, where each result item is described by a tuple $S_i = (S_i[url], S_i[title], S_i[snippet])$ with the URL, the title and the snippet[2]. Assuming that there is a logical topic structure in the result set, the output of a search result clustering algorithm is a set of labeled clusters organized in various ways such as flat partitions, hierarchies etc.

In this work, we propose a system that will by all means cluster and annotate the results of search engines having a reasonable trade-off between response time and cluster quality. We propose a novel system that exploits semantic entities of Wikipedia for categorizing the result set in different topic groups, according to the various meanings of the provided query. The proposed method utilizes a number of semantic annotation techniques using Knowledge Bases, like WordNet and Wikipedia, so as to perceive the different senses of each query term. Finally, the method annotates the extracted topics using information derived from the clusters and in following presents them to the end user.

2   Related Work

Recently, extracting knowledge from web search engine results has gained a lot of attention. In [4] they survey all the different approaches in designing clustering web search engines and provide a taxonomy of different algorithms so as to produce the clustered search results output. Current approaches fall into the following three categories:

Data-centric Methods try to label the clusters with something comprehensible to a human. More specifically, some keyword terms are selected with the highest term frequency among all document terms. The main disadvantage of this type of method, is that the output-label does not form a sentence. The most representative algorithms of this category are the Scatter/Gather [6], Lassi [19], WebCat [11] and AIsearch [24].

Description-aware Methods mainly try to resolve the disadvantage of data-centric algorithms. These methods aim to produce cluster descriptions that are understandable by humans. Clustering algorithms, in which objects are assigned to clusters primarily based on a single feature, are used by these methods as well. Then, these algorithms select the features so that they are immediately recognizable to the user as something meaningful. As a result, they can be used to assign sensible labels to the output clusters. In this category fall methods like Suffix Tree Clustering (STC) used in Grouper system [26] and SnakeT [9]. This system also makes use of continuous sentences and replaces them with approximate ones. The output of this system is a hierarchy of clusters, where a set of sentences is assigned to each cluster. Initially, the user places the query and then, SnakeT builds the clusters on-the-fly and assigns sentences of variable length as labels to them.

---

[2] A snippet is usually a short text summarizing the context in which the query words appear in the result page.

Description-centric Methods usually aim to create more meaningful label descriptions than the other two categories. These methods discard clusters that cannot be described, as they are valueless to the end user, such as Lingo [22] which is one of these methods. The main difference between Lingo and other systems is that it initially performs the label construction and then assigns documents to each cluster. Lingo is modulated in four phases: snippets preprocessing, frequent phrase extraction, cluster label induction, and content allocation, where the first two are the same as Suffix Tree Clustering (STC), except that instead of suffix trees, suffix arrays are used.

Finally, a study related to the current one is [13] in which web documents are categorized using WordNet. Primarily, more features extracted from WordNet such as hypernym, hyponym, synonym and domain are used and consequently a sense-merging algorithm to merge similar senses before grouping is employed. Moreover, in [18] an initial web search result clustering method was presented employing Wikipedia.

3   Proposed Method

The proposed method is modulated in the following steps:

3.1   Retrieving Initial Web Search Results

At first, we query an online web search engine (eg. Google), in order to process the returned web page references. Retrieving web search results from a public web search engine is possible either by using a specified API or by HTML scraping (fetch and parse the html page). The acquisition component of the search results can use regular expressions or other forms of markup detection to extract titles, snippets, and URLs from the HTML stream, served by the search engine to its end users. In our work, we collect the results of a search engine for a specific query made by the user and then we extract the URLs as well as their corresponding snippets, which constitute the input that we will afterwards use in the rest of our method. We use the procedure of HTML scraping instead of the existing search engine APIs as the latter approaches have some limitations in their use. One of these limitations concerns the number of search results returned by the search engine which is fairly small, while another has to do with the document information provided by the search engine which could be inadequate in our case because only URLs are provided. Also, some rate limits apply to specific Web Services; the APIs permit a specific number of calls made per IP address during a specific time window.

3.2   Document Representation

In the following step, document representations are produced and our system enriches the information taken from the results of search engines (URL and snippet). Furthermore, for each web page reference we download its text content (html with stripped out html tags). Then each document is processed by removing stop words and stemming the remaining terms. Thus, each document is represented as a tf/idf vector [1]. In the end, some terms of the document are annotated and mapped on senses identified in Wikipedia.

After the results are retrieved in the initial step, each result item consists of four different parts: title, URL, snippet and the content of the web page. Each element is processed by removing html tags and stop words and then stemming each remaining term with Porter stemmer. The aim of this phase is to prune the input from all characters and terms that can possibly affect the quality of group descriptions.

Consecutively, each search result is converted into a vector of terms using tf/idf weighting scheme for each one [1]. As an alternative representation, terms annotated to senses identified from Wikipedia are used. This phase is described in the following subsection.

3.3 Wikification of Retrieved Results

Furthermore, the preprocessing method enhances the texts with this extra semantic information and structure in order to exploit conceptual similarity of them. We employed a text annotation method, as initially presented in [20], that maps terms of the text to Wikipedia entities. This method also has to deal with the named entity disambiguation problem, in which it is possible for a term to have multiple Wikipedia articles as possible annotations. Disambiguation to Wikipedia entities is quite similar to the traditional Word Sense Disambiguation task, but distinct in that the Wikipedia link structure provides additional information, with which disambiguation should be compatible. Most of the existing methods for this rely on the textual context and in the collective agreement with the disambiguation of other identified spots, in order to clarify a specific spot.

Following results of [20], we employ two alternative wikipedia disambiguation methods. In the first one, the algorithm employs WordNet taking into account all the WordNet senses of the specific point, each equipped with a different weight. In particular, for a specific spot the method considers the sorted list of the WordNet candidate senses, each one of these with a weight. Then, by employing again resemblance as the text similarity metric and the glosses of the involved WordNet senses, method computes for every page a disambiguation score. The final disambiguation score is computed linearly, combining the score of relatedness weight computed and the WordNet dissambiguation score. Finally, the method opts for a given spot; the Wikipedia article with the largest disambiguation score. Then by employing an appropriate threshold distance, it collects similar in weight Wikipedia articles in order to select the one with the largest commonness.

The alternative disambiguation approach moves in the same way as the previous one. This time although, instead of using resemblance as the text similarity metric, it employs a semantic similarity metric between the WordNet senses of the target Wikipedia pages and the WordNet senses of the spot, thus moving a step further. The final disambiguation score is computed linearly combining the disambiguation score with the relatedness computed. Then, the method selects as chosen disambiguation for a given point, the Wikipedia article with the largest score by employing again an appropriate threshold distance. In detail, the method collects similar in score Wikipedia articles aiming to select the most common of all. In the case that the method cannot extract a synset from the text directly or the Wikipedia page has no corresponding synsets, it lets the sense similarity be equal to the maximum possible distance in the WordNet ontology graph.

Considering the representation of each Wikipedia page as a set of senses a possible approach would definitely be to parse the set of all the Wikipedia pages, represent their content using vector space modeling techniques, isolate the words with the dominant TF-IDF characteristics and use WordNet to locate their dominant senses. For speeding up, the method exploited knowledge already existing in available repositories which align Wikipedia with related WordNet senses, such as YAGO2 [14], could be used to have a fast and elegant representation of Wikipedia articles as set of WordNet senses. For more details refer to [20].

The set of search results along with their produced document vectors (tf/idf, wikipedia1, wikipedia2) that are extracted in this step, are given in following as input to the clustering algorithm, which is responsible for building the clusters and afterwards assigning proper labels to them.

3.4 Clustering

In following, clustering and categorization of the documents takes place. For grouping the web search results, we apply certain clustering techniques to the document representation of each item. We have used k-means [7] as common clustering algorithm. Concerning the clustering algorithm, we have used cosine similarity metric.

In each cluster that our system produces, we assign a label with various terms/senses/wikipedia articles that define each category. In the first traditional approach, we assume that the label is recovered from the feature vector of the clusters and consists of a few unordered terms. The selection is based on a set of most frequently occurring keywords in the clusters' documents. We can measure this frequency by the tf/idf scheme which we have already mentioned in document representation. More specifically, we have used the intersection of the most repeated appearing keywords of all the documents that occur in a specific cluster. An alternative is to calculate the centroid and use the most expressive dimensions of this vector.

The second approach clearly takes advantage of Wikipedia term annotations extracted in the representation phase. From these terms, we use the strongest by ranking them with the probability to be selected as a keyword in a new document. This probability is calculated by counting the number of documents where the term was already selected as a keyword divided by the total number of documents where the term appeared. Similar techniques were presented in [3,21]. What we propose here is a simplification for the sake of decreasing computational time. To incorporate Wikipedia information, we have modified the clustering methods appropriately. Clustering methods use a complex distance based the same time on the tf/idf and the Wikipedia-sense vectors in the spirit of [2], which combine different levels of term and semantic representations. Clustering procedure (eg. k-means) calculates document distances using the cosine distance of term vectors and of the Wikipedia vectors normalized at 50% respectively. So, term distance and semantic distance have the same weight in the final complex distance metric[3].

Lastly, we have employed a hybrid two step clustering algorithm. Initially, a coarse k-means clustering with 3-4 clusters using the *tf/idf* document representation, is performed. In following, each one of the initial created clusters is clustered recursively using k-means with the wikipedia document representation documents once again.

3.5   Cluster Labeling

In the ultimate step, the system assigns labels to the extracted clusters in order to facilitate users to select the desired one. In the simplest possible way, the label is recovered from the feature vector of the clusters and consists of a few unordered terms, based on their tf/idf measure. On the other hand, system uses the identified Wikipedia senses. These categories/titles serve as potential candidates for cluster labelling as shown in [3].

4   Experimental Evaluation

We have implemented our system as a web server, using Java EE and web user interfaces developed with Java Server Pages (JSP). We have utilized library JWNL[4] so as to interoperate with WordNet 2.1[5]. In order to retrieve web search results from online search engines, we employ a library web browser written in Java, called HtmlUnit[6]. This library allows high-level manipulation of websites from other Java code, including clicking hyperlinks executing Javascript etc. It moreover provides access to the structure and the details within received web pages. HtmlUnit emulates parts of browser behaviour including the lower-level aspects of TCP/IP and HTTP. For the procedure of search results' preprocessing, we use LingPipe[7]. It is a Java tool kit for working out text using computational linguistics. Stemming, stop-word methods, vector space representation and tf/idf weighting scheme, clustering algorithms etc. were implemented by making use of this library. Finally, certain steps of the system where enriched using threads like getting results from the online web search engine, parsing each document

---

[3] On our evaluation we have used the 1000 most weighted tf/idf terms and the 5 most weighted Wikipedia terms.
[4] http://sourceforge.net/projects/jwordnet/
[5] http://wordnet.princeton.edu/
[6] http://htmlunit.sourceforge.net
[7] http://alias-i.com/lingpipe/

and labeling each cluster in order to gain some speed up concerning response time to the end user. Our prototype system is deployed for testing online[8].

We have performed several runs of our service on an Intel i7@3Ghz with 6GB memory. We have used a quite ambiguated term "chicago" waiting to give results in different topics. In the following figure you can see several examples of the test queries and the produced results. We have retrieved 50 query results, used the whole tf/idf vector and also used the hybrid two step clustering algorithm described in section 3.4. We observe that system produced correctly several clusters for bank, basketball, football team etc. and a couple of larger clusters with mixed results containing university and city hall info.

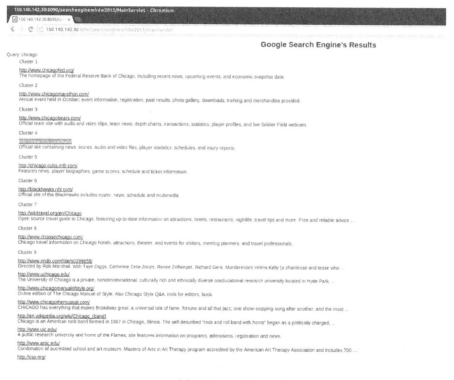

(a) Chicago Query

We evaluated our techniques by measuring the performance in terms of precision, recall and F-measure, where precision is calculated as the number of items correctly put into a cluster divided by total number of items put into the cluster; recall is defined as the number of items correctly put into a cluster divided by all items that should have been in the cluster; and F-measure is the harmonic mean of the precision and recall. The total F-measure for the entire clustering is

---

[8] http://150.140.142.30:8090/searchenginemhdw2013/

the sum for each objective category the F-measure of the best cluster (matching this category) normalized to the size of the cluster. All results presented here use the K-means clustering. In order to evaluate the efficacy of the clustering module, we have used a web/txt dataset, that is usually used in web clustering tasks, and have counted the precision of the produced web document clusters. The dataset is Reuters News Collection 21578[9], which is a collection of documents that appeared on Reuters newswire in 1987. The documents were first assembled and then indexed into categories curated by users. Thus, in Figure 4 the quality of clustering is quite well having F-measure score around 60% leading to more reasonable results than in webKB dataset. However, the utilization of the annotation process with Wikipedia senses, seems to noticeably ($\approx 25 - 30\%$) increase the results in all cases or even higher in some other.

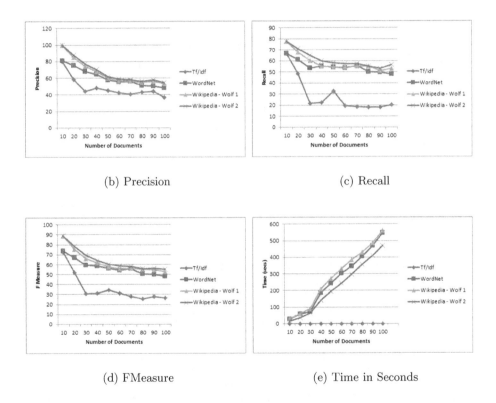

(b) Precision   (c) Recall

(d) FMeasure   (e) Time in Seconds

5   Conclusions - Future Work

The main advantage of clustering web search results is that it enables users to easily browse web page reference groups with the same meaning. In addition, it allows better topic understanding and favours systematic exploration of

---

[9] http://www.daviddlewis.com/resources/testcollections/reuters21578/

search results. This paper describes the need for query reformulation and reports the advantages of semantics in our understanding the users' search interaction. Specifically, we have formed a clustering post-search engine, as an extra feature to the standard search engine, exploiting state of the art knowledge discovery techniques.

Directions for further investigation in this line of research would be the achievement of a better trade-off between response time and clustering of web search results. Small response times are critical for user adoption of an online web search engine but not sacrificing quality of the results. Interesting would be how to distribute computational effort of how to pre-process off-line frequent patterns of queries. Another interesting topic of research would be how it could be possible to transfer a portion of the computation at client side, at user's web browser taking in mind user's preferences and previous browsing history.

Acknowledgements. This research has been co-financed by the European Union (European Social Fund-ESF) and Greek national funds through the Operational Program Education and Lifelong Learning of the National Strategic Reference Framework (NSRF)-Research Funding Program: Heracleitus II. Investing in knowledge society through the European Social Fund.

This research has been co-financed by the European Union (European Social Fund-ESF) and Greek national funds through the Operational Program Education and Lifelong Learning of the National Strategic Reference Framework (NSRF)-Research Funding Program: Thales. Investing in knowledge society through the European Social Fund.

References

1. Baeza-Yates, R.A., Ribeiro-Neto, B.A.: Modern Information Retrieval, 2nd edn. Addison Wesley (1999, 2011), http://mir2ed.org/
2. Caputo, A., Basile, P., Semeraro, G.: SENSE: SEmantic N-levels Search Engine at CLEF2008 Ad Hoc Robust-WSD Track. In: Peters, C., Deselaers, T., Ferro, N., Gonzalo, J., Jones, G.J.F., Kurimo, M., Mandl, T., Peñas, A., Petras, V. (eds.) CLEF 2008. LNCS, vol. 5706, pp. 126–133. Springer, Heidelberg (2009)
3. Carmel, D., Roitman, H., Zwerdling, N.: Enhancing cluster labeling using wikipedia. In: SIGIR 2009, pp. 139–146 (2009)
4. Carpineto, C., Osiski, S., Romano, G., Weiss, D.: A survey of Web clustering engines. ACM Comput. Surv. (2009)
5. comScore. Baidu Ranked Third Largest Worldwide Search Property (2008), http://www.comscore.com/press/release.asp?press=2018
6. Cutting, D.R., Karger, D.R., Pedersen, J.O., Tukey, J.W.: Scatter/Gather: A Cluster-based Approach to Browsing Large Document Collections. In: SIGIR 1992, pp. 318–329 (1992)
7. Dunham, M.H.: Data Mining: Introductory and Advanced Topics. Prentice Hall PTR, Upper Saddle River (2002)
8. Everitt, B.S., Landau, S., Leese, M.: Cluster Analysis, 4th edn. Oxford University Press (2001)

9. Ferragina, P., Gullì, A.: The Anatomy of SnakeT: A Hierarchical Clustering Engine for Web-Page Snippets. In: Boulicaut, J.-F., Esposito, F., Giannotti, F., Pedreschi, D. (eds.) PKDD 2004. LNCS (LNAI), vol. 3202, pp. 506–508. Springer, Heidelberg (2004)
10. Ferragina, P., Scaiella, U.: TAGME: on-the-fly annotation of short text fragments (by wikipedia entities). In: CIKM 2010, pp. 1625–1628 (2010)
11. Giannotti, F., Nanni, M., Pedreschi, D., Samaritani, F.: WebCat: Automatic Categorization of Web Search Results. In: SEBD 2003, pp. 507–518 (2003)
12. Hearst, M.A.: Search User Interfaces, 1st edn. Cambridge University Press (2009)
13. Hemayati, R., Meng, W., Yu, C.: Semantic-Based Grouping of Search Engine Results Using WordNet. In: Dong, G., Lin, X., Wang, W., Yang, Y., Yu, J.X. (eds.) APWeb/WAIM 2007. LNCS, vol. 4505, pp. 678–686. Springer, Heidelberg (2007)
14. Hoffart, J., Suchanek, F., Berberich, K., Lewis-Kelham, E., Melo, G., Weikum, G.: YAGO2: exploring and querying world knowledge in time, space, context, and many languages. In: WWW (Companion Volume) 2011, pp. 229–232 (2011)
15. Huang, J., Efthimiadis, E.N.: Analyzing and evaluating query reformulation strategies in web search logs. In: CIKM 2009, pp. 77–86 (2009)
16. Jansen, B.J., Spink, A., Blakely, C., Koshman, S.: Defining a session on Web search engines. JASIST 58(6), 862–871 (2007)
17. Jansen, B.J., Spink, A., Pedersen, J.: A temporal comparison of AltaVista Web searching. JASIST 56(6), 559–570 (2005)
18. Kanavos, A., Theodoridis, E., Tsakalidis, A.: Extracting Knowledge from Web Search Engine Results. In: ICTAI 2012, pp. 860–867 (2012)
19. Maarek, Y.S., Fagin, R., Ben-Shaul, I.Z., Pelleg, D.: Ephemeral Document Clustering for Web Applications. Tech. rep. RJ 10186, IBM Research (2000)
20. Makris, C., Plegas, Y., Theodoridis, E.: Improved text annotation with Wikipedia entities. In: SAC 2013, pp. 288–295 (2013)
21. Mihalcea, R., Csomai, A.: Wikify!: linking documents to encyclopedic knowledge. In: CIKM 2007, pp. 233–242 (2007)
22. Osinski, S., Stefanowski, J., Weiss, D.: Lingo: Search Results Clustering Algorithm Based on Singular Value Decomposition. In: Intelligent Information Systems 2004, pp. 359–368 (2004)
23. Scaiella, U., Ferragina, P., Marino, A., Ciaramita, M.: Topical clustering of search results. In: WSDM 2012, pp. 223–232 (2012)
24. Stein, B., Eissen, S.M.Z.: Topic Identification: Framework and Application. In: I-KNOW 2004, pp. 353–360 (2004)
25. Trillo, R., Po, L., Ilarri, S., Bergamaschi, S., Mena, E.: Using semantic techniques to access web data. Inf. Syst. 36(2), 117–133 (2011)
26. Zamir, O., Etzioni, O.: Grouper: A Dynamic Clustering Interface to Web Search Results. Computer Networks 31(11-16), 1361–1374 (1999)

AppendicitisScan Tool: A New Tool for the Efficient Classification of Childhood Abdominal Pain Clinical Cases Using Machine Learning Tools

Athanasios Mitroulias[1], Theofilatos Konstantinos[2], Spiros Likothanassis[2], and Mavroudi Seferina[2,3]

[1] Department of Mathematics, University of Patras, Greece
[2] Department of Computer Engineering and Informatics, University of Patras, Greece
[3] Department of Social Work, School of Sciences of Health and Care,
Technological Institute of Patras, Greece
thanasis.mitroulias@gmail.com,
{theofilk,likothan,mavroudi}@ceid.upatras.gr

Abstract. The abdominal pain is considered a very common disease during the childhood. One of the main diseases which are considered nowadays as the cause of childhood abdominal pain is the appendicitis which is very hard to be diagnosed in children. Moreover, even when it is diagnosed the doctors should decide about the type of appendicitis and take a crucial decision about the treatment (surgeon or medication). For these reasons, researchers in the last decade have focused on developing machine learning models to predict appendicitis from childhood abdominal pain clinical cases. However, most of these methods are limited to low performance and to using diagnostic factors which are not generally available. Moreover, none of them is available as a tool which could be used in practice. For all these reasons, we developed and applied a new ensemble methodology which combines the results of three machine learning models: Artificial Neural Networks, Support Vector Machines and Random Forests. The implementation is available as a standalone tool named AppendicitisScan Tool.

Keywords: Artificial Neural Networks, Support Vector Machines, Random Forests, Abdominal Pain Classification.

1   Introduction

Appendicitis is very hard to be diagnosed as many factors should be considered and examined by an expert. Moreover, this disease is too protean and cannot be effectively diagnosed by clinical instinct. This fact is even more reinforced when Appendicitis has to be diagnosed in children (below 14 years old). Even when Appendicitis is effectively diagnosed, the selection of the patients' treatment is not a straightforward procedure. Surgeries may be avoided in early stages of Appendicitis and medication treatment could be used instead. Studies [1] have indicated that upon 20% of the surgeries which are conducted for Appendicitis were not necessary as they were

L. Iliadis, H. Papadopoulos, and C. Jayne (Eds.): EANN 2013, Part II, CCIS 384, pp. 110–118, 2013.
© Springer-Verlag Berlin Heidelberg 2013

conducted to healthy people or to people suffering from Appendicitis in early stages. For these reasons computer-aided methods should be used to enable the accurate diagnosis of Appendicitis and the prediction of its appropriate treatment.

A variety of Computational Intelligence techniques have been applied so far for the computational diagnosis of Appendicitis in children. These are mainly based on Artificial Neural Networks [2], Support Vector Machines [3], Bayesian Methods [4], Decision Trees [5] and Random Forests [6]. Despite the encouraging results of the aforementioned techniques none of them provide 100 % accuracy in their prediction. Thus, they failed to have practical utilization as a false prediction may lead to patient's death or affect his life. Moreover, all these methods emphasized in the efficient diagnosis of Appendicitis without predicting the optimal treatment for the patients.

In the present paper, we present a new standalone tool named AppendicitisScan Tool which exceeds limitations of existing computational techniques. This tool deploys an ensemble method which combines three state-of-the-art computational intelligence methodologies (Artificial Neural Networks, Support Vector Machines and Random Forests) to achieve high accuracy in predicting Appendicitis in children and its appropriate treatment. In order to produce a firm result, it requires a consensus between the three independent prediction models. Otherwise, it returns a "No Prediction" response and this suggests that an expert will investigate this case with more consideration. Additionally, AppendicitisScan tool automatically applies normalization and missing values estimation to improve its usability and enable its utilization by non-computer experts.

AppendicitisScan tool was applied to a public available dataset which have already been used by other methodologies. For the problem of diagnosing Appendicitis it came up with 98% correct firm predictions and for the problem of predicting the correct treatment of Appendicities patients it came up with 75% correct firm predictions. However, all other cases were responded with a "No Prediction" answer and thus AppendicitisScan Tool did not provide any erroneous answer. To further validate its performance, we applied it to four cases provided by the Karamandaneio Children's Hospital of Patras and all of them was correctly predicted for both prediction problems.

The rest of the paper is structured as following: In section 2 the data which were used in the present study are described. In section 3 the state-of-the-art classifiers, which were used, were briefly presented and the proposed method is described in detail. In section 4 experimental results are provided while in section 5 important conclusions were made and some interesting future directions are proposed.

2   Material and Data

The datasets used in the present study were based on the data which were used in [2, 7 and 6]. These data was collected from a sample of patient records of the Pediatric Surgery Department of the University Hospital of Alexandroupolis, Hellas and it consists of 516 cases, which as indicated in Table 1 are 437 (84.69%) normal (discharge, observation, no-findings cases) and 79 (15.31%) pathogenic cases with different stages of appendicitis (focal, phlegmonous, gangrenous appendicitis and peritonitis).

The first classification which this papers deals with, is the classification of normal and pathogenic cases. To train and test, the examined machine learning models, the overall dataset was split in training and test set as indicated in Table 1. In the initial dataset 15 features (clinical and laboratorial factors) were provided to be used as inputs. These parameters are: sex, age, religion, demographic data, duration of pain, vomitus, diarrhea, anorexia, tenderness, rebound, leucocytosis, neutrophilia, urinalysis, temperature and constipation. Demographic data is one feature that limits the applicability of the extracted models as it is not available for all areas. For this reason, in the present study only the other 14 features were used as inputs.

The second classification problem which was examined in the current paper is the one of classifying the appendicitis patients to those whose cases should be confronted with medication and to others who requires a surgeon operation. The dataset splitting for the second classification problem is also described in Table 1.

Table 1. Datasets

Type of clinical Case	Total Number of Cases	Training Set' Cases	Test Set's Cases	Assigned Label for the Problem of Predicting Appendicitis	Assigned Label for the Problem of Predicting the Best Treatment For Appendicitis Cases
Discharge (Normal Appendicitis)	206	157	49	0	-
Observation (Normal Appendicitis)	186	150	36	0	-
No findings (Normal Appendicitis)	45	42	3	0	-
Focal Appendicitis	34	31	3	1	0
Phlegmonous Appendicitis	29	23	6	1	1
Gangrenous Appendicitis	8	7	1	1	1
Peritonitis	8	6	2	1	1

To further validate the performance of the AppendicitisScan tool, Karamandaneio Children's Hospital of Patras has already initiated its experimental application. From this procedure 4 more clinical cases have been extracted. Two of them were patients suffering from appendicitis on a stage that requires surgeon and two of them were children which were studied and proved to be healthy.

3   Machine Learning Methodologies

Three state of the art machine learning methods were used for the examined classification tasks in the present study. The following subsections present these classification methods, the proposed methodology and the implemented AppendicitisScan Tool.

3.1   Artificial Neural Networks

Artificial Neural Networks [8] exist in several forms in the literature and have already been used in various applications from different research domains.

The most commonly used architecture is the Multi-Layer Perceptron (MLP). A MLP consists of at least three layers of nodes: input, hidden and output layers. The network processes information starting with the input nodes which contain the value of the explanatory variables. Then, each node of the hidden layers transfers incoming information through a nonlinear activation function and provides it to the output layer. The node of every layer has weighted connections to all other nodes of the next layer.

Training a MLP Neural Network is the procedure of finding the optimal weights to enable the overall network to map the input value of the training data to the corresponding output value. This procedure starts with randomly chosen weights. The crucial point in this procedure, is to stop the training when the network has learned the training data but not the outliers and their noise. To accomplish it we deploy a methodology called early-stopping through cross validation which is considered one of the best methods for avoiding overfitting. The learning phase of an MLP is most of the times assigned to the Back Propagation (BP) Algorithm. The main problems in acquiring a near to optimal MLP is the tuning of BP algorithm's parameter and the optimization of its structure. To optimize the structure of the MLP we applied cross validation using only the training sets. To surpass the problem of tuning the parameters of the BP algorithm we used the Levenberg–Marquardt back propagation algorithm [9] which deploys an adaptive parameter tuning procedure which varies the parameters during the training procedure to balance between exploration and exploitation in the optimization phase.

3.2   Support Vector Machines

Support vector machines (SVM) belong to the wider category of Kernel Methods which are supervised learning methods that can be applied in classification and regression problems. SVMs are mainly based on the generalized algorithm developed by Vapnik [10]. And they have already been applied in many scientific problems.

The main advantage of SVMs is that they are based on a very strong mathematical background which assures that for any original separable set of two-class objects they are able to find the optimal hyperplanes that separates them providing the bigger margin area between the two hyperplanes. Moreover, they can be extended to be able to separate classes that cannot be separated with a linear classifier. This is achieved by transposing the coordinates of the objects into a feature space using nonlinear functions. The feature space in which every object is projected should be a high dimensional one where the two classes are separated with the linear hyperplane. Among the various available Kernel functions we deployed the Radial Basis Function (RBF) due to its efficiency in providing very high performance classification results.

The main problem when using RBF SVMs, is to optimize the regularization parameter C and the RBF's parameter gamma on parallel. To achieve this goal we used the findings of Keerthi and Lin at 2003 [11]. More specifically, linear kernels are a special case of RBF and this relationship could be used to tune the parameters of RBF SVMs without having to search all the C and gamma combinations. Thus, first we perform a linear search using cross validation, to locate the best value of the C parameter in Linear SVMs. Then using this value we perform a cross validation search to optimize the parameter gamma of RBF SVMs.

3.3   Random Forests

Another sophisticated machine learning method, such as the SVMs, is the Random Forest (RF) method [11]. In contrast to the simple decision trees, RFs are ensemble classifiers that create many decision trees simultaneously where each node uses a random subset of the features considered. More specifically, RFs are a combination of tree predictors such that each tree depends on the values of a random vector sampled independently and with the same distribution for all trees in the forest. The idea of growing an ensemble of trees and letting them vote for the most popular class has led to significant improvements in classification accuracy. It has already been proven that with a growing number of trees in a RF the generalization error converges to a single value. This error highly depends on the accuracy of the individual trees of the forest and the correlation between them. A cross validation search using the training set was performed to optimize the number of trees and the percentage of the initial inputs which are to be used by every tree.

3.4   AppendicitisScan Tool

Every machine learning method so far has specific advantages and disadvantages. When using only one machine learning model in a biomedical application, the uncertainty of the predictions limits its applications to research usage as the possibility of getting a wrong prediction may cost or affect a human life. For this reason in the present paper we proposed the combination of MLP NNs, SVMs and RFs to acquire an ensemble method which will provide results only when all independent models achieve consensus.

The proposed method starts with two preprocessing steps. The first preprocessing step refers to the calculation of missing values for the inputs of our models. This is very important to make the proposed tool able to have general usage and not being limited to clinical cases with all features estimated. This step is accomplished by using the KNN-Impute method [12] which calculates a missing value of a sample by using the mean value of its K nearest neighbors. In the second step categorical and continuous features are transformed to arithmetic values and they are all normalized to the interval [-1, 1]. This normalization step is used to avoid giving emphasis to features with higher absolute values.

For the problems of predicting Appendicitis and classifying its patients to ones who need medication and others who need surgeon, all the presented classification methods were trained. A cross validation procedure was used for each one of them, to locate their optimal parameters. The function described in equation 1 was used to evaluate and compare different parameter values

$$Evaluation\_Function = Accuracy + \sqrt{Sensitivity * Specificity} \quad (1)$$

This function takes into account not only the classification accuracy but also the product of Sensitivity and Specificity. This function was used as the examined data-sets were imbalanced. Thus, in order to make a fair comparison, the accuracy metric is not sufficient.

AppendicitisScan Tool returns a prediction only when all machine learning methods achieve a consensus. It has been implemented using MatlabR2011a package and it has been deployed as a standalone tool, providing a user friendly interface. This tool is available upon request.

The flowchart of AppendicitisScan Tool is presented in Figure 1.

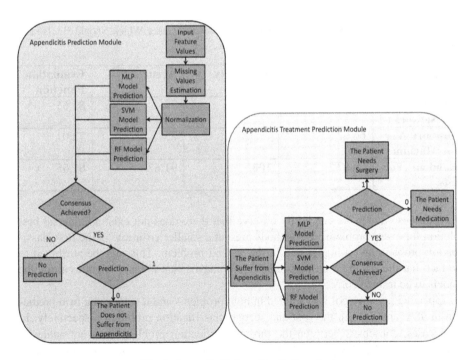

Fig. 1. AppendicitisScan Tool's flowchart

4   Experimental Results

The three presented machine learning models were applied to the tasks of prediction Appendicitis and to the task of predicting whether an Appendicitis patient requires surgeon operation or not. The datasets which were used are described in Table 1.

Due to the stochastic nature of the examined machine learning methodologies, the experiments were repeated ten times and in Tables 2 and 3 we present their results in the test sets.

Table 2. Experimental Results for Predicting Appendicitis

Method	Sensitivity	Specificity	Accuracy	Evaluation Function
Artificial Neural Networks	97,73	100	98	98,86
Support Vector Machines	96,59	100	97	98,28
Random Forests	97,73	100	98	98,86

Table 3. Experimental Results for Predicting Appendicitis Cases Which Should Be Treated With a Surgery

Method	Sensitivity	Specificity	Accuracy	Evaluation Function
Artificial Neural Networks	**100**	66,67	75	**81,65**
Support Vector Machines	66,67	77,78	75	72,01
Random Forests	66,67	**100**	**91,67**	**81,65**

From these tables we can easily observe that there does not exist a universal best solution for every problem. All methods present a smaller error rate in the first classification problem and a higher one in the second problem. This may be attributed to the fact that the second dataset is smaller and maybe is it is insufficient to allow the models to be trained correctly.

AppendicitisScan Tool was applied in both problems and it achieved a firm prediction in 98% and 75% in the first and second classification problems respectively. In other cases, consensus between the three classification problems was not achieved and thus it responded with a "No prediction" result. It is noteworthy that AppendicitisScan Tool never responded with a firm erroneous answer.

To further validate the performance of the proposed tool, it was applied to the data collected from the Karamandaneio Children's Hospital of Patras. For all 4 cases firm correct answers were extracted.

5   Conclusions and Future Work

The problems of predicting childrens' Appendicitis and its appropriate treatment method have been studied in the present paper. A new tool called AppendicitisScan was introduced. This tool overcomes the limitations of accuracy and limited functionality which other existing methodologies have. It requires the input of only 14 easily accessible feature values for every patient. Moreover, even the absence of some of these values does not constrain the proposed tool. Using simple missing values calculation methods, it is able to achieve accurate classification even if some of the 14 features are not available. The proposed tool's predictions are based on an ensemble classification method which combines Artificial Neural Networks, Support Vector Machines and Random Forests. To raise its robustness and reassure the validity of its prediction, it provides a firm prediction only if all the independent prediction models come to an agreement. Otherwise, it outputs a "No prediction" answer and the opinion of an expert is required to perform the prediction.

When AppendicitisScan was applied in public available data and in a few new cases, it achieved to avoid making a single false prediction. This finding indicates that the proposed tool is appropriate to have an auxiliary utilization together with the expert doctors. It could be used to alleviate their work load by diagnosing Appendicitis in a semi-automatics manner and predicting the correct patients' treatment for most of the cases. With AppendicitisScan Tool's usage only a few cases, for which a "No prediction" result is returned, should be really examined by experts. This will save them a lot of time.

As the utilization of AppendicitisScan rises, more cases will be available. This is very crucial for the problem of predicting the treatment of Appendicitis, as the available dataset right now is considered very small. Thus, higher prediction accuracy is expected to be achieved with more training samples for this problem. Another interesting future direction would be to apply boosting and leverage techniques [14] to technically raise the size of the dataset for the prediction of Appendicitis treatment.

Another limitation of AppendicitisScan so far is its lack of interpretability. Domain experts cannot so far extract any knowledge about the way the inputs are provided to perform the predictions, as it is a combination of three complex computational intelligence models, that do not provide any insight about their classification process. To overcome this limitation, in its next version, AppendicitisScan tool should be complemented with Fuzzy Rules extraction methods [15, 16], which are able to extract interpretable classification rules from data with a loss in performance.

References

[1] Sim, K.T., Picone, S., Crade, M., Sweeney, J.P.: Ultrasound with graded compression in the evaluation of acute appendicitis. J. Natl. Med. Assoc. 81(9), 954–957 (1989)

[2] Mantzaris, D., Anastassopoulos, G., Adamopoulos, A., Stephanakis, I., Kambouri, K., Gardikis, S.: Selective Clinical Estimation of Childhood Abdominal Pain based on Pruned Artificial Neural Networks. In: Proceedings of the 2007 WSEAS Int. Conference on Cellular and Molecular Biology-Biophysics and Bioengineering, Athens, Greece, August 26-28, pp. 50–55 (2007)

[3] Kentsis, A., Lin, Y.Y., Kurek, K., Calicchio, M., Wang, Y.Y., Monigatti, F., Campagne, F., Lee, R., Horwitz, B., Steen, H., Bachur, R.: Discovery and Validation of Urine Markers of Acute Pediatric Appendicitis Using High-Accuracy Mass Spectrometry. Ann. Emerg. Med. 55(1), 62–70 (2010)

[4] Sakai, S., Kobayashi, K., Nakamura, J., Toyabe, S., Akazawa, K.: Accuracy in the Diagnostic Prediction of Acute Appendicitis Based on the Bayesian Network Model. Methods Inf. Med. 46(6), 723–726 (2007)

[5] Ting, H.W., Wu, J.T., Chan, C.L., Lin, S.L., Chen, M.H.: Decision Model for Acute Appendicitis Treatment With Decision Tree Technology—A Modification of the Alvarado Scoring System. Journal of Chinese Medical Assiciation 73(8), 401–406 (2010)

[6] Adamopoulos, A., Ntasi, M., Mavroudi, S., Likothanassis, S., Iliadis, L., Anastassopoulos, G.: Revealing the Structure of Childhood Abdominal Pain Data and Supporting Diagnostic Decesion Making. In: Palmer-Brown, D., Draganova, C., Pimenidis, E., Mouratidis, H. (eds.) EANN 2009. CCIS, vol. 43, pp. 165–177. Springer, Heidelberg (2009)

[7] Anastasopoulos, G., Iliadis, L.: Intelligent hybrid modeling towards the prognosis of abdominal pain. International Journal of Hybrid Intelligent Systems 6(4), 245–255 (2009)

[8] Haykin, S.: Neural Networks: A Comprehensive Foundation. Prentice Hall (1998)

[9] More, J.: The Levenberg-Marquardt algorithm: Implementation and theory. In: Numerical Analysis Lecture Notes in Mathematics, vol. 630, pp. 105–116 (1978)

[10] Vapnik, V.: The Nature of Statistical Learning Theory. Springer (2000)

[11] Keerthi, S., Lin, C.J.: Asymptotic behaviours of support vector machines with Gaussian kernel. Neural Computation 15, 1667–1689 (2003)

[12] Breiman, L.: Random Forests. Machine Learning 45(1), 5–32 (2001)

[13] Batista, G., Monard, M.C.: K-Nearest Neighbour as Imputation Method: Experimental Results. Technical Report 186, ICMC-USP (2002)

[14] Meir, R., Ratsch, G.: An Introduction to Boosting and Leveraging. In: Mendelson, S., Smola, A.J. (eds.) Advanced Lectures on Machine Learning. LNCS (LNAI), vol. 2600, pp. 118–183. Springer, Heidelberg (2003)

[15] Papadimitriou, S., Terzidis, K.: Efficient and Interpretable Fuzzy Classifiers from Data with Support Vector Learning. Intelligent Data Analysis 9(6), 527–550 (2005)

[16] Papadimitriou, S., Terzidis, K., Mavroudi, S., Skarlas, L., Likothanassis, S.: Fuzzy rule based classifiers from support vector learning. WSEAS Transactions on Computers 4(7), 661–670 (2005)

Mining the Conceptual Model of Open Source CMS Using a Reverse Engineering Approach

Vassiliki Gkantouna[1], Spyros Sioutas[2], Georgia Sourla[1],
Athanasios Tsakalidis[1], and Giannis Tzimas[3]

[1] Department of Computer Engineering and Informatics, Faculty of Engineering,
University of Patras, Patras, Greece
[2] Department of Informatics, Ionian University, Corfu, Greece
[3] Department of Applied Informatics in Management & Economy, Faculty of Management and
Economics, Technological Educational Institute of Messolonghi, Messolonghi, Greece

Abstract. Model-driven engineering has become the emerging standard for software development focusing on the use of models as first-class citizens. One possible field of application of such model-driven approaches can be the open source Content Management Systems (CMS) domain. Typically, CMS are built using the source-code-oriented software development process raising issues related to usability, performance and other qualities of service in an application's lifecycle. To overcome these issues, the use of model-driven approaches in the development of CMS-based web applications (WAs) can be particular beneficial.

To this end, we propose a model-driven reverse engineering approach for automatic mining of the conceptual model of existing WAs developed using the widely used CMS Joomla! by applying data mining techniques. This methodology can be used to form the cornerstone of an evaluation framework for Joomla!-based WAs either in the design or maintenance process.

Keywords: Model-driven development, reverse engineering, web application, content management systems, Joomla!, data mining, WebML.

1 Introduction

The unprecedented growth of the World Wide Web in the last years has set the scene for the appearance of a new type of web software framework called CMSs oriented towards the management and publishing of digital content. The usage of CMSs as the base platform for the development of new WAs is rapidly increasing in the web development community as demonstrated by the plethora of CMSs available today.

CMSs are used as software products with out-of-the-box functionalities for WAs providing technical users with a standardized platform for web development and maintenance. Indeed, most CMS-based WAs can be realized with merely configuration of the CMS instead of software engineering from scratch enhancing this way the system's reusability and efficiency. Still, configuration of a CMS for the creation of the so-called data-intensive WAs can be complex and time consuming since the

L. Iliadis, H. Papadopoulos, and C. Jayne (Eds.): EANN 2013, Part II, CCIS 384, pp. 119–128, 2013.
© Springer-Verlag Berlin Heidelberg 2013

cur-rent CMS platforms have difficulties to support the design and implementation of such kind of WAs. It is a common case when business analysts have defined business requirements but they do not know how to map those to CMS configuration due to the fact that they do not have the whole picture of the WAs underlying architecture. Indeed, this is often an error-prone task, as it is highly dependent on the developer's interpretation of those requirements. The problem originates from the development nature of CMS-based WAs, typically based on the traditional source-code-oriented development process, in which source-code is the primary artifact, and design models and documentation are considered support artifacts. Therefore, it is obvious that the traditional methods used for the development of CMS-based WAs, tend to be ineffi-cient especially when the WA's complexity increases.

On the other hand, Model-Driven Engineering (MDE) is an evolving and promising approach to software development and maintenance process. The key concept of MDE approaches lies in the use of models as first-class citizens throughout the engineering lifecycle where the models used by business analysts can be traced towards more detailed models used by software engineers. This way, MDE allows developers to alleviate the complexity of platforms and express the domain concepts effectively instead of focusing on source code and implementation details.

Additionally, the integration of MDE approaches with Reverse Engineering (RE) techniques in CMS-based WAs development could help obtain high maintainability of the underlying system. RE methods support the obtainment of representations and views from an existing target system at a higher level of abstraction and the identification of its fundamental components by retrieving its constituent structures. As a result, RE approaches enable developers to focus on the overall system architecture without paying attention on the implementation details simplifying this way the maintenance, evolution and restructuring tasks.

Under this point of view, we propose a model-driven RE approach for the automatic mining of the conceptual model of existing WAs developed using the widely used CMS, Joomla! [1], by applying data mining techniques. The recovered models are specified by referring to the Web Modeling Language (WebML) [2] for the modeling of data-intensive applications. The goal of this work is to propose a methodology with the hope that it can be used as a common ground to form the cornerstone of an overall evaluation framework for Joomla!-based WAs either in the design or maintenance process.

The remainder of this paper is organized as follows: Section 2 presents the related work that is relevant to this research work, Section 3 describes the key concepts of the WebML Hypertext Model, while Section 4 presents a detailed analysis of our RE approach. Finally, Section 5 presents the tool developed to support the approach and Section 6 provides concluding remarks and discusses future steps.

2   Related Work

In this work, we are addressing a Model-Driven RE approach specifically oriented towards the class of CMS-based WAs. Many other approaches have been proposed in the literature that addresses some of the issues presented in this paper but the essential difference lies in the fact that they are not focused on CMS-based platforms

In general, WAs Model-Driven RE studies differ in the WA's aspect they focus on, the way they consider the level of abstraction of the recovered model and the formalism they adopt to represent it. Several studies [3] aim at recovering an architectural view of WA depicting its components and the relationships among them at different levels of detail. Other UML-based approaches focus on abstracting a description of WA's functional requirements utilizing UML case diagrams. In [4] the authors present an approach to recover the architecture of WAs. The approach generates UML models from existing WAs through static and dynamic techniques. UML diagrams are extracted to depict the static, dynamic and behavioral aspect of WAs.

Furthermore, other approaches are based on Ubiquitous Web Application design methodology (UWA) for the user-centered RE of the business processes implemented by a WA. In [5, 6] an approach is described for recovering user-centered conceptual models from an existing WA representing the WA's contents, relationships between contents and views on contents, as perceived by the final users of the application. Moreover, a new and upcoming trend comes with approaches [7, 8] based on WebML, a Web modeling language for specifying the content, composition, and navigation features of hypertext applications. Following this trend, we decided to utilize WebML notation for the CMS-based WAs modeling since it is a robust modeling language and supported by the WebRatio CASE tool [9]. WebML offers a set of visual primitives for defining conceptual schemas that represent the organization of the application contents and of the hypertext interface. These primitives are also provided with an XML-based textual representation, which allows specifying additional detailed properties that cannot be conveniently expressed in terms of visual notation. This is very important since it provide us the ability to model Joomla!-based WA's elements in depth.

3   The WebML Hypertext Model

In an effort to provide a comprehensive picture of WAs, the majority of the RE approaches define the development of WAs by means of three main models namely the data model, the navigational model and the presentation model. The data model specifies the application data in terms of entities and relationships representing the semantic associations between them. The navigational model describes how pages and page components are linked to form the hypertext and finally the presentation model expresses the layout and the graphic appearance of webpages.

Towards this end, WebML defines the Data model, the Hypertext model and the Presentation model [2]. In this work, we focus mainly on the Hypertext model which specifies the composition and the navigation of WAs reflecting this way the WA's structure and behavior. Regarding the Data model, it was not feasible to obtain a representation of the WA's database schema as most WAs do not allow the access to the underlying database. Nevertheless, we manage to capture the data distribution by applying semantic similarity methods on the content published by the HTML pages.

The Hypertext model is composed of two sub-models namely the composition and the navigation model. The composition model specifies which pages compose the hypertext, and which content units make up a page. The key ingredients of WebML are pages, units, and links, organized into modularization constructs called areas and site views. Units are the atomic pieces of publishable content used to publish the information described in the data model and they are the building blocks of pages. Five types of units are predefined in WebML to compose pages: data, multi-data, index (and its variants multichoice and hierarchical), entry, and scroller. Each unit is associated to one underlying entity, from which the content of the unit is computed. These five basic types of content units can be combined to represent Web pages of arbitrary complexity. Pages are typically built by assembling several units of various kinds, and they are the actual interface elements delivered to the users. Page and units do not stand alone, but are linked to form a hypertext structure. Links express the possibility of navigating from one point to another one in the hypertext, and the passage of parameters from one unit to another unit, which is required for the proper computation of the content of a page. The navigation model expresses how pages and content units are linked to form the hypertext. A detailed description of the Hypertext model is provided in [2].

4   The Reverse Engineering Process

This section describes the model-driven RE approach that we have defined to automatically mine the conceptual model of a target existing Joomla! WA utilizing data mining techniques. The whole process is based on the following two considerations: firstly, the composition model can be recovered from Joomla!-based WAs by identifying specific structural elements (particular types of Joomla! extensions) in HTML pages and mapping these elements to their corresponding WebML representation. These elements are the components and modules and they are indicative of the HTML page's layout structure. Secondly, the navigation model can be extracted by identifying the hyperlinks between HTML pages as well as the possible navigation flows between the structural components of the WA created by Joomla! navigation modules such as menus and breadcrumbs, and mapping the identified links to the corresponding WebML representation.

4.1   Modeling Joomla! Elements to WebML Units

From the aforementioned reasons, it seems necessary to define an appropriate WebML representation for WA's pages and page components. Based on the default Joomla! installation, we noticed that the webpages have a fixed structure composed by five parts: the central part containing the main content and the peripheral parts namely the top, left, right and bottom containing atomic blocks of content (Fig.1.a). Furthermore, we noticed that the layout of the content published in the central part is defined by the particular webpage's component whereas the content displayed on the peripheral parts is defined by the published modules. Based on these observations, we defined the WebML representation of a Joomla! WA's webpage which consists of two conjunctive nested pages (AND subpages) namely the Modules subpage containing all the modules that lie within the page and the Layout subpage modeling the way the particular component of the webpage displays content (Fig.1.b).

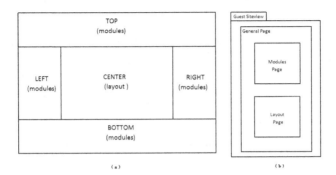

Fig. 1. (a) Joomla! HTML page structure (b) WebML representation of a webpage

In the same fashion, we define WebML representations for the Components and Modules extensions. Components produce the major content of WAs and have one or more "views" that control how the content is displayed in the webpage's central part. The basic categories of components are: content, contact, weblinks, news feed, user, search and administrator, each one having a number of subcategories. For example, the content component includes the following subcategories: single article, article categories, article category blog, article category list, featured articles and archived articles. Based on the way they publish content, we resulted in WebML representations consisting of a combination of content units, for every type of the available components. Some indicative examples are shown in Figure 2 (Fig. 2).

Modules are small blocks of content that can be displayed in various positions on the peripheral parts of a webpage. The core of Joomla! includes twenty four modules ranging from login to search to random image. The basic module categories are: navigation, content, user, display and utility, each one having corresponding subcategories. For example, the content module includes the most read articles, news flash, latest articles, archived articles, related articles, article categories and article category. Again, we defined proper WebML representations for every module reflecting the module's functionality and the way it delivers the content to the user. Some repre-sentative examples are shown in n the Figure 3(Fig. 3).

4.2   The Reverse Engineering Approach

Driven by the considerations described in the previous section, we have developed a RE approach that is able to mine the hypertext model from existing Joomla!-based WAs. The proposed approach is made up of two phases that enable the mining and the recovery of the following models: the constitution and the navigation model which in turn synthesize the WA's hypertext model.

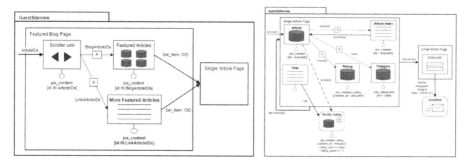

Fig. 2. Featured blog and single article WebML representations

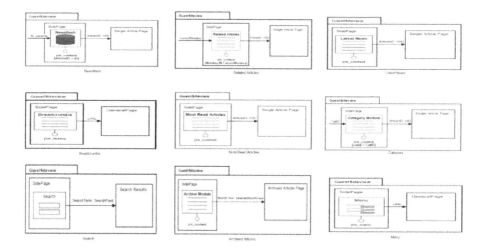

Fig. 3. Modules WebML representations

As a first step, the HTML pages of the Joomla!-based WA to be modeled are captured by using a web crawler in order to perform source code analysis-parsing. More specifically, having in mind which and how Joomla! elements will be modeled to WebML units, the parsing aims to find such items in the source code of a webpage that would strongly imply the presence of those elements in that particular page. We have noticed that the values of the class attribute of an HTML page can characterize the various structural elements facilitating this way the identification of the published components and modules. We refer to these values as keywords. Based on this fact, we decided to build our mining technique on the identification of these keywords. We have carefully collected the keywords used for the identification of all components and modules of the default Joomla! installation after a thorough study of their HTML specification. As a result, by parsing the HTML code of a page, we can identify the keywords it contains and this way to find out the component and modules made up the structure of that particular page. For instance, the keyword describing a single article is the word "item-page" and the keyword for a login module is the word "userdata".

Another aspect of WAs that must be modeled is the implying functionalities performed as a result of potential navigational flows. In WebML, this is described by using operation units [2]. For this reason, we have examined the possible navigation "scenarios" that could imply a potential operation and modeled them to the appropriate WebML operation units. For the identification of those scenarios, we searched again for specific keywords in the source code of webpages. For example, when a page contains an email icon is modeled as a page having a link to a sendmail unit. As a result, given the HTML code of a page, based on these keywords we can mine the component used and all the displayed modules abstracting a view of the page's layout structure and behavior.

In the following diagram, the RE process for mining of the composition and navigation model is presented.

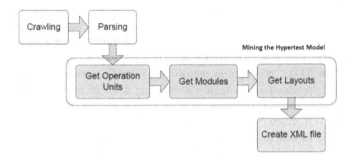

Fig. 4. The Reverse Engineering methodology for mining the WA's Hypertext model

4.3   Mining the Composition Model

In the above diagram we can distinguish three main tasks, corresponding to the WebML modeling concepts that the process is able to identify: the operation units, the content units modeling the modules and the content units modeling the components displaying in HTML pages of the WA. The identification of the operation units is carried out by searching for specific keywords in the HTML code of the WA as mentioned before. The possible scenarios currently modeled are: user registration and registration tasks, login task, reset and remind password tasks, content vote and content rating tasks and send email task. The specification of their WebML representation was based mainly on the functionality they perform. Similarly, the identified components and modules in the parsing phase had to be modeled to their WebML representation. Based on the particular nature of each module and component, at this step we examine their publishing properties in order to achieve the most representative WebML representations. For example, in the case of a menu we compute its depth in order to decide if an index or a hierarchical index unit models it best.

4.4   Mining the Navigation Model

Units and pages do not exist in isolation, but must be connected through links to form the hypertext structure of the WA. Links abstract and generalize the fundamental notion of hypertexts, the concept of the HTML anchor. The following practical cases, referred to an HTML-based hypertext, are all examples of what can be considered an anchor: an HTML anchor tag with an href attribute that refers to another page, an HTML anchor tag with an href attribute that refers to the same page and the confirm button of an HTML form used for searching.

To capture the above cases in the case of Joomla!-based WAs, during the parsing phase, we identified all the links (by recording all the occurrences of the keyword "href" in source code) between pages as well as links created by modules like the ones displayed by a breadcrumb module. Especially in the case of the captured operation units, the links are modeled as OK and KO-links expressing the concept of operation success and failure, respectively.

4.5   Creation of the XML File

Based on the recovered composition and navigation model, we finally compose the hypertext model of the target WA in its textual Extensible Markup Language (XML) syntax. The XML file follows a specific structure dictated by the WebML Document Type Definition (DTD) provided in the official WebML website [10]. The Joomla!-based WA is represented as a siteview containing a set of pages modeling the real world WA webpages and operation units modeling the possible operations implemented by the WA. Each page is comprised from two subjunctives nested pages, the Modules and the Layout pages modeling the modules and the component respectively identified in the specific page. Links connecting pages or content units model the WA navigation behavior.

We mentioned above that we built our mining technique based on specific keywords corresponding to Joomla! elements supported by the Joomla! default installation. It must be stressed that this assumption does not restrict the proposed methodology to be applied only to WAs based on the template provided in default installation. Instead, in the case of an alternative template, the tool (http://alkistis.ceid.upatras.gr/research/modeling) supporting the proposed methodology takes as input an XML file containing the mapping of the additional modules and components supported by the alternative template to the ones of the default template. Furthermore, since WAs do not allow access to the underlying database, we could not abstract the WA's structural schema. However, we have used some semantic similarity methods on the content published by the HTML pages in order to capture the data distribution. More specifically, we utilize semantic similarity measurement techniques based on the Wordnet lexical database [11] and we apply them in the content displayed on each page. This way, we obtained a realistic perception of the data distribution among web pages.

5   Tool Support

In the context of this work, a tool supporting the proposed RE methodology has been developed. It can be accessed at http://alkistis.ceid.upatras.gr/research/modeling. The tool takes as input the Uniform Resource Locator (URL) of the Joomla!-based WA and produces as output an XML file describing the WA's conceptual model according to the WebML textual XML syntax. Having this URL as a starting point, the crawling phase begins and records WA's pages. The next phase is the parsing of the HTML code of the identified pages. To perform the parsing, we utilize the Html Agility Pack [12], an agile HTML parser for .NET applications. During parsing the HTML code of a page, the tool searches for the occurrence of the specific keywords within the code, which are indicative of the component and the modules enclosed in that particular page as we mentioned previously. After the iterative application of this step to all the identified HTML pages, the Composition model is abstracted. Then, for the mining of the navigation model, the HTML code of the pages is parsed, searching for the occurrences of the href attribute which implies the presence of hyperlinks. Finally, the identified component and modules for each page, which are actually the stepping stones for the composition and navigation model, are interpreted into their corresponding WebML textual syntax composing in turn the WA's hypertext schema. The general architecture of the tool is presented in the following figure.

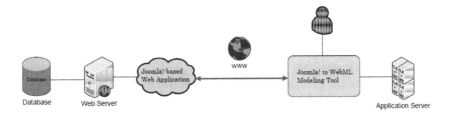

Fig. 5. The general architecture of the tool

6   Conclusions and Future Work

In conclusion, we present a novel RE approach for the automatic mining of conceptual models from existing WAs developed by Joomla! referring to the WebML notation. Starting from the WA's source code, we extract the composition and navigation model which eventually are combined to form the WA's hypertext model. A tool supporting   the   proposed   methodology   can   be   accessed   at http://alkistis.ceid.upatras.gr/research/modeling. Now, having obtained the conceptual model of Joomla!-based WA, the next logical question that arises is the following: in which ways it can be utilized. As we mentioned above, our vision is to form an overall evaluation framework for Joomla!-based WAs. Currently, our top priority is to extend the proposed methodology in order to integrate techniques for the discovery of design patterns or potential recurrent design solutions that lie in the recovered hypertext model of Joomla!-based WAs. By applying the approach to a large number of Joomla!-based WAs, we believe that we can identify templates for the Joomla! framework for specific domains and discover new design patterns. Actually, the recovered models can be exploited by a previous work [13] of some of the authors oriented to the discovery of re-usable design solutions in Web conceptual schemas.

In the future, we are planning to extend the proposed methodology to model WAs developed by other types of CMS. Another interesting usage of the recovered model is to use it as a common ground for the specification of a meta-model that can capture the key concerns required to model and implement Joomla!-based WAs. Moreover, it can serve as the base of new modeling languages specified within the Joomla! domain.

References

1. Joomla! Content Management System, http://www.joomla.org/
2. WebML modeling language, http://www.webml.org/webml/page1.do
3. Di Lucca, G.A., Fasolino, A.R., Tramontana, P.: Reverse engineering web applications: the ware approach. Journal of Software Maintenance and Evolution 16, 71–101 (2004)
4. Weijun, S., Shixian, L., Xianming, L.: An approach for reverse engineering of web applications. In: International Symposium on Information Science and Engieering, vol. 2, pp. 98–102 (2008)
5. Bernardi, M.L., Lucca, G.A.D., Distante, D.: The re-uwa approach to recover user centered conceptual models from web applications. International Journal on Software Tools for Technology Transfer 11(6), 485–501 (2009)
6. Bernardi, M.L., Cimitile, M., Distante, D.: Web applications design recovery and evolution with RE-UWA. Journal of Software: Evolution and Process (2012)
7. Katsimpa, T., Panagis, Y., Sakkopoulos, E., Tzimas, G., Tsakalidis, A.K.: Application modeling using reverse engineering techniques. In: SAC 2006, pp. 1250–1255 (2006)
8. Faliagka, E., Rigou, M., Sirmakessis, S., Tzimas, G.: A Tool for Extracting Model Clones From a Conceptual Schema. In: IASTED Conf. on Software Engineering 2006, pp. 39–44 (2006)
9. The WebRatio Development Environment, http://www.webratio.com/
10. WebML Document Type Definition,
    http://www.webml.org/webml/page33.do?UserCtxParam=0&GroupCtx Param=0&ctx1=EN
11. WordNet-based semantic similarity measurement,
    http://www.codeproject.com/Articles/11835/WordNet-based-semantic-similarity-measurement
12. The Html Agility Pack, http://htmlagilitypack.codeplex.com/
13. Panagis, Y., Sakkopoulos, E., Sirmakessis, S., Tsakalidis, A.K., Tzimas, G.: Discovering Re-usable Design Solutions in Web Conceptual Schemas: Metrics and Methodology. In: Lowe, D.G., Gaedke, M. (eds.) ICWE 2005. LNCS, vol. 3579, pp. 545–556. Springer, Heidelberg (2005)

Representation of Possessive Pronouns in Universal Networking Language

Velislava Stoykova

Institute for Bulgarian Language, Bulgarian Academy of Sciences,
52, Shipchensky proh. str., bl. 17, 1113 Sofia, Bulgaria
vstoykova@yahoo.com

Abstract. The paper[1] presents a complex approach to multilingual representation of possessive pronouns information using statistically-based knowledge discovery techniques of Sketch Engine (SE) software. It analyses semantics, grammar features and related ideas, principles and problems for formal representation of possessive pronouns in light of multidisciplinary complexity of the task. The analysis of representation of possessive pronouns for three different languages in frameworks of Universal Networking Language (UNL) is presented with respect to linguistic motivation used for representations and for machine translation.

Keywords: Knowledge Discovery, Linguistic Data Mining, Knowledge Representation, Semantic Networks, Machine Translation.

1 Introduction

Possessive pronouns exist in almost all European languages and they share similar semantics and grammar features. Recently, the statistically-based techniques and semantic networks were successfully used to represent both lexical and grammar information with multilingual applications. The Sketch Engine (SE) software was successfully applied to extract semantic relations whereas semantic networks like DATR language for lexical knowledge representation [4], WordNet [5] and Universal Networking Language (UNL) [14] were used for formal representation.

For all of them (as for example SE [6] and DATR [3,10,13]) there are programming applications made for lots of languages which suggest various approaches and techniques that can be successfully used for multilingual application in machine translation. Further, we are going to present analysis of semantic and grammar features of possessive pronouns and related approaches for UNL formal multilingual representations.

2 The Sketch Engine Statistical Approaches

The Sketch Engine (SE) [7,8] software for processing electronic text corpora allows use of combined statistical approaches for semantically similar words extraction and comparison of results between several corpora which allows multilingual application. It performs search for keywords, concordances, collocations and co-occurrences for a

---

[1] The research is supported by the project BG051PO001-3.3-05/0001 "Science and Business" of the Ministry of Education, Youth and Science of Bulgaria.

L. Iliadis, H. Papadopoulos, and C. Jayne (Eds.): EANN 2013, Part II, CCIS 384, pp. 129–137, 2013.

related word. The keywords are evaluated on the base of word frequency lists and can be generated by using statistical search. Collocations and co-occurrences are words which are most probably to be found with the related keyword. We have used techniques of $T - score$, $MI - score$ and $MI^3 - score$ incorporated in the SE. For which the following terms are used: N - corpus size, f_A - number of occurrences of the keyword in the whole corpus (the size of the concordance), f_B - number of occurrences of the collocated keyword in the whole corpus, f_{AB} - number of occurrences of the collocate in the concordance (number of co-occurrences). The related formulas are as follows:

$$\text{MI-Score} \quad \log_2 \frac{f_{AB}N}{f_A f_B}$$

$$\text{T-Score} \quad \frac{f_{AB} - \frac{f_A f_B}{N}}{\sqrt{f_{AB}}}$$

$$\text{MI}^3\text{-Score} \quad \log_2 \frac{f_{AB}^3 N}{f_A f_B}$$

The SE software allows processing of large-scale electronic text corpora.

2.1 Extracting Semantic Relations

Generally, semantic conceptual relations are regarded to be horizontal and vertical.

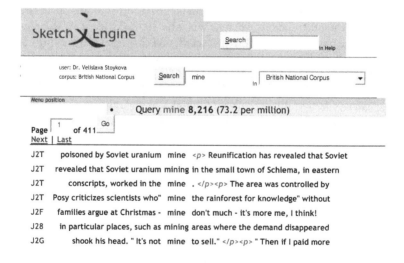

Fig. 1. The concordance of English word *mine* from BNC

The horizontal semantic relations are those of synonymy, anthonymy and meronymy. They can be evaluated by estimation of statistical semantic similarity or distance. The vertical semantic conceptual relations are those of hierarchy like hyperonymy and hyponymy.

All types of semantic relations are defined by generating related word contexts through extraction of keywords, word concordances and collocations by the use of different statistical approaches based on retrieval and clustering of statistically similar words.

2.2 Corpora Search

For our research, we have used the British National Corpus (BNC) [6] - approximately 112 million words as well as Bulgarian National Corpus (BulNC) [6] - approximately 26 million words.

The main task of possessive pronouns extraction is considerd as disambiguation task. Fig. 1 shows a concordance of English possessive pronoun *mine* from BNC and presents disambiguation of *mine* as a pronoun instead *mine* as a verb or a noun. For that, we have used statistically-based SE technique to evaluate collocations (Fig. 2).

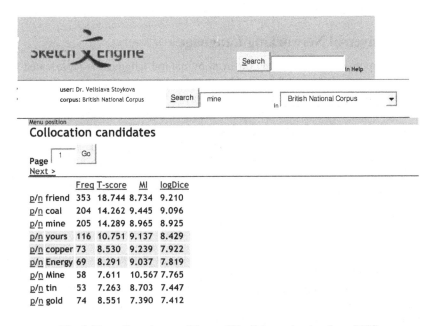

Fig. 2. The collocation candidates of English word *mine* from BNC

As a result, we have found that the most frequent words which are collocated with word *mine* are *friend, coal, mine, yours, copper*, etc. Thus, the most frequent use of *mine* is as a pronoun. The analysis improve that collocation extraction is a reliable techniques to resolve grammar features disambiguation tasks.

It is also a reliable method for extracting semantic relations like in a $friend - of - mine$ or $mine - yours$ however, the process require further human competence analysis. Moreover, for the oppostie task of knowledge representation a detailed analysis of semantics and grammar features is needed.

Similar results have been obtained from BulNC for related Bulgarian possessive pronoun which suggest a common complex approach to formal multilingual representation of possessive pronouns using semantic networks application of UNL.

3   The Semantics and Grammar Features of Possessive Pronouns

The traditional description of possessive pronouns given at academic grammar works is presented with respect to their semantics, grammar features, functions and usage. Generally, the semantics of pronouns is connected to their roles to substitute, determine, relate, and agree both with other words in the sentence or within the whole text. However, these roles are realized by using subsequent grammar features of person, number, and gender (for some languages also case, definiteness, etc.) They can be used for formal representations of pronouns with respect to computational applications.

It is important to note that related grammar features are universal for all types of pronouns and are used for multilingual purpose which can result in successful machine translation applications. Moreover, in suggested approach they are used to present both grammar and lexical information by underlay special linking hierarchical mechanism.

4   The Universal Networking Language

In the UNL approach, information conveyed by natural language is represented as a hypergraph composed of a set of directed binary labeled links (referred to as "relations") between nodes or hypernodes (the "Universal Words"(UWs)), which stand for concepts. UWs can also be annotated with "attributes" representing context information [14].

Universal Words (UWs) represent universal concepts and correspond to the nodes to be interlinked by "relations" or modified by "attributes" in a UNL graph. They can be associated to natural language open lexical categories (noun, verb, adjective and adverb). Additionally, UWs are organized in a hierarchy (the UNL Ontology), are defined in the UNL Knowledge Base and exemplified in the UNL Example Base, which are the lexical databases for UNL. As language-independent semantic units, UWs are equivalent to the sets of synonyms of a given language, approaching the concept of "synset" used by the WordNet.

Attributes are arcs linking a node onto itself. In opposition to relations, they correspond to one-place predicates, i.e., function that take a single argument. In UNL, attributes have been normally used to represent information conveyed by natural language grammatical categories (such as tense, mood, aspect, number, etc). Attributes are annotations made to nodes or hypernodes of a UNL hypergraph. They denote the circumstances under which these nodes (or hypernodes) are used. Attributes may convey three different kinds of information: (i) The information on the role of the node in the UNL graph, (ii) The information conveyed by bound morphemes and closed classes, such as affixes (gender, number, tense, aspect, mood, voice, etc), determiners (articles and demonstratives), etc., (iii) The information on the (external) context of the utterance. Attributes represent information that cannot be conveyed by UWs and relations.

Relations, are labeled arcs connecting a node to another node in a UNL graph. They correspond to two-place semantic predicates holding between two UWs. In UNL, relations have been normally used to represent semantic cases or thematic roles (such as agent, object, instrument, etc.) between UWs.

UNL-NL Grammars are sets of rules for translating UNL expressions into natural language (NL) sentences and vice-versa. They are normally unidirectional, i.e., the enconversion grammar (NL-to-UNL) or deconversion grammar (UNL-to-NL), even though they share the same basic syntax.

In the UNL Grammar there are two basic types of rules: (i) Transformation rules - used to generate natural language sentences out of UNL graphs and vice-versa and (ii) Disambiguation rules - used to improve the performance of transformation rules by constraining their applicability.

The UNL offers universal language-independent and open-source platform for multilingual applications [2]. The UNL application for English language is available but some applications for other languages like Russian [1] and Bulgarian [11,9] are available as well.

4.1   Representing Possessive Pronouns in UNL

The UNL specifications offer types of formal grammar rules particularly designed to present pronouns grammar features of person, number and gender which are capable to interpret inflectional features (like case and definiteness) as well. The inflection can be represented with respect to prefixes, suffixes, infixes, and to the sound alternations taking place during the process of inflection.

However, the semantics of possessive pronouns includes various relationships like: possession (depending whether it is an object or a subject of possession), relational, etc. It is presented in different languages by different grammar features but generally, is expressed by the features of case (syntactic or morphological) or by agreement in gender and number (at morphological level). Thus, multilingual representation of possessive pronouns require formal models which allow development of both syntactic and morphological rules.

Thus, UNL allows two types of transformation inflectional rules: (i) A-rules (affixation rules) apply over isolated word forms (as to generate possible inflections) and (ii) L-rules (linear rules) apply over lists of word forms (as to provide transformations in the surface structure). Affixation rules are used for adding morphemes to a given base form. They are used for generating inflections or derivations. There are two types of A-rules: (i) simple A-rules involve a single action (such as prefixation, suffixation, infixation and replacement), and (ii) complex A-rules involve more than one action (such as circumfixation).

There are four types of simple A-rules: (i) prefixation, for adding morphemes at the beginning of a base form, (ii) suffixation, for adding morphemes at the end of a base form, (iii) infixation, for adding morphemes to the middle of the base form, (iv) replacement, for changing the base form. Further, we are going to analyse the way possessive pronuns are represented in the lexical database or UNL dictionary for English, Bulgarian and Russian by comparing examples of personal, possessive and reflexive pronouns.

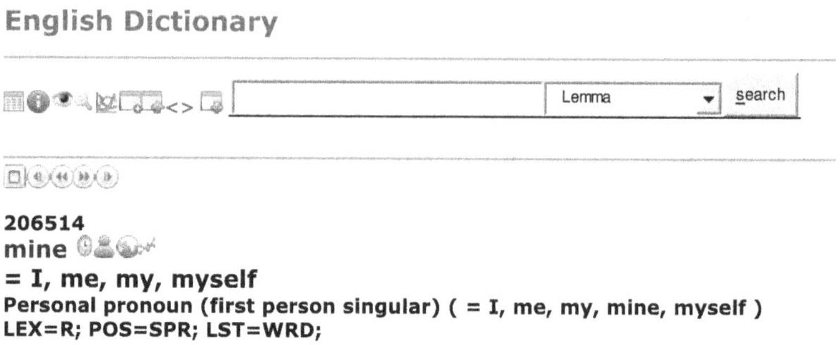

English Dictionary

206514
mine
= I, me, my, myself
Personal pronoun (first person singular) (= I, me, my, mine, myself)
LEX=R; POS=SPR; LST=WRD;

Fig. 3. The English possessive pronoun *mine* in UNL representation

The English pronouns UNL representation uses mostly rules for syntactic transformations. However, the UNL dictionary uses grammar feature of person to link different types of pronouns by means of synonymic hierarchical lexical relations. For example, it relates personal pronoun (for subject) *I* to personal pronoun (for object) *me* and to possessive adjective *my* and to possessive pronoun *mine* and reflexive pronoun *myself*.

Fig. 3 shows UNL dictionary entry for possessive pronoun *mine* where lexical definition relates grammar information like the type of pronoun and the grammar feature of person. It also presents grammar information through specifications LEX=R and POS=SPR which are used by syntactic grammar rules.

Fig. 4 presents UNL dictionary entry for reflexive pronoun *myself* which includes the same types of information structured the same way.

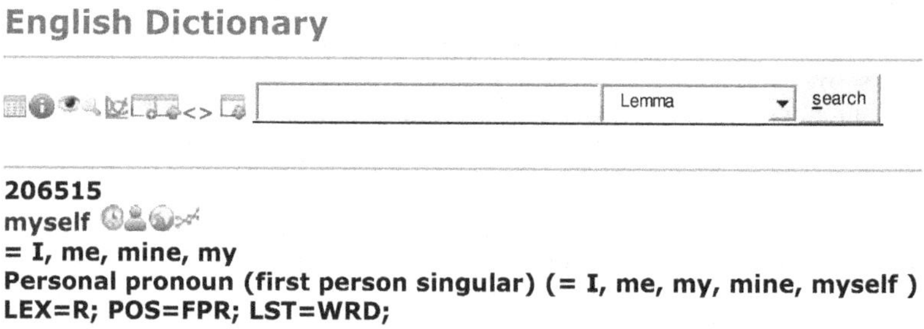

English Dictionary

206515
myself
= I, me, mine, my
Personal pronoun (first person singular) (= I, me, my, mine, myself)
LEX=R; POS=FPR; LST=WRD;

Fig. 4. The English reflexive pronoun *myself* in UNL representation

However, there are languages like Bulgarian and Russian which present syntactic information for definiteness or case by using inflection. For them possessive pronouns are presented additionally by applying inflectional rules.

The formal representation of Bulgarian possessive pronouns is a part of inflectional morphology application [11,12] made within the framework of the project 'The Little Prince Project' of the UNDL Foundation aimed to develop UNL grammar and lexical resources for several European languages based on the book 'The Little Prince'. It offers interpretation of inflectional morphology which uses A-rules.

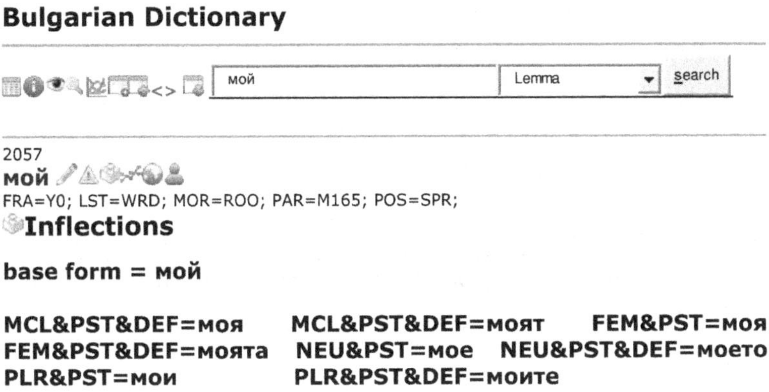

Bulgarian Dictionary

🔳🔵ℹ️🔍📝🔧<> 🔲	мой	Lemma ▾	search

2057
мой ✏️⚠️🖐️🔖🔴👤
FRA=Y0; LST=WRD; MOR=ROO; PAR=M165; POS=SPR;
🔹**Inflections**

base form = мой

MCL&PST&DEF=моя	MCL&PST&DEF=моят	FEM&PST=моя
FEM&PST&DEF=моята	NEU&PST=мое	NEU&PST&DEF=моето
PLR&PST=мои	PLR&PST&DEF=моите	

Fig. 5. The Bulgarian possessive pronoun moj in UNL representation

Fig. 5 shows UNL dictionary entry for Bulgarian possessive pronoun moj. It presents both lexical and grammar information. The grammar information is given by both syntactic and inflectional rules. The specification POS=SPR is used by syntactic grammar rules and relates Bulgarian to English lexical entry.

In general, the UNL interpretation of Bulgarian inflectional morphology offers a sound alternations account mostly by the use of A-rules. The inflectional rules are defined without using hierarchical inflectional representation even they define related inflectional types. The sound alternations and the irregularity are interpreted within the definition of the main inflectional rule. The information about inflection is given through the specifier PAR=M165 which assign related inflectional type consisting of inflectional grammar rules (given at the Appendix).

The Russian interpretation of possessive pronouns in UNL combines both hierarchical lexical representation and grammar representation. Thus, it directly relates (Fig. 6) possessive pronoun moj to $mnoj$, mne, $menja$ and ja, and to English I, me, my, $mine$ and $myself$ through grammar feature of person (Personal pronoun (first person singular)) giving in this way its possible translation. Additionally, the syntactic relations are presented by using specifier POS=SPR which correlates that lexical entry to its Bulgarian and English counterparts and allows use of syntactic transformation rules.

The grammar information also includes representation of case relations given by inflectional rules for generation of inflected forms of pronouns. Fig. 6 shows UNL lexical entry for the Russian pronoun moj and also presents all generated case inflected forms.

Generally, the Russian representation of possessive pronouns combines approaches used to represent pronouns in English (by defining both lexical and grammar relations) and in Bulgarian (by defing inflectional rules) using common UNL frameworks.

The UNL application, also, represents a web-based intelligent information and knowledge management system which allows different types of semantic search with respect to various search criteria.

Russian Dictionary

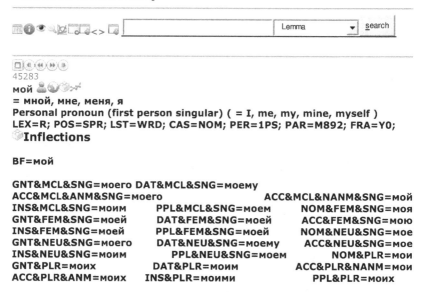

Fig. 6. The Russian pronoun *moj* in UNL representation

5   Conclusions

The analyzed formal representations of English, Bulgarian and Russian possessive pronouns use semantic networks representation. They underlay common formal framework based on the use of grammar feature of person to relate pronouns for different languages. Also, they use the UNL inflectional rules representation scheme to generate and relate different inflected forms. Additionally, they present syntactic information by using common specifiers (like POS=SPR) to relate syntactic rules through linking.

The UNL pronouns representation scheme is capable to offer adequate representation not only for related languages (like Bulgarian and Russian) but also for non-related languages (like English and Bulgarian or English and Russian) and is successfully used for machine translation. The application is open for further improvement and development by introducing additional grammar rules and by enlarging database.

References

1. Boguslavsky, I.: Some lexical issues of unl. In: Cardeosa, J., Gelbukh, A., Tovar, E. (eds.) Universal Network Language: Advances in Theory and Applications. Research on Computing Science, vol. 12, pp. 101–108 (2005)
2. Boitet, C., Boguslavskij, I.M., Cardeñosa, J.: An evaluation of unl usability for high quality multilingualization and projections for a future unl++ language. In: Gelbukh, A. (ed.) CICLing 2007. LNCS, vol. 4394, pp. 361–373. Springer, Heidelberg (2007)
3. Corbett, G., Fraser, N.: Network morphology: a datr account of russian nominal inflection. Journal of Linguistics 29, 113–142 (1993)

4. Evans, R., Gazdar, G.: Datr: A language for lexical knowledge representation. Computational Linguistics 22(2), 167–216 (1996)
5. Fellbaum, C.: WordNet: An Electronic Lexical Database. MIT Press (1998)
6. Kilgarriff, A.: The sketch engine: Sketch engine on-line database. Tech. rep. (2013), http://www.sketchengine.co.uk/
7. Kilgarriff, A., Rundell, M.: Lexical profiling software and its lexicographic applications: a case study. In: Proceedings from EURALEX 2002, pp. 807–811 (2002)
8. Kilgarriff, A., Rychly, P., Smrz, P., Tugwell, D.: The sketch engine. In: Proceedings from EURALEX 2004, pp. 105–116 (2004)
9. Noncheva, V., Stancheva, Y., Stoykova, V.: The little prince project - encoding of bulgarian grammar. Tech. rep. UNDL Foundation (2011), www.undl.org
10. Stoykova, V.: The definite article of bulgarian adjectives and numerals in datr. In: Bussler, C.J., Fensel, D. (eds.) AIMSA 2004. LNCS (LNAI), vol. 3192, pp. 256–266. Springer, Heidelberg (2004)
11. Stoykova, V.: Bulgarian inflectional morphology in universal networking language. In: Kay, M., Boitet, C. (eds.) Proceedings of 24th International Conference on Computational Linguistics (COLING 2012): Demonstration Papers, pp. 423–430. ACL Anthology (2012)
12. Stoykova, V.: The inflectional morphology of bulgarian possessive and reflexive- possessive pronouns in universal networking language. In: Procedia Technology, vol. 1, pp. 400–406. Elsevier (2012)
13. Stoykova, V.: Formal representations of bulgarian possessive and reflexive-possessive pronouns. In: Recent Advances in Computer Engineering Series, vol. 9, pp. 144–149 (2013)
14. Uchida, H., Zhu, M., Senta, T.D.: Universal Networking Language (2005)

Appendix

Bulgarian Grammar

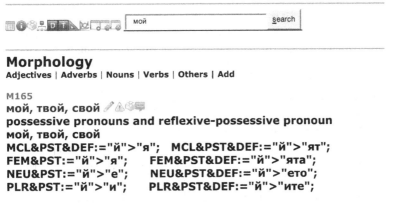

Fig. 7. The inflectional rules definitions for Bulgarian possessive pronoun *moj*

Sleep Spindle Detection in EEG Signals Combining HMMs and SVMs

Iosif Mporas, Panagiotis Korvesis, Evangelia I. Zacharaki,
and Vasilis Megalooikonomou

Department of Computer Engineering & Informatics,
University of Patras, GR-26500 Patras, Rio, Greece
imporas@upatras.gr, korbesis@ceid.upatras,
ezachar@upatras.gr, vasilis@ceid.upatras.gr

Abstract. In this paper we present a combined SVM-HMM sleep spindle detection scheme. The proposed scheme takes advantage of the information provided from each of the two prediction models in decision level, in order to provide refined and more accurate spindle detection results. The experimental results showed that the proposed combined scheme achieved an overall detection performance of 90.28%, increasing the best-performing SVM-based model by 2% in terms of absolute performance.

Keywords: sleep spindles, EEG, support vector machines, hidden Markov models.

1   Introduction

Over the last decades sleep medicine is studying sleep for the purpose of sleep disorders treatment. Electroencephalographic (EEG), eye movement (EOG) and electromyographic (EMG) signals of the subject are recorded throughout the sleep cycle. The analysis of those signals and the detection of specific patterns offers information related to sleep disorders. One such pattern is the sleep spindle.

Sleep spindles are characteristic transient oscillations that appear on the EEG during non-rapid eye movement (non-REM) sleep. Sleep spindles, also referred to as "sigma bands" or "sigma waves", may represent periods where the brain is inhibiting processing to keep the sleeper in a tranquil state. Along with K-complexes they are characteristic indicators of the onset of stage 2 sleep [1]. They are characterized by progressively increasing, then gradually decreasing waveforms, are affected by medication, aging and brain pathology and may be involved in learning processes [2]. Sleep spindles are important for the classification of the NREM sleep and the evaluation of the degree of arousal.

The amount and the distribution of the sleep spindles is essential for describing the morphology of the sleep EEG, thus the assessment of the distribution of sleep spindles over a whole sleep cycle is needed [3]. The low amplitude of some spindles, compared to the background EEG activity, renders the detection of sleep spindles

L. Iliadis, H. Papadopoulos, and C. Jayne (Eds.): EANN 2013, Part II, CCIS 384, pp. 138–145, 2013.
© Springer-Verlag Berlin Heidelberg 2013

difficult even for sleep experts. Therefore, in such cases the visual detection of spindles would not be accurate. Moreover, manual annotation of the spindle occurrences for a whole sleep cycle recording is time consuming and tedious for the sleep technician. Additionally, manual annotations are affected by the subjectivity of the sleep technician and the degree of his/her experience. Therefore, automatic detection of sleep spindles is essential for reducing the workload and allowing the processing of large enough amount of data.

In this paper we present a scheme for the automatic detection of sleep spindles, which is based on the combination of discriminative and statistical machine learning methods. The rest of the paper is organized as follows. In section 2 we describe the background of the sleep spindles in EEG signals and the state of the art in automatic detection of sleep spindles. In Section 3, the architecture of the proposed scheme is presented. The experimental protocol and the evaluation results are described in Sections 4 and 5, respectively.

2   Background

Sleep spindles are bursts of oscillatory brain activity visible on EEG that occur during stage 2 of the sleep cycle [2]. Spindles consist of 12–14 Hz waves, however the frequency-domain information of the sleep spindles can be extended to 10-20 Hz [1, 4].Typical time duration for sleep spindles is 0.5 - 2.5 seconds [1]. Sleep spindles are used together with other EEG features, such as K-Complexes, to classify the sleep stages [5] and are considered as crucial objective indicators for neurodegenerative disorders [6]. An example of sleep spindle captured from the CZ EEG electrode is shown in Fig. 1.

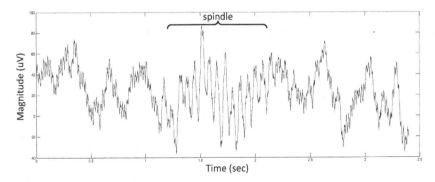

Fig. 1. Example of sleep spindle captured from the CZ electrode of EEG signal recordings

Several publications related to detection of sleep spindle are found in the literature. The most commonly reported performance metrics are the sensitivity (true positive rate), the specificity (true negative rate) and the overall accuracy. In [7], the authors proposed spindle detection based on teager energy operator (TEO) and wavelet transform (WT) and reported accuracy equal to 93.7%. In [8], short time Fourier transform (STFT) and WT were used and the achieved true detection rate was 92%. In [9] the authors presented a sleep spindle detection method, which takes into account several types of artifacts (such as saturations, unusual increase of EEG, abrupt transitions, movement artifacts, interference and unusual low-frequency waveforms). Using variable thresholds of the statistical properties of the signal a sensitivity of 76.9% for a specificity of 90% is achieved. In [3] multilayer perceptron neural networks (MLP) and support vector machines (SVMs) using autoregressive modeling based features were used. The reported accuracy was 93.6% for the MLP and 94.4% for the SVM classifier. MLP was also used in [10] for spindle detection in band-pass (10.5-16 Hz) filtered EEG signals, without feature extraction other than that produced by the band-pass filtering. The reported sensitivity varied from 79.2% to 87.5% and the specificity ranged from 3.8% to 15.5%. In [11] an automatic sleep spindle detection system, using MLP was evaluated in detecting spindles of both healthy controls, as well as Mild Cognitive Impairment (MCI) and Alzheimer's disease (AD) patients. The authors reported that the sensitivity of the detector was 81.4%, 62.2%, and 83.3% and the false positive rate was 34%, 11.5%, and 33.3%, for the control, MCI, and AD groups, respectively. Decision trees were used in [12] for spindle detection in EEG signals, where three methods were followed (STFT, multiple signal classification and TEO). The reported sensitivity was equal to 96.17% for specificity equal to 95.54%. In [13] two sleep spindle detectors were presented, based on the Short Time Fourier Transform and the wave morphology of sleep spindles. The two detectors were combined by applying the AND logical operator on the detection results. The best reported performance was 93% sensitivity when the combination of the detectors was used.

3   Automatic Detection of Sleep Spindles

For the detection of sleep spindles we relied on the combination of discriminative and statistical models. Specifically, the support vector machines [14] and the hidden Markov models (HMMs) [15] were selected due to their advantageous performance in similar signal processing tasks [3, 11, 10]. The sleep spindle detection is performed in two stages. In the first stage the signal is pre-processed, parameterized and processed independently from the discriminative (SVMs) and the statistical (HMMs) models. In the second stage the output recognition results from each model are combined by a fusion method in order to provide the final sleep spindle detection results. The block diagram of the proposed scheme is illustrated in Fig. 2.

As shown in Fig. 2 an EEG signal $X = \{x_k\}$, with $x_k \in \mathbb{R}$ and $1 \le k \le N$, is introduced to the system. The signal is pre-processed with framing of the signal to blocks of L samples. Each frame is time-shifted by its preceding one by s samples, where $s \le L$, thus resulting to M overlapping frames w_i, with $w_i \in \mathbb{R}^L$ and $1 \le i \le M$. At the feature extraction block for each of the computed frames w_i a parametric vector $V_i \in \mathbb{R}^k$ is computed by a signal parameterization algorithm p, i.e. $V_i = p(w_i)$, with $1 \le i \le M$.

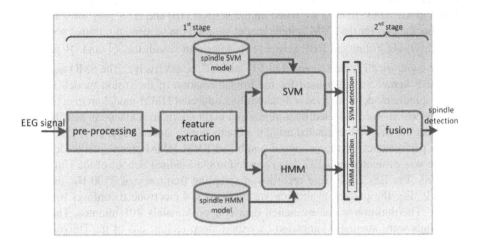

Fig. 2. Block diagram of the combined SVM-HMM sleep spindle detection scheme

The output of the feature extraction block is forwarded in parallel to each of the two models, i.e. the discriminative SVM-based and the statistical HMM-based sleep spindle models. Each of the two models C_j, $1 \le j \le 2$ estimates whether the i-th incoming feature vector corresponds to sleep spindle or not, i.e. providing binary classification results d_i^j with the corresponding recognition score for each of the two classes, i.e. $d_i^j = C_j(V_i)$, $d_i^j \in \mathbb{R}^2$, $1 \le i \le M$ and $1 \le j \le 2$.

The second stage of the sleep spindle detection scheme exploits the recognition results of the two models, i.e. the SVM-based and the HMM-based, in order to combine them and provide a final decision for each feature vector V_i. Specifically, the recognition results d_i^j, estimated at the first stage by each of the two models, are concatenated in a single vector $D_i \in \mathbb{R}^4$ with $1 \le i \le M$ as shown in Fig. 2. A fusion model f utilizes the SVM-based and HMM-based predictions, which are included in the D_i vector, in order to provide the final decision y_i for the i-th frame of the EEG signal, i.e. $y_i = f(D_i)$. In the present work the fusion model f was implemented with the SVM algorithm.

4   Experimental Setup

The sleep spindle detection scheme shown in Fig. 2 combines the recognition results produced by the SVM-based and the HMM-based model. For the implementation of the SVM and the HMM models we relied on the WEKA [16] and HTK [17] software tool-kits. Specifically, for the SVM spindle model we used the sequential minimal optimization (SMO) algorithm and RBF kernel [14]. After grid search the C and γ parameters were empirically selected equal to 1.0 and 0.01, respectively. The SMO algorithm with RBF kernel was also used for the implementation of the fusion model. For the HMM spindle models, we used a 3-state fully connected HMM model architecture, the states of which were modeled by a mixture of eight continuous Gaussian distributions. HMM parameters were estimated using the Baum-Welch algorithm [18].

The performance of the proposed combined SVM-HMM sleep spindle detection scheme was evaluated on EEG data recorded at the Medical School of the University of Patras. The EEG data were recorded at sampling frequency of 2500 Hz, using 64 channels. For the present evaluation we used the CZ electrode recordings from one subject. The duration of the evaluated data is approximately 401 minutes. The sleep recordings were manually annotated by expert sleep technicians of the University of Patras. The evaluated data include 1228 occurrences of sleep spindles and no overlap between training and test data subsets existed.

For the decomposition of the EEG signal to feature vectors it initially was frame blocked to overlapping frames of 0.5 seconds with time-shift equal to 0.1 seconds. For each frame the power spectral density (PSD) was estimated for the frequency range [10, 20] Hz, with accuracy 1 Hz, i.e. PSD estimations at {11, 12, ..., 19, 20} Hz resulting to 11-dimentional vectors. The PSD vectors were expanded to their derivative (delta) coefficients, thus resulting to feature vectors of dimensionality equal to 22. The sequence of computed feature vectors was used as input for both the SVM and HMM based sleep spindle models.

5   Experimental Results

We firstly investigated the performance of each of the SVM-based and HMM-based sleep spindle models separately. The performance of the SVM-based spindle models implemented with the sequential minimal optimization algorithm is shown in Table 1.

Table 1. SVM-based sleep spindle detection performance (in percentages)

Recognized as →	Spindle	Non-spindle
Spindle	88.15	11.46
Non-spindle	11.53	88.47

As can be seen from the confusion matrix of Table 1, the SVM-based model achieved sleep spindle recognition accuracy of approximately 88%. Both the false positive and the false negative rate were found equal to approximately 11.5%.

The overall recognition performance for the SVM-based sleep spindle model was found equal to 88.22%.

The performance of the HMM-based spindle model, described in the previous section, is shown in Table 2 in terms of confusion matrix.

Table 2. HMM-based sleep spindle performance (in percentages)

Recognized as →	Spindle	Non-spindle
Spindle	85.00	15.00
Non-spindle	12.63	87.37

As can be seen in Table 2 the hidden Markov model offered slightly lower recognition performance than the support vector machines. In detail, sleep spindles were detected with accuracy equal to 85%, i.e. approximately 3% lower than the SVM-model. The decrease in recognition of non-spindles was approximately 1%, while the false negative and false positive rates were found equal to 15% and 12.63% respectively. The overall HMM-based sleep spindle detection performance was found equal to 87.2%.

The slight decrease in spindle detection performance when the statistical model (HMM) was used can be attributed to the limited amount of the available annotated data combined with the feature vectors length, which did not result to the precise estimation of the HMM parameters, in contrast to the SVM discriminative method which does not suffer from the curse of dimensionality and always converges to the global minima. It is worth mentioning that in 23.62% of the cases the SVM and the HMM spindle models resulted to different decisions, thus indicating the need for combination of the two methods.

In a second step we combined the results of the discriminative SVM and the statistical HMM models, using an SVM-based fusion function. The sleep spindle detection performance is shown in Table 3.

Table 3. SVM-HMM sleep spindle performance (in percentages)

Recognized as →	Spindle	Non-spindle
Spindle	88.54	11.85
Non-spindle	03.41	96.59

As can be seen in Table 3 the combination of the discriminative SVM model results with the statistical HMM based results slightly improved the spindle accuracy by approximately 0.5%, while the false negative rate did not significantly change. However, the use of the fusion scheme significantly improved the both the true and false negatives. Specifically, the true negative rate reached approximately 96.5% and the false negative rate reduced to approximately 3.5%. The overall performance of the SVM-HMM combination scheme was found equal to 90.28%. In Table 4, for the purpose of direct comparison we present the accuracy of the SVM-based spindle model, the HMM-based spindle model and the proposed combined SVM-HMM scheme.

Table 4. Comparison of sleep spindle detection schemes performance (in percentages)

Detection Scheme	Accuracy
SVM	88.22
HMM	87.20
SVM-HMM	90.28

As can be seen in Table 4, the proposed combined SVM-HMM scheme outperformed the best-performing SVM-based model by approximately 2%, in terms of absolute performance. This is in agreement with [13], where also slight improvement was found when applying the logical AND operator on the detection results of two models. The advantageous performance of the proposed combined scheme is owed to the exploitation by the fusion model of the underlying information of both the SVM and HMM detection results, which in 23% of the cases was not used when each of the models operated alone.

6   Conclusion

In the present work we evaluated a scheme for automatic detection of sleep spindles from the CZ EEG electrode. The proposed scheme combines the detection results produced by a discriminative (SVM) and a statistical (HMM) based spindle detection model, in order to provide more accurate results. The experimental results showed that the fusion of the two detection models outcome achieved accuracy of 90.28% improving the overall accuracy by approximately 2% in terms of absolute performance, by exploiting the underlying information in the 23% of the cases where the two models did not result to the same decision.

Studies on scalp-EEG have showed topographic distinction between two major sleep spindle categories, namely slow and fast spindles. Slow spindles present spectral peak frequency at around 12 Hz and are more pronounced over frontal scalp electrodes. Fast spindles present spectral peak frequency at around 14 Hz and exhibit mainly parietal and central scalp distribution [19]. We presume that different detection models have different detection ability on different types of spindles. Thus the proposed scheme is valuable in exploiting the prediction ability of different models.

Acknowledgement. The research reported in the present paper was partially supported by the ARMOR Project (FP7-ICT-2011-5.1 - 287720) "Advanced multipaRametric Monitoring and analysis for diagnosis and Optimal management of epilepsy and Related brain disorders", co-funded by the European Commission under the Seventh' Framework Programme. Project web-site: http://armor.tesyd.teimes.gr/.

The authors wish to acknowledge the contribution of Dr. A. Koupparis, Dr. V. Kokkinos, and Prof. G.K. Kostopoulos from the Medical School at the University of Patras, who supported the collection and annotation of the medical data.

References

1. Sanei, S., Chambers, J.A.: EEG Signal Processing. John Wiley & Sons, Ltd. (2007)
2. De Gennaro, L., Ferrara, M.: Sleep spindles: an overview. Sleep Medicine Reviews 7(5), 423–440 (2003)
3. Gorur, D., Halici, U., Aydin, H., Ongun, G., Ozgen, F., Leblebicioglu, K.: Sleep Spindles Detection Using Autoregressive Modeling. In: Proc. of ICANN/ICONIP (2003)
4. Huupponen, E., Gomez-Herrero, G., Saastamoinen, A., Varri, A., Hasan, J., Himanen, S.-L.: Development and comparison of four sleep spindle detection methods. Artificial Intelligence in Medicine 40, 157–170 (2007)
5. De Gennaro, L., Ferrara, M., Bertini, M.: Effect of slow-wave sleep deprivation on topographical distribution of spindles. Behavioral Brain Research (116), 55–59 (2000)
6. Ktonas, P.Y., Golemati, S., Xanthopoulos, P., Sakkalis, V., Ortigueira, M.D.: Time–frequency analysis methods to quantify the time-varying microstructure of sleep EEG spindles: possibility for dementia biomarkers. Journal of Neuroscience Methods, no 1(185), 133–142 (2009)
7. Ahmed, B., Redissi, A., Tafreshi, R.: An automatic sleep spindle detector based on wavelets and the teager energy operator. In: Proc. of Annual International Conference of the IEEE Engineering in Medicine and Biology Society, pp. 2596–2599 (2009)
8. Duman, F., Erogul, O., Telatar, Z., Yetkin, S.: Automatic sleep spindle detection and localization algorithm. In: Proc. of EUSIPCO 2005, pp. 2003–2006 (2005)
9. Devuyst, S., Dutoit, T., Didier, J.F., Meers, F., Stanus, E., Stenuit, P., Kerkhofs, M.: Automatic Sleep Spindle Detection in Patients with Sleep Disorders. In: Proc. of the 28th IEEE EMBS Annual International Conference, New York City, USA (2006)
10. Ventouras, E., Monoyiou, E., Ktonas, P., Paparrigopoulos, T., Dikeos, D.: Sleep Spindle Detection Using Artificial Neural Networks Trained with Filtered Time-Domain EEG: A Feasibility Study. Computer Methods and Programs in Biomedicine 78(3), 191–207 (2005)
11. Ventouras, E., Economou, N., Kritikou, I., Tsekou, H., Paparrigopoulos, T., Ktonas, P.: Performance evaluation of an Artificial Neural Network automatic spindle detection system. In: Proc. of the Conf. IEEE Eng. Med. Biol. Soc., pp. 4328–4331 (2012)
12. Duman, F., Erdamar, A., Erogul, O., Telatar, Z., Yetkin, S.: Efficient sleep spindle detection algorithm with decision tree. Expert Systems with Applications 36(6), 9980–9985 (2009)
13. Costa, J., Ortigueira, M., Batista, A., Paiva, T.: Sleep Spindles Detection: a Mixed Method using STFT and WMSD. International Journal of Bioelectromagnetism 14(4), 229–233 (2012)
14. Keerthi, S.S., Shevade, S.K., Bhattacharyya, C., Murthy, K.R.K.: Improvements to Platt's SMO algorithm for SVM classifier design. Neural Computation 13(3), 637–649 (2001)
15. Rabiner, L.R.: A tutorial on Hidden Markov Models and Selected Applications in Speech recognition. Proceedings of the IEEE 77(2) (1989)
16. Witten, H.I., Frank, E.: Data Mining: practical machine learning tools and techniques. Morgan Kaufmann Publishing
17. Young, S., Evermann, G., Gales, M., Hain, T., Kershaw, D., Liu, X., Moore, G., Odell, J., Ollason, D., Povey, D., Valtchev, V., Woodland, P.: The HTK Book (for HTK Version 3.4). Cambridge University Engineering Department (2006)
18. Baum, L.E., Petrie, T., Soules, G., Weiss, N.: A Maximization Technique Occurring in the Statistical Analysis of Probabilistic Functions of Markov Chains. Annals of Mathematical Statistics 41(1), 164–171 (1970)
19. Ventouras, E.M., Ktonas, P.Y., Tsekou, H., Paparrigopoulos, T., Kalatzis, I., Soldatos, C.R.: Independent Component Analysis for Source Localization of EEG Sleep Spindle Components. In: Computational Intelligence and Neuroscience, vol. 2010 (2010)

Classifying Ductal Trees Using Geometrical Features and Ensemble Learning Techniques

Angeliki Skoura[1,*], Tatyana Nuzhnaya[2], Predrag R. Bakic[3], and Vasilis Megalooikonomou[1,2]

[1] Department of Computer Engineering and Informatics, University of Patras, Patras, Greece
{skoura,vasilis}@ceid.upatras.gr
[2] Data Engineering Laboratory, Center for Data Analytics and Biomedical Informatics, Temple University, PA, USA
tatyana.nuzhnaya@temple.edu
[3] Department of Radiology, University of Pennsylvania, Philadelphia, USA
predrag.bakic@uphs.upenn.edu

Abstract. Early detection of risk of breast cancer is of upmost importance for effective treatment. In the field of medical image analysis, automatic methods have been developed to discover features of ductal trees that are correlated with radiological findings regarding breast cancer. In this study, a data mining approach is proposed that captures a new set of geometrical properties of ductal trees. The extracted features are employed in an ensemble learning scheme in order to classify galactograms, medical images which visualize the tree structure of breast ducts. For classification, three variants of the AdaBoost algorithm are explored using as weak learner the CART decision tree. Although the new methodology does not improve the classification performance compared to state-of-the-art techniques, it offers useful information regarding the geometrical features that could be used as biomarkers providing insight to the relationship between ductal tree topology and pathology of human breast.

Keywords: Feature Extraction, Classifier Ensembles, Breast Imaging.

1   Introduction

Geometrical and structural variability of the breast ductal tree is associated with abnormalities in patients with nipple discharge such as papilloma, breast cancer and atypia [1]. Mammary malignancies typically develop from the ductal epithelial cells, and spread within the ducts (ductal carcinoma in situ) and outside of the ducts (infiltrating ductal carcinoma). Consequently, the ductal tree structures are an important area of investigation of both normal breast development and malignant breast transformation. Automated frameworks for the structural analysis of breast ducts have been proposed to provide information related to early cancer detection and cancer risk estimation. Among the challenging issues in designing frameworks for analysis of anatomical structures of tree topology is the selection of descriptive features and

---

* Corresponding author.

L. Iliadis, H. Papadoulos, and C. Jayne (Eds.): EANN 2013, Part II, CCIS 384, pp. 146–155, 2013.
© Springer-Verlag Berlin Heidelberg 2013

patterns which can discriminate patients from normal subjects. However, the task of analysis is hindered as it is affected by imaging modality's ability to capture the maximum level of tree branching from its complex surroundings as none of the available imaging methods has succeeded in tracking the full ductal tree [2].

In this study, we investigate whether a set of features which capture the geometrical variability of ductal trees could be considered to be important descriptors with respect to the risk of the breast cancer. The proposed methodology consists of three basic parts; ductal tree tracing, feature measurement and classification process. Furthermore, we explore three ensemble learning schemes to boost the classification performance. For evaluation, the methodology was applied to a dataset of 71 clinical galactograms. The experimental results appear promising, improve our understanding of the relation between the ductal tree topology and the underlying pathology and our framework could be potentially used to facilitate medical diagnosis.

2   Background

Existing methodologies to characterize tree structural variability of human ductal trees with respect to the risk of breast cancer include two basic components; feature extraction and classification process. Regarding the extracted features, quantification schemes for topological characteristics such as tree asymmetry [3] and fractal dimension have been proposed. The branching patterns have been investigated using tree encoding and text mining techniques [4]. According to [4], encoding schemes such as the Prüfer code and the depth-first string encoding were used to obtain a symbolic representation of the tree topology and employing them as signatures the problem of comparing trees was reduced to sequence comparison and it was handled using text mining techniques. Moreover, the branching patterns have been analyzed in terms of Sholl analysis which quantifies the spatial distribution of branching points and in terms of ramification matrices using the Strahler ordering which represents the probabilities of branching at various tree levels [5].

Although several approaches have been proposed for the analysis of breast ductal trees, the simple k-nn classifier and its variation, the weighted k-nn, was usually employed as the main objective of the former studies was the evaluation of the feature extraction scheme. However, the potential of ensemble learning schemes which are of the most important developments in classification methodology have not been explored in this field of medical images. Among ensemble learning schemes, boosting is the most popular method of ensemble learning. Boosting works by sequentially applying a classification algorithm to reweighted versions of the training data and takes a weighted majority vote of the sequence of classifiers that are produced. Although the idea to employ multiple classifiers and combine their predictions is challenging, no successful method has been reported so far to apply boosting to k-nn [6].

3   Methodology

The core steps of the proposed framework are image preprocessing, feature extraction and classification process. Initially, ductal tree structures need to be traced and segmented from the background of original medical images. Using ductal tree outlines, we explore a set of quantitative features derived from the ellipse that has the same second central moment as the tree structure, the tree asymmetry index and the number of tree nodes. Finally, classification is applied using three variants of boosting algorithms where as weak learners were considered decision trees.

3.1   Preprocessing

At this stage, each galactogram is segmented into two distinct components: image background and centerline of the breast ductal tree. In the application presented here, segmentation and skeletonization was performed manually by clinician experts, using nipple as the root (Fig. 1b, 1d). Although automated techniques for image segmentation have been implemented, the precision of such techniques is low in the case of galactograms due to high levels of noise fluctuations throughout the original galactograms. The resulting image after preprocessing is a binary image depicting the centerline of a ductal tree. We focus on the skeletons of ductal trees because the centerlines characterize more precisely the geometry of ductal trees as the margins of ductal trees are usually unclear due to the high levels of noise contained in the original galactograms (Fig. 1a, 1c). Typical galactograms and the corresponding tracings of the ductal trees are shown in Figure 1.

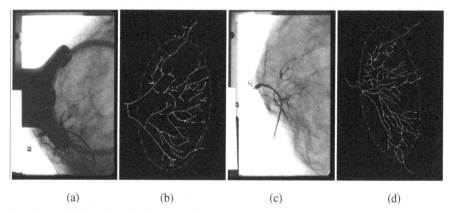

(a)	(b)	(c)	(d)

Fig. 1. (a, c) Original medical images of galactograms, (b, d) The corresponding outlined ductal trees magnified and the fitted ellipses

3.2   Feature Selection

Concerning the large variation of geometry among ductal trees (Fig.1), we considered a hypothetical ellipse that best fits each ductal tree to describe its geometrical properties. More formally, an ellipse is computed that has the same second central moment (also known as moment of inertia of plane area, area moment of inertia, or second area moment) as the traced tree structure. In general, the second central moment of a binary object is a geometrical property which reflects how its points are distributed with regard to an arbitrary axis. In the case of binary image, the second central moment is the covariance matrix of the image. Considering a matrix $n \times 2$ containing the coordinates of the points of a tree structure, a 2-D normal distribution is fitted to the data and the contour line of such a distribution is an ellipse. Figures 1b and 1d illustrate the computed ellipses for two different ductal trees.

Our objective is to explore the geometrical profile of ductal trees and towards this direction, the following shape descriptors are computed based on the fitted ellipse; Major Axis Length, Minor Axis Length, Eccentricity and Orientation. These elliptical descriptors are computed as follows:

1) Major Axis Length (MAxis) is a scalar number specifying the length (in pixels) of the major axis of the ellipse.
2) Minor Axis Length (mAxis) is a scalar number specifying the length (in pixels) of the minor axis of the ellipse.
3) Eccentricity is a scalar number that specifies the eccentricity of the ellipse. The eccentricity is computed as $Ecc = \sqrt{\left(1 - \left(\frac{MA}{mA}\right)^2\right)}, Ecc \ni [0,1]$. An ellipse whose eccentricity is 0 is actually a circle, while an ellipse whose eccentricity is 1 is a line segment.
4) Orientation is a scalar number that corresponds to the angle $[-90, 90]$ formed by the horizontal reference axis and the major axis of the ellipse.

These four features have been already used for recognition system in other research fields such as segmentation and characterization of brain tumors and recognition of sign language [7]. However, here for the first time these descriptors are applied for the characterization of ductal trees in galactograms. Apart from the above descriptors, the following features of a ductal tree are measured:

5) Solidity is a scalar number specifying the proportion of the pixels in the convex hull that are also in the ductal tree. The convex hull is the smallest convex polygon that contains the ductal tree. Solidity is a measure of the compactness of the ductal tree to the convex hull and for a solid object solidity is 1, while the value is lower for an object having a rough perimeter or an object which contains holes.
6) Tree Asymmetry (TA) is a numerical index for the topological structure of a binary tree. For the computation of TA, reconstruction of the tree nodes needed and was performed manually using nipple as the root. The reconstruction of the ductal structures aimed at identifying points of branching and resolving potential ambiguities such as anastomoses. Generally, the values of TA range from zero for fully symmetric trees (only possible when the number of tree's terminal nodes is a power of two) to a maximum value for fully asymmetric trees, which value approaches one as the asymmetric tree becomes larger. For a complete analysis on tree asymmetry index, the reader could refer to [8]. In [8], the authors proved that the tree-asymmetry measure appears to be sensitive to topological differences and it is almost independent of the size of the trees.
7) Number of tree nodes (nNodes) is a scalar number specifying the total number of tree nodes i.e. the sum of branching nodes and leave nodes of the ductal tree.

The set of the above seven properties capturing structural features of the ductal trees are used to characterize, compare and classify ductal trees. Note that all seven features are invariant under operations of rotation, reflect and translation, thus no preprocessing step of image registration is required.

3.3   Classification

For the classification process, we explore three ensemble learning techniques based on the boosting technique; Real AdaBoost [9], Gentle AdaBoost [10] and Modest AdaBoost [11]. Let $X \ni \mathbb{R}^n$ denote the data space and Y be the set of possible class labels. In our application, there are two classes $Y = \{-1, +1\}$. The goal of classification is to build a mapping function $F: X \rightarrow Y$ that given the feature vector $x \in X$ calculates the correct class label $y \in Y$. Also, we consider a sequence of labeled data for training $S = \{(x_1, y_1), \dots, (x_N, y_N)\}, x_i \in X; y_i \in Y$.

In all schemes, as weak learner is considered the Classification and Regression Tree (CART) decision tree [12]. In CART, as in any decision tree algorithm, the leaves represent the classification result and nodes represent features. CART algorithm searches all variables and for all possible values in order to find the best split and constructs a node following the procedure:

- For each feature, find the threshold θ that separates the training set S with least error
- Choose feature x_i with least error, construct the node with predicate $x_i > \theta$ and associate the two leaves derived from this node with respective class labels

After determining the root node, the leave with the largest error is chosen and the above procedure is then repeated (using only those training samples that are associated with the chosen leaf) until a predefined number of nodes, $CARTnodes$, is constructed.

Here, three variants of AdaBoost are employed to build sequentially a linear combination of base CART classifiers that focus on difficult examples. The AdaBoost algorithm, learns a combination of the output of M (weak) classifiers $H_m(x)$ in order to produce the final decision of classification given by

$$H(x) = sign(\sum_{m=1}^{M} a_m \cdot H_m(x))$$

where a_m is the weight (contribution) of each classifier. The weak classifiers are trained sequentially and the weight distribution of the training set instances x_i is updated between iterations according to the accuracy of classification of the previous classifiers. The weight of the misclassified instances is increased for the next iteration, whereas the weight of the correctly classified instances is decreased. The next classifier is trained with a re-weighted distribution. The amount of change on the weight of each instance is proportional to the classification error of the instance. This allows for the current classifier to be focused on the most difficult instances, that is, the ones that were not well classified by the previous classifiers [13].

The AdaBoost based techniques can be considered as a greedy optimization method for minimizing exponential error function [Survey]:

$$E = \sum_{i=1}^{N} e^{-y_i \cdot H(x_i)}$$

where $H(x)$ is the constructed ensemble of classifiers.

A brief presentation of the three variants of AdaBoost follows. Real AdaBoost [13] is the first variant of AdaBoost and the term "Real" derives from the fact that the classifiers produce a real value which is the probability that a given input instance belongs to a class, considering the current weight distribution for the training set. Thus, the Real AdaBoost algorithm can be considered as a generalization of the basic AdaBoost algorithm. Gentle AdaBoost produces a more stable ensemble model. Although, the Real AdaBoost algorithm performs exact optimization with respect to H_m, at each iteration, this algorithm improves it, using adaptive Newton stepping, and uses weighted least-squares regression to minimize the function E. Modest AdaBoost is known to have less generalization error and higher training error, as compared to the previous variants. Modest AdaBoost scheme favors weak classifiers that have maximally decorrelated error with each other [13].

To calculate the three different boosting schemes we used the GML AdaBoost Toolbox for Matlab developed at Moscow State University which is available on-line at [14].

4   Experimental Results

4.1   Dataset

The proposed methodology is applied and evaluated on a dataset containing 71 X-ray galactograms acquired at Thomas Jefferson University Hospital and at the Hospital of the University of Pennsylvania. From these images, 31corresponded to women with no reported galactographic findings (class NF) and 40 to women with reported findings (class RF). The images represented a large variation in topology among the visualized ductal trees. Initially, galactograms were segmented and ductal trees were outlined manually by physician experts. After the phase of feature extraction, the seven features were calculated for each image was obtained. The values of average and the standard deviation of the seven features for both classes are presented in Table1.

Table 1. Average and standard deviation of the extracted geometrical features regarding the two classes of ductal trees

Geometrical Features of Ductal Trees							
		MAxis	mAxis	Eccentricity	Orientation	Solidity	TA	nNodes
NF	mean	2187.00	978.39	0.84	-17.35	0.065	0.65	86.25
class	std	809.83	474.90	0.12	48.32	0.022	0.22	57.86
RF	mean	2136.75	959.19	0.88	-0.91	0.099	0.72	66.85
class	std	1047.54	667.06	0.07	46.93	0.117	0.11	43.83

4.2   Classification Results

A sequence of base CART classifiers was built and experiments were performed for different number of nodes of the CART model ($CARTnodes$). For classification, the training set was randomly selected as 2/3 of the total dataset and the test set consisted of the remaining instances. Since the original dataset included similar number of ductal trees for each class, the two classes were considered as balanced. The parameter M

which represents the number of base classifiers was set to 70; this value ensured stable evaluation results. The number of CART nodes ranged from 1 to 15. The distribution of weights for the instances of the training set was initialized for all boosting schemes $w_i = 1/N$, $i \in \{1, ..., N\}$, where N is the cardinality of the training set. For the evaluation of the proposed classification methodology, the performance was measured in terms of $Sensitivity = TP/(TP + FN)$, $Specificity = TN/(TN + FP)$ and $Accuracy = (TP + TN)/(TP + TN + FP + FN)$, where TP, TN, FP and FN are the number of True Positive, True Negative, False Positive and False Negative cases of ductal trees after validation using the test set. The performance of the three variants of Ada-Boost for $CARTnodes = \{1, 2, ..., 15\}$ is presented in Fig. 2-4. Also, the performance of the weak learner, i.e. the single CART decision tree, is presented for comparison.

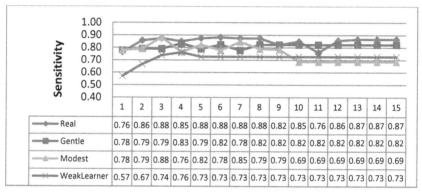

Fig. 2. The obtained results of sensitivity using the weak learner and the three variants of AdaBoost

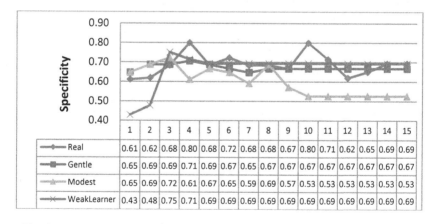

Fig. 3. Results of specificity for the three variants of AdaBoost and the weak learner

Regarding sensitivity (Fig.2), the Real algorithm outperformed other variants of AdaBoost achieving 88% sensitivity for various numbers of CART nodes $CARTnodes = \{3,5,6,7,8\}$. The percentage of sensitivity converged to 87% when the number of nodes is larger than 13, whereas the sensitivity of Gentle algorithm converged to 82%. Although the Modest algorithm presented the lowest sensitivity (69%) compared to other algorithms after convergence (even lower than weak learner), it achieved high levels of sensitivity 85% and 88% when the number of CART nodes was 7 and 3 correspondingly.

Regarding specificity (Fig.3), the Real AdaBoost achieved the best performance (80%) compared to other algorithms. Although the weak learner achieved specificity lower than 50% when the number of nodes is 1 or 2, it presented similar results to these of Real AdaBoost after convergence.

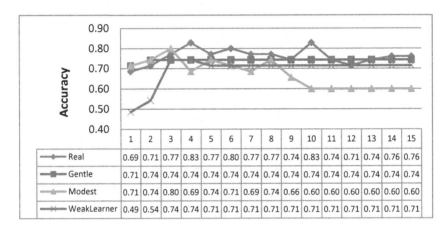

Fig. 4. Comparing accuracy for the three variants of AdaBoost and the weak learner

According to physician experts, accuracy is the most important evaluation metric for the application of classification of galactograms. Regarding accuracy, Real Ada-Boost outperformed the other two variants and the weak learner; the best percentage of accuracy is 83% which was achieved when $CARTnodes = \{5,10\}$. The trade-off between sensitivity and specificity is reflected onto the accuracy metric, thus the Modest algorithm which presented by far the lowest specificity presented also the lowest accuracy. The Gentle algorithm converged to 74% accuracy whereas the weak learner converged to 71% accuracy.

Focusing on the features that were utilized for node construction of CART trees, we observed that the sequence of features remained unchanged as the user-defined parameter $CARTnodes$ was increased more than 8. Specifically for any value of $CARTnodes \geq 8$, the resulted CART tree was composed of eight nodes with root node representing the feature nNodes, one node representing the feature mAxis, four nodes representing the feature of eccentricity and two nodes representing the feature of tree asymmetry. The selection of only four features (nNodes, mAxism, eccentricity, tree asymmetry) offers an interpretable description of features that discriminate the two classes providing insight to the relation of geometry and underlying pathology of ductal trees.

Compared to state-of-the-art techniques for classification of galactograms which employ encoding techniques to obtain a symbolic representation of ductal trees and afterwards use text mining techniques to compute the similarity among ductal structures [3,4], the proposed methodology presented lower performance. Although such a comparison is not fair since the proposed methodology and these previous works use different classification schemes; the proposed method includes ensemble learning techniques whereas included the weighted k-nn classifier, we present the results of sensitivity, specificity and accuracy for two previous methodologies on the dataset of 71 galactograms used in this study (Table 2). The results of state-of-the-art techniques were higher than the proposed ones but suffered from instability regarding the values of parameter k. Moreover, there is no effective way to boost multiple k-nn classifiers into an ensemble learning scheme as k-nn is fairly stable with respect to resampling [6]. Finally, although the proposed scheme did not improve the performance of classification, indicated geometric features related to the risk of breast cancer which could be combined with previous methods to increase the performance of classification.

Table 2. Evaluation results of two state-of-art techniques

	Prüfer tree encoding and text mining methodology [4]			Official tree encoding and text mining methodology [3]		
	Sensitivity	Specificity	Accuracy	Sensitivity	Specificity	Accuracy
$k=2$	0.86	0.84	0.85	0.87	0.75	0.81
$k=3$	0.83	0.84	0.84	0.81	0.71	0.76
$k=4$	0.84	0.88	0.86	0.85	0.81	0.83
$k=5$	0.82	0.84	0.83	0.75	0.66	0.71

5   Conclusion

In this paper, a methodology based on extraction of a new set of geometrical features and ensemble learning techniques is proposed for the classification of ductal trees. The experimental results enhance our understanding regarding the anatomy of human ductal trees and contribute to a comprehensive analysis of the geometrical properties of ductal trees that could be potentially be used as biomarkers of breast cancer risk.

Acknowledgments. This research has been co-financed by the European Union (European Social Fund – ESF) and Greek national funds through the Operational Program "Education and Lifelong Learning" of the National Strategic Reference Framework (NSRF) - Research Funding Programs: Heracleitus II and Thales, Investing in knowledge society through the European Social Fund.

References

1. Guray, M., Sahin, A.A.: Benign Breast Diseases: Classification, Diagnosis, and Management. The Oncologist 11(5), 435–449 (2006)
2. Eyal, E., Furman-Haran, E., Degani, H.: 3-D tracking of the mammary ductal tree using diffusion tensor MR imaging. In: Proceedings of International Society for Magnetic Resonance in Medicine (ISMRM), pp. 588–590 (2008)

3. Skoura, A., Barnathan, M., Megalooikonomou, V.: Classification of ductal tree structures in galactograms. In: Proceedings of 6th IEEE Int. Symposium on Biomedical Imaging (ISBI), pp. 1015–1018. IEEE Press (2009)
4. Megalooikonomou, V., Barnathan, M., Kontos, D., Bakic, P.R., Maidment, A.D.: A Representation and Classification Scheme for Tree-like Structures in Medical Images: Analyzing the Branching Pattern of Ductal Trees in X-ray Galactograms. IEEE Trans. on Medical Imaging 28(4), 487–493 (2009)
5. Bakic, P.R., Albert, M., Maidment, A.D.: Classification of galactograms with ramification matrices: preliminary results. Academic Radiology 10, 198–204 (2003)
6. Garcia-Pedrajas, N., Ortiz-Boyer, D.: Boosting k-Nearest Neighbor Classifier by Means of Input Space Projection. Technical Report, Computational Intelligence and Bioinformatics Research Group (2008)
7. Rambol, R.K., Ahmad, N., Deepak, A.: An Effective Security Management of Database through DNA Fingerprinting Recognition using Geometric Parameters. International Journal of Computer Applications and Information Technology 1(2), 37–38 (2012)
8. Pelt, J.V., Uylings, H.B., Verwer, R.W., Pentney, R.J., Woldenberg, M.J.: Tree asymmetry – a sensitive and practical measure for binary topological trees. Bulletin of Mathematical Biology 54(5), 759–784 (1992)
9. Schapire, R.E., Singer, Y.: Improved boosting algorithms using confidence-rated predictions. Machine Learning 37(3), 297–336 (1999)
10. Friedman, J., Hastie, T., Tibshirani, R.: Additive logistic regression: A statistical view of boosting. The Annals of Statistics 38(2), 337–374 (2000)
11. Vezhnevets., A., Vezhnevets, V.: Modest AdaBoost – teaching AdaBoost to generalize better. In: Graphicon (2005)
12. Breiman, L., Friedman, J.H., Olshen, R.A., Stone, C.J.: Classification and Regression Trees. Wadsworth, Belmont (1984)
13. Ferreira, A.: Survey on Boosting Algorithms for Supervised and Semi-supervised Learning, Instituto de Telecomunicacoes. Technical Report (2007)
14. Vezhnevets, A.: Moscow State University, MSU Graphics & Media Lab., Computer Vision Group,
http://graphics.cs.msu.ru/en/science/research/
machinelearning/adaboosttoolbox

Medical Decision Making via Artificial Neural Networks: A Smart Phone-Embedded Application Addressing Pulmonary Diseases' Diagnosis

George-Peter K. Economou[1] and Vaios Papaioannou[2]

[1] Visiting Professor, Hellenic Open University
[2] Teacher of Informatics

Abstract. A prototype Medical Decision Making System (MDMS) based on feed forward Artificial Neural Networks (fANNs) and its software implementation as an application of 'smart phone' devices, is the topic of this article. An MDMS, generally structured to cover categories of distressed body organs, is presented, focusing on the full spectrum of Pulmonary Diseases (PDs). The fANNs that compose this MDMS have been taught by using real world patients' clinical data. The fANNS have been taught on a powerful PC and their weights ported to be used as a part of a wide range 'smart phone' devices' applications software.

Keywords: Medical Decision Making Systems, Artificial Neural Networks, Pulmonary Diseases, 'smart phones'.

1   Introduction

R&D activities have emphasised into merging human understanding, reasoning and expertise in systems that would treat information in a not 'algorithmic' fashion. Decision Making Systems [2, 15] can be the platforms of such an approach, while MDMS aim to assist doctors of medicine (MDs), guide trainees and encourage medical experts in their diagnoses [5, 8, 9]. In addition, MDs who serve on remote locations, can utilize MDMS in order to judge more accurately upon a particular not so familiar disease.

A creative MDMS, was developed by a team of medical and technical experts. Guidelines and specific data flow were set to fulfil the diagnosis and treatment needs of human diseases. PDs were implemented first of a series of reconfigurable MDMS, due to the raising exhibition of Tuberculosis and Lung Cancer [8, 9].

Real world clinical data were used to instruct its ANNs' layers and preliminary and more detailed experiments showed its great capabilities of making correct classification of symptoms and PDs (about 83% out of 1'000 possible non learnt PD clinical cases). 'Smart phone' applications [12] is the means utilized to port the proposed MDMS to a great variety of such devices, due to their portability and huge acceptance.

L. Iliadis, H. Papadopoulos, and C. Jayne (Eds.): EANN 2013, Part II, CCIS 384, pp. 156–163, 2013.
© Springer-Verlag Berlin Heidelberg 2013

2   Artificial Intelligence and ANNs

Decision Making Systems (DMS) are mostly built by using Artificial Intelligence (AI) techniques, either to accomplish their inference engine or as the means to implement both a knowledge base and their induction rules [14, 18, 24]. Further, the ever expanding use of MDMS in medical centres all over the world has begun to show their value in real world cases, whereas new methods are altogether posed to give the necessary evaluation criteria to compare their performances [6, 13, 18]. In short, ANNs' architectures seem to be a head off than other AI-based or more 'algorithmic' procedures.

The promotion of ANNs towards the formation of MDMS advances vigorously and aims to different goals [3, 8, 9, 14, 18-21]. Assets that AI and medical experts seek in MDMS can be found in ANNs being a part of their mechanism. Parallel searching, dynamic data storage, robustness, generalization virtues and their improved speed factor are inherent to them. ANNs may be implemented purely in software, on general-purpose platforms or on microprocessors, or via custom hardware / VLSI chips [25].

On the other hand, ANNs' vast application domain and general tasks' accomplishment, furnish a sound ground of exploitation. To enhance this end, a large number of ANNs were simulated on a PC and were taught by means of a variety of learning algorithms to favour the best. Finally, the most known and the most severe Pulmonary Diseases' (PDs) symptoms were integrated into their artificial synapse weights. The teaching process was mostly based on supervised learning techniques.

3   Organization of Inputs

The building of knowledge based environments for assisting MDs, is a complex task. MDs' handling of certain medical data (*MD*) tends to differentiate, whereas the importance of all and each *MD* is a fact. Then again, ANNs categorize and generalize *MD* into new patients' cases by building their synapse weights based on individual symptoms. Besides, any MDMS should follow step by step the Clinical Differential Diagnosis Methodology (CDDM) [13, 14] and ANNs suit best [1, 9]. CDDM requires all *MD* to be weighed separately, intermediate results to be judged at the end, and all analysis data to be processed from the more generic to the more specific. A mapping of patients' symptoms to the classes of possible diseases, is thus achieved.

Clinical experts in PDs, established the boundaries of the project. A definite number of questions/inputs were set, the same ones that MDs ask when examining patients. They contain related findings of each one of PDs' symptoms, i.e. Cough, Sputum, Haemoptysis, Fever, Dyspnoea, Wheezing, Chest Pain and historical as well as data obtained from physical exams. Moreover, those *MD* were fed to a large number of ANNs [16, 17, 22] and related to both a sum of thirty five (35) PDs and they related twelve (12) major PDs' classes by separate levels of implementation.

MD were fed by referencing symptoms' findings in possible PDs. Major influences, as the significance to determine certain PDs, multiple PDs' interference in a diagnosis and resulted PDs' ordering on a higher fitness basis, were left to the ANNs to learn. Still, for lethal PDs a patient could suffer, it was made certain that they were excluded or confirmed by the MDMS when evaluating new cases, by using suitable input patterns. Consequently, the same procedure can apply for other diseases. The mapping of major symptoms' classes to possible diseases is similarly performed. The infrastructure of the ANNs, remains the same throughout the procedure.

4   Proposed Composition of fANNs

The feed forward ANN proved to be the most suitable to form the basis of the suggested MDMS. It is arranged in a three layers formation, following *MD*'s time propagation sequence (Fig. 1). Back propagation [17] and Kalman filtering of back propagation equations [22] learning algorithms were forwarded to teach the fANNs. The latter, however, utterly swayed; its performance required less initialization steps and presented better learning speed, increased accuracy, higher convergence rates and data handling. Learning parameters include fANNs with binary inputs, floating point arithmetic, 30 - 44 nodes, ~300 input patterns, 3 slabs each (Neural Layers, NL) and a requested learning accuracy of an average error level of 0.5%. Learning times spread between 20' - 30', i.e. 2,000 - 3,000 learning cycles, and involved mostly multiplications.

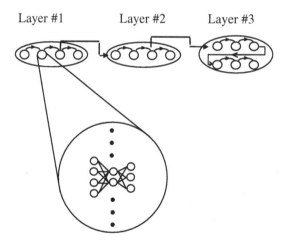

Fig. 1. Layers:ANNs' Composition

4.1   First Layer

A fANNs' formation of four levels, three slab structured, is used. Inputs to the first level are fed separately for each one of the major symptoms (subjective *MD*) and in random order. Also, two other identical structured fANNs treat historical and physical examinations data (objective *MD*). The outputs of those fANNs, are the general classes of the possible PDs (percentages of similarity and compatibility to their learnt patterns).

These outputs are subsequently weighed by eight three-slab fANNs in the second level by forming pairs of symptoms/historical and physical examinations data, enforcing the former as the CDDM imposes. Outputs will be again the classes of PDs. On the third level, the outputs of the second are propagated in another three-slab ANN which, again, have the same outputs as the previous ones, combining all *MD*.

Up to this point, all *MD* given are pondered aloof and interactively considered finally only (step by step implementation of the CDDM). Thus, the end results show the general tendency of the possible present PDs. In the fourth level, though, clinical examinations are suggested by a two-slab ANN. Fig. 2 depicts first Layer.

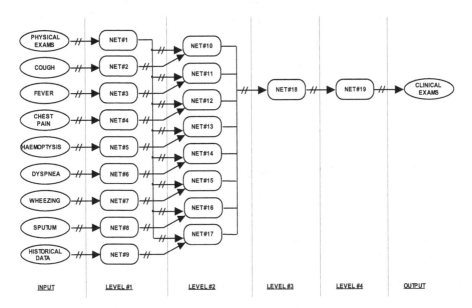

Fig. 2. Layer #1's Structure

This induction methodology scheme is so far submitted, due to its transparency to the intermediate results that are generated. An expert is able at every level of the process to intervene and select the most crucial diagnosis' facets, according to his own opinion. The fANNs, altogether, do not inhibit but, instead, offer percentages of possible PDs' existence, letting the expert to make the final decision. However, MDMS can be let to offer different diagnoses' scenarios and to prune out not possible PDs, too.

4.2   Second Layer

First level's scheme given, another four-levelled, three-slab fANNs' formation was used. These fANNs operate exactly as already stated, but they handle PDs in all their outputs, intermediate and final (and not their classes). Also, some new inputs are added: those that relate the final PDs' percentages of fitness, the third level of the first layer have already computed. This way, a strong positive feedback will be exercised in the second layer fANNs to enhance the final diagnosis. Preliminary and more elaborate results have shown a great increase in outputs' accuracy (25% - 38%).

4.3 Third Layer

Two three-levelled fANNs' formations are also used. These handle all data the former ones did, plus clinical examinations' conclusions and all the new data that could result from the PD(s)' eventual progress (sometimes, elongated time passes for an examination to be done). Still, the final outputs, are scheduled to be the necessary medication potions as well as their dosage and time schedule.

As the reader may note, this novel architecture is built in such a way that assures friendliness, transparency and efficiency. As for the latter quality, *MD* and clinical experts' assistance help to make the system achieve a good performance. So far results have shown an overall performance of nearly 83% of success as the already taught fANNs of this MDMS were left to generalize into their inputs (1'000 total new cases).

This performance does not mean that in the 17% of the new cases presented to the MDMS, PDs are classified incorrectly, thus resulting in a patient's maltreatment. One has to bear in mind that the first priority of MDs is to provide for the correct clinical examinations. By the use of the examinations' results, it can be said that the overall performance of the proposed MDMS is expected to be well over the achieved one.

5 'Smart Phone' Application

Based on nowadays technology evolution and the widespread use of mobile devices, an android version of the MDMS was developed. Thus, MDs can access diagnosis data via 'smart phone' mobile devices. Android is the first free, open source and fully customizable mobile platform, a software stack for mobile devices that includes an operating system, middleware and, key mobile applications (Fig. 3). The android Software Development Kit provides for free the tools and Application Programming Interface necessary to develop applications and port the MDMS [4].

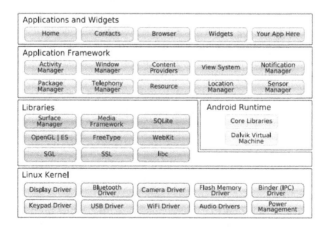

Fig. 3. The Android system architecture

Android applications are written in the Java programming language, and are divided into four types: Activities, Services, Content providers and, Broadcast receivers. The graphical user interface for an Android application is built by using a hierarchy of View and ViewGroup objects. The Android SDK and the Eclipse IDE tools were used for porting the MDMS [10]. Fig. 4 shows this programming hierarchy.

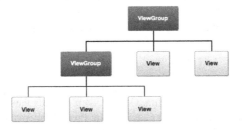

Fig. 4. How ViewGroup objects form branches in the layout and contain other View objects

Most of the MDMS source code, written in the C programming language, was ported directly to the android platform via the Android NDK toolset that allows a user to applications parts of by using native code languages, i.e. C and C++ (see Fig. 5).

Fig. 5. Snapshot of the application User Interface Development

The user can choose which diagnosis data sets to load and perform an instant diagnosis either by using the 'smart phone' MDMS or as a means to upload the *MD* data set to a server for faster diagnosis calculation via a cloud computing model [11].

In terms of performance, a typical desktop PC might achieve around 50GFlops whereas a typical high end 'smart phone' about is 1GFlop [7]. Most Android devices bear ~1GB of RAM by typical Secure Digital cards. I/O performance can reach transfer rates up to 832Mbit/sec, whilst SATA III disks can re ach up to 6,000 Mbit/s. Hence, these performance results are more than adequate when compared to most PCs.

6   Conclusions

The proposed mapping of an ANNs' composition, towards the formation of a powerful general purpose MDMS, in a 'smart phone' application based system was proven to provide for very prominent results. This composition was used for the structuring on an efficient MDMS to assist MDs in human diseases' diagnosis. Special care was given to make it re-targetable and achieve great performance results.

The precursory along to the ultimate results are very stimulating. However, the intensifying of this MDMS through the augmentation of its data, is a necessary next step. The presentation of a general purpose MDMS to be the basis of both other medical diseases induction diagnosis and form the mapping in other areas of expertise, is our major goal. Also, steps are being taken towards using the scanning 'smart phone' devices capabilities in order to insert low level picture *MD* in these powerful tools.

References

1. Ahsan, M.R., Ibrahimy, M.I., Khalifa, O.O.: A Step towards the Development of VHDL Model for ANN based EMG Signal Classifier. In: Proc. of International Conference on Informatics, Electronics & Vision, pp. 542–547 (2012)
2. Al-Absi, H.R.H., Abdullah, A., Hassan, M.I., Shaban, K.B.: Hybrid Intelligent System for Disease Diagnosis Based on Artificial Neural Networks, Fuzzy Logic, and Genetic Algorithms. In: Abd Manaf, A., Zeki, A., Zamani, M., Chuprat, S., El-Qawasmeh, E. (eds.) ICIEIS 2011, Part II. CCIS, vol. 252, pp. 128–139. Springer, Heidelberg (2011)
3. Bounds, D.G., et al.: A Multi Layer Perceptron Network for the Diagnosis of Low Back Pain. In: Proc. Int. Conf. on NN, San Diego, pp. 481–489 (1988)
4. Chang, G., et al.: Developing Mobile Applications on the Android Platform, pp. 264–286. Springer, Berlin (2010)
5. Dhawan, A.P.: An Expert System for the Early Detection of Melanoma Using Knowledge-Based Image Analysis. Anal., Quant. Cyt. and Hist. (1988)
6. Distante, F., et al.: Mapping NN onto a Massively Parallel Architecture: A Defect-Tolerance Solution. In: Proc. of the IEEE, pp. 444–460 (1991)
7. Dongarra, J.J., et al.: The LINPACK Benchmark: Past, Present, and Future. Concurrency Computat.: Pract. Exper., 803–820 (2003)
8. Economou, G.-P.K., et al.: Decision Support Systems for Tele-Medicine Applications. Research Studies Press Ltd., Hertfordshire (2004)
9. Economou, G.-P.K., et al.: Enhanced Applications of a Tele-Medicine Decision Support Platform. WSEAS Trans. on Softw. App. 4, 707–715 (2006)
10. http://developer.android.com
11. http://en.wikipedia.org/wiki/Cloud_computing
12. http://en.wikipedia.org/wiki/Smartphone
13. Hart, A., Wyatt, J.: Evaluating Black-Boxes as Medical Decision Aids: Issues Arising from a Study of NN. Med. Inf., 229–236 (1990)
14. Henson-Mack, K., et al.: Integrating Probabilistic and Rule-Based Systems for CDD. In: IEEE Proc. Southeastcon 1992, Birmingham, AL, USA, pp. 699–702 (1992)
15. House, W.C.: Decision Support Systems: A Data-Based, Model-Oriented, User-Development Discipline. Petrocelli Books Inc., McGraw Hill (1991)
16. Hush, D.R., Horne, B.G.: Progress in Supervised NN. IEEE Sig. Proc. Mag., pp. 8–39 (1993)
17. Lippmann, R.P.: An Introduction to Computing with NN. IEEE ASSP Mag., 4–22 (1987)
18. Marozas, V., Jurkonis, R., Kazla, A., Lukosevicius, M., Lukosevicius, A., Gelzinis, A., Jegelevicius, D.: Development of Teleconsultations Systems for e-Health. Transformation of Health Care with Information Technologies. IOS Press, Holland (2011)

19. Mulsant, B.H.: A NN as an Approach to Clinical Diagnosis. M. D. Comp., 25–36 (1990)
20. O' Leary, T.J., et al.: Computer-Assisted Image Interpretation: Use of a NN to Differentiate Tubular Carcinoma from Sclerosing Adenosis. Mod. Path., 402–405 (1992)
21. Poli, R., et al.: An NN Expert System for Diagnosing and Treating Hypertension. IEEE Comp., 64–71 (1991)
22. Scalero, R.S., Tepedelenlioglu, N.: A Fast New Algorithm for Training Feedforward NN. IEEE Trans. on Sig. Proc., 202–210 (1992)
23. Sibai, F.N.: A Fault Tolerant Digital Artificial Neuron. IEEE Design & Test of Computers, 76–82 (1993)
24. Umbaugh, S.E., et al.: Applying Artificial Intelligence to the Identification of Variegated Coloring in Skin Tumors. IEEE Eng. in Med. and Biol. (1991)
25. Watanabe, T., et al.: A Single 1.5-V Digital Chip for a 10^6-synapse NN. IEEE Trans. on NN, 387–393 (1993)

A Simulator for
Privacy Preserving Record Linkage

Alexandros Karakasidis and Vassilios S. Verykios

Hellenic Open University, Patras, Greece
{a.karakasidis,verykios}@eap.gr

Abstract. Nowadays, there is an abundance of data produced by human activities. Integrating these data for processing, raises privacy issues. Privacy Preserving Record Linkage is an area of research, developed in the last few years, aiming at integrating personal data without compromising the privacy of the people these data refer to. To solve this problem, many methods have been recently developed. To be able to assess the performance and the overall behaviour of these methods, we have built a Privacy Preserving Record Linkage Simulator which we present in this paper.

1 Introduction

In the last few years, we have all witnessed the results of the information explosion. Large volumes of data describing individuals are collected by public or private organizations. This is either a result of tracking human behavior and condition, or a result of people making available their personal data by themselves. In the first case, examples of data collection include human actions, like shopping, interacting with public agencies, or admission to healthcare facilities. In the second case, we mostly refer to data made available to social networks.

Being able to interconnect and analyze this independently stored information would provide benefits to both businesses and governments. For example, being able to interconnect data stored by distinct health organizations could prevent epidemics outbreaks or help us better understand the mechanisms behind certain diseases. However, linking heterogeneous data is not a trivial task. First of all, since data are stored separately by independent organizations, there is no unique identifier for linking them. As such, they should be linked using personal or private information. Furthermore, these data tend to suffer from low quality, exhibiting typos, abbreviations, misspellings etc.

Most of these issues, comprising the classical version of the Record Linkage problem, have been resolved [3]. However, classical record linkage approaches do not take any particular measures to protect the privacy of the subjects described by these data. Considering this fact together with the need for privacy assurance, when processing personal information, we are led to a shift to the problem of record linkage so that it takes privacy into account. As such, Privacy Preserving Record Linkage (PPRL) is the problem where data from two or more

L. Iliadis, H. Papadopoulos, and C. Jayne (Eds.): EANN 2013, Part II, CCIS 384, pp. 164–173, 2013.

heterogeneous data sources are integrated in such a way that after the integration concludes, the only extra knowledge that each source gains relates to the records which are common among the participating sources.

Privacy Preserving Record Linkage consists of two main phases, Privacy Preserving Matching (PPM) and Privacy Preserving Blocking (PPB). Privacy Preserving Matching is the process of accurately matching datasets using elaborate algorithms which do not compromise data privacy. Since these algorithms are elaborate, they manage to achieve high matching performance but they also exhibit low time performance. Time performance is an important aspect of the problem, since we usually need to integrate large volumes of data. As such, Privacy Preserving Blocking techniques have been introduced which aim at reducing Privacy Preserving Matching times. These techniques aim at putting aside unlikely to match candidate pairs.

To explore the behavior and the performance of Privacy Preserving Matching and Privacy Preserving Blocking algorithms, we have built a simulator which materializes many of the techniques falling in the aforementioned categories and which is under constant development. In this paper we present the architecture of our Simulator, the algorithms we have materialized and some experimental results received by using our tool.To the best of our knowledge, this is the first work in the literature presenting a Privacy Preserving Record Linkage Simulator. In addition, our approach features more than one Privacy Preserving Matching and Privacy Preserving Blocking technique.

The rest of this paper is organized as follows. In Section 2 we discuss previous approaches to the problem and how they are related to our work. Section 3 contains a description of the architecture of our Simulator and Section 4 provides some experimental results from our Simulator. Finally, in Section 5 we discuss some conclusions and some thoughts for the future development of the PPRL Simulator.

2   Related Work

In this Section we briefly present existing techniques for Privacy Preserving Blocking and Privacy Preserving Matching. Starting with PPB, the first attempt was made by Al-Lawati et al [1]. They reduce matching complexity by proposing various levels of blocking in exchange of privacy. Next, there are the works of Inan and his colleagues [6,7]. In [6] they use tree-like structures, called value generalization hierarchies, for reducing matching candidates. In [7] they propose the use of an approach based on differentially privacy for performing PPM. In both of their papers the use homomorphic encryption for privately matching numeric attributes. In [9], Karakasidis et al. propose the use of legacy phonetic encoding as PPB method based on the inherent privacy characteristics of the phonetic algorithms. These properties were explored in [13]. Recently, the use of reference texts was examined for PPB [11,10,12]. In [11] the use of reference text clustering was proposed for PPB. This was the first attempt to use reference sets for PPB. In [11] the authors introduced a method for multidimensional PPB in combination with reference text clustering. Finally, in [12] there is a sorted neighborhood method suitable for encrypted fields.

Regarding PPM, there have been attempts from many different perspectives. The Bloom Bigrams method suggested in [18] is considered as a state-of-the-art method for privately linking alphanumeric identifiers. Substrings of string identifiers are hashed into bit vectors. However, as indicated in [14], under certain conditions, this method is susceptible to constraint satisfaction cryptanalysis. Another method for string identifiers was suggested in [16]. This method uses a reference table and exploits the triangular inequality to assess similarity between actual and reference data. Scannapiecco et al. [17] achieve privacy preserving data matching using embedding spaces. They also suggest a privacy preserving schema matching technique based on a global common schema. In [8], phonetic encoding is used for private record matching using a secure multiparty computation protocol. Finally, The most recent work in the area is the one of Kuzu et al. [15], where an SMC protocol for matching both numeric and is presented based on a Paillier encryption scheme.

3   The Simulator

The aim of this Section is to provide a description of the architecture of the PPRL Simulator we have implemented. First of all, we describe the simulator' s architecture and discuss certain architectural choices. Next, we present how the Simulator is configured. Finally, we briefly present the algorithms that are currently available in the PPRL Simulator.

3.1   Considerations Taken When Designing a PPRL Simulator

The process of building a PPRL Simulator is very similar to the situation of building an actual PPRL system. The reason is that, in any case, the building blocks of the simulator should exhibit as much as possible the same behavior with the actual system. in order to be able to extract as safe conclusions as possible about a method's performance and behavior. We chose Java to build our system. There were many practical reasons for this decision. First of all, Java 's ability of *Write Once Run Anywhere*, provides the benefit of machine independence. Additionally, there is an abundance of out-of-the-box libraries and data structures and many more built by the community [4]. Finally, due to Java's architecture, it is easy to incorporate the functionality of the classes implemented in the Simulator in a future real world system. Without any additional modification to these classes, this system could be a web service, a standalone, a web, a cloud, or even a mobile application.

Having these in mind, we made a particular effort to limit the assumptions when designing the architectural components of the PPRL Simulator to elements that would not affect an actual system and its behavior. In specific, the only assumptions we have made are the following. First of all, at present, all operations are executed sequentially. In many PPB or PPM algorithms there are functions that require the parallel operation of the matching parties. In our simulator this is simulated by a processing queue. Secondly, we currently assume that communication links propagate data without delay. As such we do not measure communication times.

A modular design was utilized, separating the simulation functionality from the actual methods implemented. This also allows for its extendability. As we may observe from Figure 1, first of all, there is the basic simulator module. This module hosts simulator nodes. A simulator node may either be a *Data Source* node (DS), or a *Trusted Third Party* node (TTP).

The basic simulator module actually executes the simulator. It hosts the simulator nodes and it is also connected to the results module, where results are processed. Each simulation may be executed multiple times with different parameters. These parameters are stored in a manually configured *Properties File*.

3.2 PPRL Simulator Architecture

The role of a DS node is self evident. It holds data in order to privately link them with the data held by another DS node. However, in many PPRL algorithms, an auxiliary node is assumed, the TTP. The operation of this node is usually to enhance the privacy provided by each method by hiding part of the exchanged information from the DSs. As such, it usually behaves differently than the DSs. In our simulator, we distinguish these two types because depending on the algorithm we examine, any number of DS nodes may be created by the user, having a minimum of two of course. However, a TTP is automatically created, when this is required. In any case, all of them may be configured when necessary, as we will see later in this section. Regardless of their type, all nodes communicate with each other through a shared message queue. At this point we should mention that all DS nodes and TTP nodes obligatory follow the same protocol.

Data Source Nodes. A DS node, as illustrated in Figure 1, should necessarily contain a *Data Connection Module* providing the necessary data so as to act as a data source. Next, it should also contain a *PPM Module*. This module is again mandatory. Finally, it may also contain a *PPB Module*. A PPB Module is not mandatory, since a PPRL system might not feature a PPB technique.

Trusted Third Party Nodes. There is no necessary module for a TTP node. This is because the PPM or PPB algorithms used may not require a Trusted Third Party. In such a case, the simulator does not create a node of this kind. However, a TTP Module may only consist of a PPM Module, or of a PPB Module, or of both Modules. A Data Connection Module is not mandatory but may be used as well, since for some algorithms the TTP needs to generate or retrieve external data. [16,12].

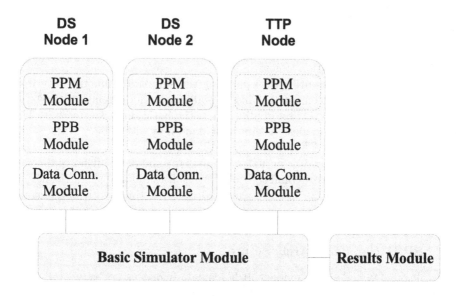

Fig. 1. Architecture of the PPRL Simulator. In this Figure we assume two DS nodes and one TTP node. Modules with a dashed border are optional.

3.3   The Results Module

The results calculated by the PPRL Simulator are stored into two csv files. The first csv file, which we call the *Matching Output File* contains the record ids of the records calculated as matching by the PPM algorithm used. This file is what we actually expect from a PPRL system. It may be used either for manual review or for debug purposes. The second csv file created, the *Measurements File* contains measurements and statistics for the experiment. This file holds the number of true positives, the number of false positives, the precision, recall and F-score of the method, the time required for database input/output the time required for PPB and the time required for PPM.

When the datasets held by the DS nodes have the same unique identifiers (for instance when corrupted versions of the same dataset are used), the PPRL automatically calculates metrics based on the common *PrimaryKey*, as illustrated in Listing 1.1. In the case however that the datasets used by the DS nodes do not share common primary keys, if we want automatic calculation of the measures, we should use a matching csv file containing the matches between different primary keys. When the simulator performs more than one runs of the same algorithm, then an equal number of Matching Output Files is created, while all the measurements are appended to the same Measurements File.

3.4   PPRL Simulator Configuration

One of the goals set when designing the PPRL Simulator was to be highly configurable. Therefore, we use a Java properties file. As we may see from Listing

1.1, each simulator element has its own configuration section. However, this is only for display reasons, since even in the case that the configuration commands were mixed the file would be still parsed correctly.

Listing 1.1. A Sample Configuration File

```
# Basic Simulation Module
DB = sqlite
NoExecutions = 3
DSNodes = 2
MatchMethod = BLOOMBIGRAMS   # Schnell et. al
BlockMethod = SNEF      # Karakasidis et. al

# Common Parameters for All Nodes
# PPB Parameter
SNEFWindowSize = 20
SNEFRSSize = 400
SNEFKanon = 3
SNEFRefTextTable = master

# PPM Parameters
MatchingThreshold = 0.45
BloomFilterSize = 1000
HashFunctions = 4

# First DS Node parameters
1_TableName = voters
1_PrimaryKey = NCID
1_MatchVariables = LAST_NAME;FIRST_NAME;MIDL_NAME;STREET_NAME
1_BlockVariables = FIRST_NAME;MIDL_NAME

# Second DS Node parameters
2_TableName = census
2_PrimaryKey = REC_ID
2_MatchVariables = SURNAME;GIVEN_NAME;MIDL_NAME;ADDRESS
2_BlockVariables = GIVEN_NAME;MIDL_NAME
```

Configuration of the Basic Simulation Module. We could divide the configuration of the basic simulation module in two parts. The first part has to do with experiment execution parameters. The second part has to do with configuration parameters specific for each algorithm. In specific, for the experiment configuration section of the basic simulation module, the following parameters of the simulation may be set. First of all, the PPM algorithm that is used is configured. This setting is universal for all types of nodes and, depending on the node type, it attaches the appropriate class to the PPM module. Next, the PPB algorithm to be used, if any. Again, we follow the same rationale with the PPM module. The next parameter to be configured has to do with the database type to be used. Currently three database types are available: mySQL, SQLite and comma delimited csv files. The selection of the database is universal again. However, each node uses a separately configured table.

Next, the number of executions of the same experiment is configured. This is useful in two cases. First, when time performance is assessed and we need multiple executions of the same experiment in order to ensure that the measurements are accurate. Second, when probabilistic PPM or PPB methods are used. In such cases, the matching performance may vary among executions. The last parameter that is set, has to do with the number of DS nodes that will

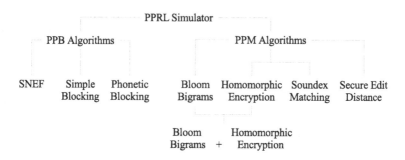

Fig. 2. Algorithms currently implemented in the PPRL Simulator

participate in the simulation. In the case that a TTP node is required by either
the PPM or the PPB algorithm, it is automatically created without having to
explicitly declare it.

Regarding the case of PPM and PPB parameters, these are setup by the
basic simulation module. This is because these parameters should be common
to all nodes so as they correctly adhere to the specific protocol. In the sample
configuration file, illustrated in Listing 1.1, the SNEF method [12] is used for
PPB while the Bloom Bigrams method [18] is used for PPM.

Configuration of the DS Nodes. The first parameter of a node that has to
be set, regardless of the algorithm used, is the *TableName*, designating the table
where the specific node retrieves its data from. As we can see from Listing 1.1,
the *TableName* parameter is prepended by a number followed by an underscore.
This designates the node that the specific parameter refers to. We follow this
pattern with the next parameters as well.

Next, there is the *MatchVariables* parameter. This one tells the simulator
which parameters will be used as matching fields. This parameter is independently configured, since nodes may hold databases with different schemas. In
any case, the fields are corresponded by the order they are referenced. For example, in Listing 1.1, field *LAST_NAME* of DS 1 will be checked against field
SURNAME of DS 2. The same logic holds in the case that PPB is used.

3.5   Implemented Algorithms

The PPB and PPM algorithms currently implemented in the PPRL Simulator
are illustrated in Figure 2. Beginning with the implemented PPB Algorithms,
the first one is the Sorted Neighborhood on Encrypted Fields (SNEF) algorithm
[12], a privacy enhanced version of the Sorted Neighborhood algorithm introduced by Hernandez and Stolfo [5]. The next blocking algorithm implemented is
the Simple Blocking algorithm introduced by Al Lawati et al. [1]. This method
incorporates both Privacy Preserving Blocking and Matching. Next, there is
Phonetic Blocking which uses Soundex's inherent privacy characteristics [9].

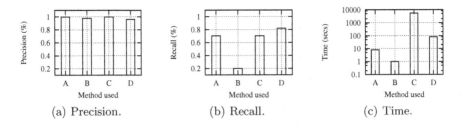

(a) Precision.   (b) Recall.   (c) Time.

Fig. 3. Algorithm comparison results using the PPRL Simulator. A: SNEF+Bloom Bigrams, B: Simple Block, C: SNEF+Bloom Bigrams+Homomorphic, D: Bloom Bigrams.

Regarding PPM algorithms, the first one on the left, as we can see in Figure 2 is the Bloom Bigrams method [18], currently considered as a reference point by the PPRL community. Next, there is the Homomorphic Encryption method, used for numerical fields in [6,7]. The fast Soundex Matching method [8] follows, using Soundex's inherent privacy characteristics for PPM. To continue, there is the Secure Edit Distance method, also introduced in [9] securing Edit Distance through the use of Bloom Filters. We may also see that there is a combination of the Homomorphic Encryption and the Bloom Bigrams method. This uses the implementations of these two methods and it is applicable in the cases that we use both numeric and alphanumeric matching fields.

4   Simulation Results

In this Section we present some experimental results that we retrieved using our Simulator. We provide results of the Simple Blocking technique proposed by Al Lawati et al. [1]. We also use a combination of Homomorphic Encryption [7] and Bloom Bigrams [18] as a PPM method and SNEF [12] as a PPB blocking method. We employ the Bloom Bigrams method for PPM and we examine it both in its plain form and when it is enhanced with the SNEF Privacy Preserving Blocking Method.

We have conducted all of our experiments using a Core i3 PC with 4GB of RAM. For our setup we assumed having two DS nodes and one TTP node. We experimented using a synthetic dataset produced by FEBRL [2] consisting of 5000 original and 5000 duplicate records. We have used $\{FirstName, Suburb\}$ as blocking variables and $\{FirstName, Surname, Address\}$ as matching variables. In the case where homomorphic encryption is used, our matching variables are $\{FirstName, Surname, Age\}$. The results were normalized having as a basis the performance of the Bloom bigrams method. The results are illustrated in Figure 3. For Precision and Recall, in Figures 3(a) and 3(b), the vertical axes are linearly scaled, while the vertical axis for Time Performance in Figure 3(c) is logarithmically scaled.

Observing these results, we are led to the following conclusions. The first conclusion is that Al Lawati's method is very fast, while homomorphic encryption performs poorly in terms of time, despite SNEF blocking. On the contrary, when homomorphic encryption is not used, the combination of SNEF and Bloom Bigrams is fast and efficient. Moreover, while Al Lawati's method is fast, it is not accurate, as it scores poorly in terms of recall. On the other hand, precision and recall, when using homomorphic encryption, is high. The SNEF + Bloom Bigrams combination scores similarly in terms of precision and recall with the case of using homomorphic encryption, requiring nevertheless a very small fraction of execution time.

5   Conclusions

In this paper we have presented a modular, highly configurable PPRL Simulator. We have also presented some results from our simulations. Our next steps include the implementation and incorporation of more methods into our simulator. Next, we plan to extend the basic simulator module so that it performs experiments using a list, or a range, of values for each parameter. Our ultimate aim however is, based on the accumulated experience we will gain from evolving the PPRL Simulator, to build a real world PPRL system for production use.

References

1. Al-Lawati, A., Lee, D., McDaniel, P.: Blocking-aware private record linkage. In: IQIS, pp. 59–68 (2005)
2. Christen, P.: Febrl-: an open source data cleaning, deduplication and record linkage system with a graphical user interface. In: Proceeding of the 14th ACM SIGKDD International Conference on Knowledge Discovery and Data Mining, pp. 1065–1068. ACM (2008)
3. Christen, P.: Data Matching. Data-Centric Systems and Applications. Springer (2012)
4. Cohen, W.W., Ravikumar, P., Fienberg, S.E.: A comparison of string distance metrics for name-matching tasks. In: IIWeb, pp. 73–78 (2003)
5. Hernández, M.A., Stolfo, S.J.: Real-world data is dirty: Data cleansing and the merge/purge problem. Data Min. Knowl. Discov. 2(1), 9–37 (1998)
6. Inan, A., Kantarcioglu, M., Bertino, E., Scannapieco, M.: A hybrid approach to private record linkage. In: ICDE, pp. 496–505 (2008)
7. Inan, A., Kantarcioglu, M., Ghinita, G., Bertino, E.: Private record matching using differential privacy. In: EDBT 2010, pp. 123–134. ACM, New York (2010)
8. Karakasidis, A., Verykios, V.S.: Privacy preserving record linkage using phonetic codes. In: Proceedings of the 4-th Balkan Conference of Informatics, pp. 101–106 (2009)
9. Karakasidis, A., Verykios, V.S.: Secure blocking + secure matching = secure record linkage. JCSE 5(3), 223–235 (2011)
10. Karakasidis, A., Verykios, V.S.: A highly efficient and secure multidimensional blocking approach for private record linkage. In: 2012 IEEE 24th International Conference on Tools with Artificial Intelligence, pp. 428–435 (2012)
11. Karakasidis, A., Verykios, V.S.: Reference table based k-anonymous private blocking. In: SAC 2012: Proceedings of the 27th Annual ACM Symposium on Applied Computing, pp. 859–864. ACM, New York (2012)

12. Karakasidis, A., Verykios, V.S.: A sorted neighborhood approach to multidimensional privacy preserving blocking. In: 2012 IEEE 12th International Conference on Data Mining Workshops, pp. 937–944 (2012)
13. Karakasidis, A., Verykios, V.S., Christen, P.: Fake injection strategies for private phonetic matching. In: Garcia-Alfaro, J., Navarro-Arribas, G., Cuppens-Boulahia, N., de Capitani di Vimercati, S. (eds.) DPM 2011 and SETOP 2011. LNCS, vol. 7122, pp. 9–24. Springer, Heidelberg (2012)
14. Kuzu, M., Kantarcioglu, M., Durham, E., Malin, B.: A constraint satisfaction cryptanalysis of bloom filters in private record linkage. In: Fischer-Hübner, S., Hopper, N. (eds.) PETS 2011. LNCS, vol. 6794, pp. 226–245. Springer, Heidelberg (2011)
15. Kuzu, M., Kantarcioglu, M., Inan, A., Bertino, E., Durham, E., Malin, B.: Efficient privacy-aware record integration. In: Proceedings of the 16th International Conference on Extending Database Technology, EDBT 2013, pp. 167–178. ACM, New York (2013)
16. Pang, C., Gu, L., Hansen, D., Maeder, A.: Privacy-Preserving Fuzzy Matching Using a Public Reference Table. In: McClean, S., Millard, P., El-Darzi, E., Nugent, C. (eds.) Intelligent Patient Management. SCI, vol. 189, pp. 71–89. Springer, Heidelberg (2009)
17. Scannapieco, M., Figotin, I., Bertino, E., Elmagarmid, A.K.: Privacy preserving schema and data matching. In: SIGMOD Conference, pp. 653–664 (2007)
18. Schnell, R., Bachteler, T., Reiher, J.: Privacy-preserving record linkage using bloom filters. BMC Medical Informatics and Decision Making 9(1), 41+ (2009)

Development of a Clinical Decision Support System Using AI, Medical Data Mining and Web Applications

Dimitrios Tsolis[1], Kallirroi Paschali[2], Anna Tsakona[2],
Zafeiria-Marina Ioannou[2], Spiros Likothanassis[2], Athanasios Tsakalidis[2],
Theodore Alexandrides[3], and Athanasios Tsamandas[4]

[1] University of Patras, School of Business Administration,
Cultural Heritage Management and New Technologies DPT, 30100 Agrinio, Greece
dtsolis@upatras.gr
[2] University of Patras, School of Engineering, Computer Engineering and Informatics DPT,
26504 Patras - Rio, Greece
tsak@ceid.upatras.gr
[3] University of Patras, School of Medicine, Internal Medicine DPT, 26504 Patras - Rio, Greece
thalex@med.upatras.gr
[4] University of Patras, School of Medicine, Pathology DPT, 26504 Patras - Rio, Greece
tsa@med.upatras.gr

Abstract. The need of an advanced hospital information system is imminent as it supports electronic patient record management and use of decision support leading to effective diagnosis and treatment. Data mining algorithms and techniques are playing a key role to this process, enhancing access to critical data by the medical personnel and optimizing functionality for the decision support services. In addition, web services make access to critical information feasible from any place, at any time and from any device. In the current paper, DEUS, a clinical decision support system is proposed and presented which combines efficient data mining, artificial intelligence and web services so as to support diagnosis and treatment planning. The system is tested throughout two case studies a) thyroid cancer and b) hepatitis.

Keywords: Clinical Decision Support Systems, Medical Data Mining, Artificial Intelligence, Expert Systems, Thyroid Cancer, Hepatitis.

1 Introduction

Clinical decision support (CDS) systems provide clinicians, staff, patients, and other individuals with knowledge and person-specific information, intelligently filtered and presented at appropriate times, to enhance health and health care [1]. The international institutes and organizations of Medicine have long recognized problems with health care quality worldwide, and for more than a decade have advocated using health information technology (IT), including electronic CDS, to improve quality [2]. Since 2004, the importance of electronic medical records (EMRs), was promoted there has been a slow but increasing adoption of health IT [3]-[5].

L. Iliadis, H. Papadopoulos, and C. Jayne (Eds.): EANN 2013, Part II, CCIS 384, pp. 174–184, 2013.

It must be remembered, though, that these health IT applications are a means to improve health care quality, not an end in themselves. Further, although EMRs can improve accessibility and legibility of information, it is unlikely that there will be major improvements in the quality and cost of care from the use of health IT without proper implementation and use of CDS [6].

In this framework, this paper presents an advanced clinical decision support system (called DEUS) which supports EMRs management and use of decision support and data mining methods, leading to effective diagnosis and treatment. The system is under implementation and will be tested in two case studies a) thyroid cancer and b) hepatitis.

The next sections present the state of the art in CDS systems, the originality of the proposed CDS, its current implementation phase and future perspectives.

2   State of the Art

Early CDS systems were derived from expert systems research, with the developers striving to program the computer with rules that would allow it to "think" like an expert clinician when confronted with a patient [7]. From this early research there was growing recognition that these systems might be useful beyond research, that they could be used to assist clinicians in decision making by taking over some routine tasks, warning clinicians of potential problems, or providing suggestions for clinician consideration [8].

Common features of CDS systems that are designed to provide patient-specific guidance include the knowledge base (e.g., compiled clinical information on diagnoses, drug interactions, and guidelines), a program for combining that knowledge with patient-specific information, and a communication mechanism—in other words, a way of entering patient data (or importing it from the EMR) into the CDS application and providing relevant information (e.g., lists of possible diagnoses, drug interaction alerts, or preventive care reminders) back to the clinician [9]. CDS can be implemented using a variety of platforms (e.g., Internet-based, local personal computer, networked EMR, or a handheld device). Also, a variety of computing approaches can be used. These approaches may depend on whether the CDS is built into the local EMR, whether the knowledge is available from a central repository (possibly outside the local site and accessed and incorporated locally when needed), or whether the entire system is housed outside the local site and is accessed, but not incorporated into the local EMR. In principle, any type of CDS could utilize any of these underlying computational architectures, methods of access, or devices. The choices among these elements might depend more on the type of clinical systems already in place, vendor offerings, workflow, security, and fiscal constraints than on the type or purpose of the CDS [10].

Many of the technology differences described need not be apparent to the user. The following factors may be more relevant to the clinician user or those assisting with implementation: (1) the primary need or problem and the target area of care for which the CDS is being considered (e.g., improve overall efficiency, identify disease early, aid in accurate diagnosis or protocol-based treatment, or prevent dangerous adverse events affecting the patient); (2) to whom and how the information from the CDS will be delivered; and (3) how much control the user will have in accessing and responding to the information [11].

Based on the above analysis, current implementations on CDSs are mainly focusing on the clinician user and do not focus on assisting the patient to be actively involved in the treatment procedure. It is efficient to involve the patient at various stages to the procedure as she / he plays a key role in diagnosis and therapy. The proposed CDS will support the patient, by informing and alarming him, to optimize the result of the proposed therapy.

In addition more CDS do not alarm the doctor or the patient for critical condition in the procedure. The proposed CDS-IS will produce alarms for certain cases considered critical by the decision support tools.

Finally the existing CDSs' are not using data mining techniques to improve access to critical information and its use to the clinical decision process.

It is proven that a CDS can also potentially lower costs, improve efficiency, and reduce patient inconvenience. The proposed CDS could address all three of these areas simultaneously for example, by alerting clinicians to potentially duplicative testing. For more complex cognitive tasks, such as diagnostic decision making, the aim of the proposed CDS will be to assist, rather than to replace, the clinician. The CDS may relieve the clinician of the burden of reconstructing orders for each encounter. The CDS may offer suggestions, but the clinician must filter the information, review the suggestions, and decide whether to take action or what action to take.

Finally, the proposed CDS will broaden the type of devices accessing its services including tablets and smartphones. Current CDSs have not yet expanded their usability features towards this goal.

The next table provides examples of CDS that address a range of target areas and depicts the state of the art [12]:

Target Area of Care	Example
Preventive care	Immunization, screening, disease management guidelines for secondary prevention
Diagnosis	Suggestions for possible diagnoses that match a patient's signs and symptoms
Planning or implementing treatment	Treatment guidelines for specific diagnoses, drug dosage recommendations, alerts for drug-drug interactions
Follow-up management	Corollary orders, reminders for drug adverse event monitoring
Hospital, provider efficiency	Care plans to minimize length of stay, order sets
Cost reductions and improved patient convenience	Duplicate testing alerts, drug formulary guidelines

3   Deus - Originallity of the Approach

DEUS, the proposed Clinical Decision Support system design and implementation is based on a combination of technologies and basic principles.

At first, digitization technologies for medical data and the patient's electronic file is being used. The digitization process involves the use of international technological standards and good practices aiming at an interoperable di gital medical content which could be further used to the expert system.

For data access and exchange the use of metadata standards is supported in the expert system. The metadata standards used are focusing on the following three categories:

a) Patient identity - data elements about a patient, which includes a patient's full name, previous names with associated date ranges (as an optional element), date of birth, address, zip code and one type of patient identification (ID) data along with the origin of that ID; The metadata standards adopted include Health Level 7 version 2 (HL7 V2) messages and Health Level 7 Clinical Document Architecture Release 2 (HL7 CDA R2) [13];

b) Provenance - data elements about the source of the clinical data, which provides information on the "who, what, where and when" and includes a tagged data element (TDE), a time stamp, and digital signatures used to ensure the data has not been altered since its creation; the HL7 CDA R2 and the Clinical Data Interchange Standards Consortium (CDISC) standards are used to provide the information on provenance.

c) Privacy - data elements include a privacy policy pointer and content elements descriptions such as data type and sensitivity. The Clinical Document Architecture Release 2: and Patient Consent Directives (CDA R2 PCD) standards are used for privacy.

At a second stage, mapping of the metadata already being collected to international metadata standards for bio and medical informatics is implemented. The mapping process is being implemented in a fully automated way within the databases used [14]. The database design is primarily based on the aforementioned metadata standards. The database is structured and implemented in a way which supports the full documentation of the patient's data in line with the international metadata standards. The number of entries is until now strictly restricted as the system is at its early implementation stages. The entries selected represent a typical set of data preserved, accessed and exchanged for its patient. The digitization, metadata ingestion and management process as well as the mapping process ensuring interoperability has been successfully completed.

DEUS is an expert system which uses Artificial Intelligence and contributes to the diagnosis and therapy of each individual disease. The component for automated diagnosis will be built, based on artificial intelligence techniques, such as artificial neural networks, decision trees, probabilistic neural networks, association rules, expert systems, in the aim to predict and classify different therapeutic patterns for each case. Association rules will be also used to discover the occurrence of implications with a certain confidence degree, in a database. The system will use both current and history patient data to support doctor's decisions. Depending on the data flow analyzed in the next sections the system uses: A neural network algorithm based on a Bumptree Network [15] which combines the advantages of a binary tree with an advanced classification method using hyper ellipsoids in the pattern space instead of lines, planes or curves. The arrangement of the nodes in a binary tree greatly improves both learning complexity and retrieval time. A fuzzy logic algorithm, which if necessary, is used to determine or predict the maximum or minimum values of certain biometric data under examination [16], [17]. Web user interfaces for the doctor and medical personnel, which allow data insertion, search, editing and support through the diagnosis and treatment stages of the thyroid cancer. The services can be offered through the internet, at a distributed and geographically dispersed way.

Both neural network and fuzzy logic algorithms are being implemented as black boxes for input data and output results. They are transformed from programming languages to dynamic link libraries which can be integrated to the whole information system. The web user interfaces which are implemented in web scripting languages are able to can and execute the algorithms based on the data inserted by the final user. The results are also integrated to the web applications. The integrated system support doctors and medical personnel to at first diagnose effectively the thyroid cancer and at a second stage to offer personalized treatment to the patient [18]. The decision support subsystem is in the phase of its mid- implementation. In the next stage, the decision support subsystem should be fully completed and evaluated. The evaluation should focus on algorithmic performance measures for the decision making process, statistical data and an evaluation study of the whole performance of the expert system. Based on the evaluation study the expert system should be finally optimized.

Furthermore, since in the long term the system collects an enormous amount of data generated by thousands of patients, data mining techniques will be used towards two (2) directions efficient access to critical data by the medical personnel and use of the critical data to the Decision Support subsystem. The key in the use of data mining methods lies on the extraction of patient specific rules from the clinical data collected, which are further used to support the clinical decision process. The medical data include medical history, current diseases, symptoms, lab results, allergies, treatment plans etc. The analysis of this data is a complex and difficult process which requires data mining methods for the extraction of the most useful and critical – per – case information. In the proposed Clinical Decision Support System data mining is mainly used (1) to find human-interpretable patterns and associations in set of data so as to improve access to critical data and (2) patterns of prediction, which aim to foretell the outcome of specific point of interest and will be applied to the construction of the decision models for support the procedure of diagnosis and treatment planning. Once evaluated and confirmed the clinical decision models will be embedded to the clinical information system [26].

The proposed system presents considerable originality and provides an important innovation regarding relevant CDSs. The main innovations are:

a) Development of a specific Therapies and Evidence Based Medicine Knowledge based unique in international level.
b) Holistic management of diagnosis, treatment, long term therapy.
c) Possibility to use updated clinical parameters assuring a more precise and correct therapeutic approach, based on a data recording process.
d) Structure the doctor-patient interactivity, even in mobility

e) Collecting data and using data mining methods which can be further used for the access and decision support and constantly provide epidemiologic indications and an analytic census of the assisted population, in order to plan and define other prevention interventions.

f) Producing integrated services to be accessed by various devices as support between doctor and the patient.

The proposed CDS will be configured directly by doctors and medical personnel, through its pilot testing with two case studies presented in the next sections. This feature is considered essential to produce strict protocols able to generate therapeutic information on users', caregivers' and health care workers' smartphones. The Decision Support subsystem configured as specified above, will also produce alerts based on Green, Yellow and Red codes regarding the different cases under treatment. This feature is not yet covered by the state-of-the-art CDS systems in medical informatics and applications. The project will also produce a Knowledge Base for specific Therapies and a database for Evidence Based Medicine.

The data gathered for targeted therapies will be based on international metadata protocols and standards so as at the next step the data flow diagrams to be produced for the selected therapy and treatment case. The production of the data flow diagrams is essential for the effective operation of the Decision Support System. The data flow diagrams are considered as graphical presentations of "flow" of data which derive from external bodies and are applied in the information system. They finally present the way in which data entered affects the overall system's decisions (Yes or No, Min / Max values in critical parameters, etc.), regarding the diagnosis, treatment and final reporting to the final user. The data flow diagrams are already implemented and presented in following sections.

The system will be pilot tested for the diagnosis and therapy of two (2) case studies a) thyroid cancer and b) hepatitis. The next sections briefly present the work fulfilled regarding encoding the 2 patience and producing the necessary data flow diagrams to be embedded to the decision support subsystem.

Furthermore, since it collects and analyzes with data mining techniques an enormous amount of data on one hand, it allows to investigate cause effect relations between the therapy assigned and the benefits verifiable on a patient. Vice-versa, according to the observation of clinical data generated by thousands of patients, it points out phenomena that otherwise would not be identified, such as, for example, the impact of a pathology within a specific geographical area or the identification of patients with a greater improvement trend in order to point out the improvement causes, and therefore, use them to evolve medical protocols, etc.

4   Encoding the Thyroid Cancer Therapy

The Clinical Decision Support System will support diagnosis and treatment of Thyroid Cancer as a case study. To achieve this goal the process of encoding Thyroid

Cancer so as to produce data flow diagrams is necessary and has been completed. The data flow diagrams consist of decision trees, values and value limits (min / max) which will be used by the decision support system help the doctor to diagnose and treat the Thyroid Cancer.

The thyroid gland is located in the neck, in front of the trachea between the cricoid cartilage and ypersternikis (jugular) incision. The thyroid is composed of two lobes that are connected through an isthmus. The normal rate is 12 με 20 gr, vascular and soft texture. Four parathyroid glands, which produce parathyroid hormone are in the rear area of each pole of the thyroid gland and should be recognized during thyroid surgery to prevent paralysis of the vocal cords [19]-[24].

The treatment of thyroid cancer created is depicted in the next data flow diagram

TREATMENT OF THYROID CANCER

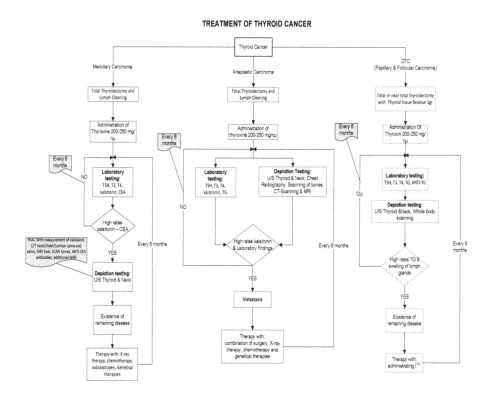

5   Encoding Hepatitis Diagnosis and Therapy

The liver is the largest organ, accounting for approximately 2% to 3% of average body weight. Located in the right upper quadrant of the abdominal cavity beneath the right hemidiaphragm and on top of the stomach, right kidney, and intestines, it is protected by the rib cage and maintains its position through peritoneal reflections, referred to as ligamentous attachments.

Hepatitis is a medical condition defined by the inflammation of the liver and characterized by the presence of inflammatory cells in the tissue of the organ. Hepatitis varies in severity from a self- limited condition with total recovery to a life- threatening or life-long disease. The most common types of Hepatitis are: Hepatitis B, and Hepatitis C. However, hepatitis can also be caused by alcohol and some other toxins and infections, as well as from our own autoimmune process [25].

The Clinical Decision Support System will support diagnosis and treatment of all the types of Hepatitis. To achieve this goal the process of encoding Hepatitis in a machine understandable format was fulfilled. Based on this process the following data flow diagrams were developed. These diagrams consist of decision trees, values and value limits (min / max) which will be used by the decision support system so as to foster the medical personnel through diagnosis and treatment. The next figures present a small part the data flow diagrams constructed.

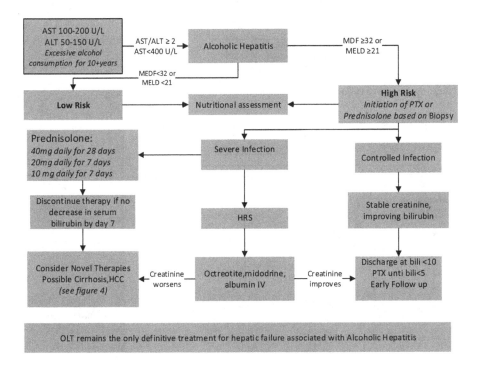

Fig. 1. Alcoholic Hepatitis Diagnosis and Treatment Algorithm

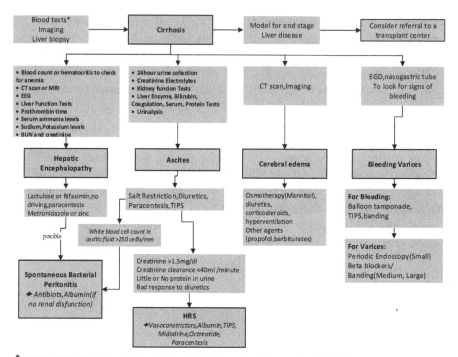

*: ALT,AST, GGT,ALP, Bilirubin,Albumin,Prothrombin Time, Special Tests to confirm viral causative factor,MRI,CT,Ultrasound,

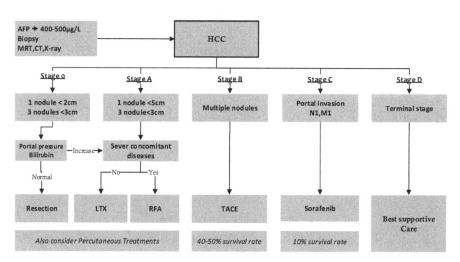

Fig. 2. Cirshosis and HCC Treatment Algorithms

6   Conclusions

An important factor in the medical's personnel correct diagnosis and treatment with a minimum error percentage is the creation of clinical decision support system which medical professionals can utilize and which will lead them to clinical decision making. The reliability and efficiency of the decision support part can be optimized with the analysis of the data flow diagrams produced and by applying data mining techniques to the medical data gathered by the system on the long term.

For the system to be effective, it must act intelligently, be user friendly and have immediate and effective data entry, while it is equally important to have quick access at any time, everywhere and from any device. Based on these principles DEUS the proposed clinical decision support systems, which are under implementation, will contribute highly to achieve the foreseen goals and provide to the medical personnel efficient services regarding diagnosis and therapy.

The system is currently under implementation. The following milestones have been achieved:

1. Digitization of the patient records.
2. Database design, implementation and testing
3. Metadata sets definition, collection and ingestion.
4. Encoding of the illnesses consisting the case studies for the systems pilot testing.
5. Primary implementation of web services for metadata access and management.

The future objectives are focusing mainly on the implementation of the Decision Support subsystem with the data mining methods embedded and web services for accessing critical information through mobile devices.

References

1. Osheroff, J.A., Teich, J.M., Middleton, B.F., et al.: A roadmap for national action on clinical decision support. American Medical Informatics Association (June 13, 2006), http://www.amia.org/inside/initiatives/cds/
2. Dick, R., Steen, E., Detmer, D.E.: The computer- based patient record: An essential technology for health care, revised edition. The National Academies Press, Washington, DC (1997)
3. DesRoches, C.M., Campbell, E.G., Rao, S.R., et al.: Electronic health records in ambulatory care—a national survey of physicians. N. Engl. J. Med. 359(1), 50–60 (2008)
4. Menachemi, N., Saunders, C., Chukmaitov, A., et al.: Hospital adoption of information technologies and improved patient safety: a study of 98 hospitals in Florida. J. Healthc. Manag. 52(6), 398–409 (2007)
5. Hsaio, C., Burt, C., Rechtsteiner, E., et al.: Preliminary estimates of electronic medical records use by office-based physicians. Health E-Stat National Center for Health (2008)

6. Glaser, J.P., Davenport-Ennis, N., Robertson, R.M., et al.: AHIC April 2008 meeting: clinical decision support recommendation letter. American Health Information Community (April 22, 2010)
7. Osheroff, J.: Improving medication use and outcomes with clinical decision support: a step- by-step guide. The Healthcare Information and Management Systems Society, Chicago (2009)
8. Sim, I., Gorman, P., Greenes, R.A., et al.: Clinical decision support systems for the practice of evidence-based medicine. J. Am. Med. Inform. Assoc. 8(6), 527–534 (2001)
9. Payne, T.H.: Computer decision support systems. Chest 118(2 suppl.), 47S–52S (2000)
10. Berner, Ed.D.E.S.: Clinical Decision Support Systems: State of the Art", 1, Agency for Healthcare Research and Quality, U.S. Department of Health and Human Services.
11. Garg, A.X., Adhikari, N.K.J., McDonald, H., et al.: Effects of computerized clinical decision support systems on practitioner performance and patient outcomes. JAMA 293(10), 1223–1238 (2005)
12. Berlin, A., Sorani, M., Sim, I.: A taxonomic description of computer-based clinical decision support systems. J. Biomed. Inform. 39(6), 656–667 (2006)
13. Warwick, C.: Metadata: an overview (retrieved January 6, 2010)
14. Nielsen, F.: Neural Networks-Algorithms and Applications. Niels Brock Business College (2008)
15. Von Altrock, C.: Fuzzy logic and Neurofuzzy Applications explained. Prentice Hall PTR, Upper Saddle River (1995)
16. McNeil, D., Freiberger, P.: Fuzzy Logic: The Revolutionary Computer Technology that is Changing our World, New York (1994)
17. Jeanette, L.: Introduction to Neural Networks. California Scientific Software Press (1994)
18. Health Level Seven International, "Introduction to HL7 Standards" (2012), http://www.hl7.org
19. Harrison, L.J.: Endocrinology. Parisianou A.E., Greece (2007)
20. Dr. Vainas, I., Dr. Chrisoulidou, A.: Thyroid Cancer. In: 2nd Postgraduate Symposium of Endocrine Oncology, Thessaloniki (2006)
21. Ain, K.B.: Papillary thyroid carcinoma. Etiology, assessment, and therapy. Endocr. Metab. Clin. North Am. 24, 711–760 (1995)
22. Mazzaferri, E.L.: Thyroid carcinoma: papillary and follicular. In: Mazzaferri, E.L., Samman, N. (eds.) Endocrine Tumors, p. 278. Blackwell Scientific Publication Inc., Cambridge (1993)
23. Dalles, K., Athanasiou, K.: Thyroid Cancer. Archives of Hellenic Medicine 24(3), 250–264 (2007)
24. Schlumberger, M., Pacini, F.: Thyroid tumors, 2nd edn., p. 216. Nucleon, Paris (2003)
25. Boon, N.A., Colledge, N.R., Walker, B.R., Hunter, J.: Davidson's Principles & Practice of Medicine, part2, ch. 23, 20th edn., pp. 936–938,971–972. Churchill Livingstone, UK (2006)
26. Aleksovska-Stojkovska, L., Loskovska, S.: Data Mining in Clinical Decision Support Systems. In: Recent Progress in Data Engineering and Internet Technology. LNEE 2013, pp. 287–293 (2013)

Supporting and Consulting Infrastructure for Educators during Distance Learning Process: The Case of Russian Verbs of Motion

Oksana Kalita[1], Alexander Gartsov[1], Georgios Pavlidis[2], and Photis Nanopoulos[2]

[1] Peoples' Friendship University of Russia, Moscow, Russia
[2] Department of Computer Engineering and Informatics, Faculty of Engineering,
University of Patras, Patras, Greece

Abstract. This paper presents a supporting and consulting infrastructure (methods and tools) assisting educators during the distance learning process. The main focus is on the information sketching a student profile. Having identified all possible behaviors, the proposed infrastructure allows the development of probabilistic distribution models of the emerging educational events. The determination of probabilities is estimated by applying data mining techniques and then using the method of maximum entropy. As an example, the process of learning the semantic structure of motion verbs in Russian and Greek language is taken. Emphasis is given to those cases where verbs structures match entirely, partially overlap, as well as those whose semantic sizes are specific, characterize either the Russian, or solely the Greek language.

Keywords: distance learning, supporting and consulting infrastructure, Russian and Greek verbs of motion, data mining, maximum entropy method.

1 Introduction

The benefits as well as the problems of distance learning (DL) have been studied extensively in literature [1, 2, 3, 4]. Many researchers believe that the learning attitudes and performance is determined to a great extent by factors related to either students' characteristics, or more specific aspects of the learning process. Factors, like students' socio-economic background, culture and mentality also play an important role in DL.

Certainly, the above factors are not acting independently, as more of the output performance depends on the interaction of several factors. These interactions increase the complexity and the amount of the information that educators (a team of professors, pedagogues, IT engineers, administrators, etc.) have to manage, analyze and interpret correctly for selecting adaptive educational strategies. Those strategies restructure the content, mode and tempo of the educational procedures. Unfortunately, existing electronic platforms for DL have not so far the necessary infrastructures (methods and tools) to process this kind of information and to provide support and intelligent guidance/advising for educators.

L. Iliadis, H. Papadopoulos, and C. Jayne (Eds.): EANN 2013, Part II, CCIS 384, pp. 185–192, 2013.
© Springer-Verlag Berlin Heidelberg 2013

The purpose of this paper is to present our main ideas towards the development of a Supporting and Consulting Infrastructure (SCI) for assisting trainers, based on the accumulated information lying in existing DL platforms. To this end, an IT infrastructure will be developed to support the proposed approach. The methodology to collect and maintain the training material of the SCI lies on the well-proven Statistical and Information Management methods which:

a)   facilitate the structuring of the content lying in existing DL systems, aiming to the creation of common metadata systems,

b)   provide strategies for data analysis techniques based on well proven estimations as well as data mining methods the discovery of potential correlative structure, and

c)   enable the development of   decision support tools for educators.

In addition, these strategies and tools will help to dynamically adapt the educational process and behaviors by producing real-time probabilities of the events that shape the students' profile and help educators adjusting their strategies. By using such kind of infrastructure, the students' behaviors in the DL process can be specified or interpreted and educators can have the opportunity to develop scenarios to efficiently manage "unexpected" behaviors.

An indicative example of the possible benefits of adopting the proposed SCI is the case of learning the Russian language by Greek students, focusing on the specific case of the "motion related verbs". The choice of this narrow area is dictated by the particular characterization of the semantic content of words expressing motion in the two languages. The example topic offers several types of futures due to the comparative analysis of the two languages including not only the semantic similarities and dissimilarities but also the mental and cultural background of the students. Indeed, the choice of Russian and Greek languages offers considerable complexity, thus stressing out the importance of our ideas.

2   Current Status

Problems during the DL of a foreign language arise when they become concrete. That is about the case of learning a particular foreign language by specific students in a specific technological environment. In [5, 6] the problems arising in the process of learning the Russian language by students having French, English, Chinese, etc.as a native language have been studied and extensively analyzed . Based on these studies, specific ethno-oriented methodologies have been developed [7].

However, ethno-oriented methodologies for learning the Russian language by Greek students have not been proposed yet, despite the fact that the first studies date from 1828 [8]. In the context of this work, a particular field of application of the proposed SCI, as a small contribution to the positive development of a related ethno-oriented methodology, will be the DL of the Russian verbs of motion by Greek stu-dents.

3   Greek Verbs Corresponding to Russian Verbs of Motion

It is well known that the first steps towards the creation of an educational model for a specific subject in DL is the analysis of the basic semantic elements that com-pose it, and the study of the apparent mental paths that students use to learn the relevant concepts. The basic concepts are divided to sub-concepts and tools while tests, activities, abstractions, dynamic representations etc., are created to assist students in their effort to thoroughly understand the meaning and the problems that this concept arises.

From this perspective, the analysis of the Russian verbs of motion with or without prefixes and the development of the relevant concepts, sub-concepts and tools for the Greek students is a complex task. The complexity is due to the fact that apart from the cultural and mentality barriers, there are rules whose semantic meaning is specific only for the Russian or the Greek language, or even coincide between the two languages.

Russian verbs of motion without prefix and the corresponding Greek verbs differ because the first ones (a) form correlated pairs, for instance {idti-hodit, ehat-ezdit}, and (b)also have a number of peculiarities in the way they are used to express their semantic content and features (Fig.1).

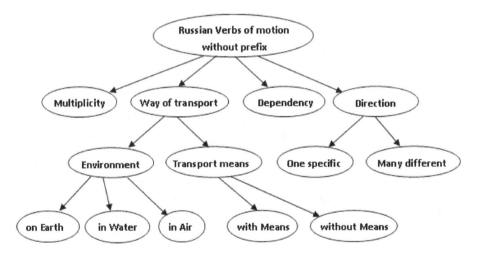

Fig. 1. Part of semantic features of Russian verbs of motion without prefix

The pairs of the Russian verbs of motion without prefix have imperfective form either they are of a particular direction or of multiple directions.

In the case of the Russian motion verbs with prefix, the "way to move" is represented in the root of the verb, while the more abstract concept "direction" is usually expressed through the prefix, but only the union of the two components completes semantically a verb of motion with prefix. For example <prefix> U <root> = {u-hodit, pri-begat, za-letat}. Verbs with a prefix expressing a particular direction have a perfective form whereas those expressing many different directions have imperfective form, for example {priidti-prihodit, uehat-uezjat}.

Correlated pairs of verbs of motion without prefix and the association of <prefix> with <root> is elementary logic for the teachers of the Russian language, but it is practically difficult learning material to be adopted by the Greek students. As result, a limited number of Russian verbs can correspond to one Greek verb and vice versa. This type of situations increases the complexity of the educational process and imposes:

(a)   the development of a modern ethno-oriented methodology for learning the Russian language by Greek students,
(b)   the transfer of this methodology in platform for DL,
(c)   the use of data mining techniques and,
(d)   the evaluation of the results extracted by the SCI by the educator.

4   Student Relevant Information and Profile

In addition to the educational material for the needs of the proposed SCI, the information that describes the student plays a very important role. This information consists of the following three parts:

<Student profile>
(Part-1) = <identity information>
(Part-2) = <social-economic information>
(Part-3) = <education-relevant information>
<End>

The first part consists of the standard identification information: name, address, registration number, birth, marital status, family, education, etc. The second part is quasi-stationary and covers the socio-economic background and cultural interests of the student. It also contains possible changes in the professional field. All this infor-mation concerns the interaction of the student with the changes in her/his direct or indirect environment. These changes may, during the training process, affect behav-ior, because the student gradually matures and redefines her/himself over time.

The third and final part of information on a student is related to her/his involve-ment, interaction and performance within the education environment. It is fully dynamic and includes the results from the specific learning process: tests, proposals, collaborations, aspirations, initiatives, questions, etc. The information containing all the three parts of the student profile shows the differences between students on an individual level and the range of the learning capacity. It depends on the quality (relevance, completeness, accuracy, etc.) of the full information package for the SCI to produce pertinent conclusions on the process of maturity and self-determination of the student.

We must point out here that one of the objectives of this work is to build a classification system (nodes and tree structures) the most generic possible in order to allow better and deeper analysis through a standard classification of "events" combining student characteristics with educational subjects parameters.

5   System Architecture

Due to space limitations, in what follows we do not provide a detailed description of the architecture of the SCI. A modular representation of the system architecture is depicted in Figure 2.

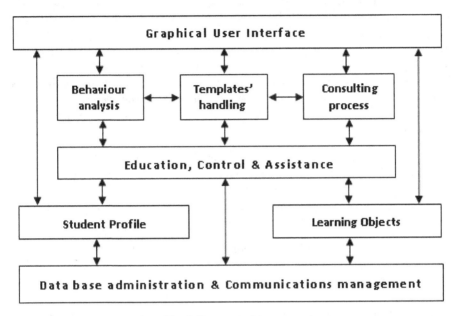

Fig. 2. System Architecture

During the training process, the SCI analyses the behavior of the student, namely:

- what is the concept of the learning subject, including the level of com-plexity,
- how long, which way, when in time and where in space,
- after how many attempts,
- under what kind of conditions, exploration freedom, help offered, etc was the educational material collected and assimilated.

When a student encounters difficulties and expresses "elementary questions", they are answered and marked either by the assistant or by the DL platform "automatically" (as FAQ). If the student makes questions with increased complexity, they are going to be forwarded to the educator who will evaluate them. On one hand the educator will give the necessary clarifications and answers and on the other hand, he will record and use them to improve the educational process in terms of the structure, flow and content. Obviously, along with the clarified questions, SCI will enter in results using the available tools, for example results of tests conducted at regular and predefined intervals.

6   Probabilistic Aspects

The SCI logs in real time all the facts concerning all the subjects and the events intervening in a DL platform. This concerns changes in the state of the elements that characterize the student, the educator, the learning objectives and the educational materials. The three sample populations namely the students (U1), the educators (U2) and the educational materials (U3) are combined into one set of the possible situations in which the individual events occur. They are governed by the (unknown) probability measure P. So, we have the probability space (U, E, P) with U=U1xU2xU3 as sample space, E is an algebra generated by all described events, and P is the (unknown) probability measure, being the basis upon which we build models of the probabilistic distributions allowing the estimation of the probabilities of specific events.

In the context of recording incoming information, a filtering process which consists of three stages follows:

1. seeking for changes concerning the parameters and characteristics of the student,
2. searching for changes related to the conduct, control and support required of the student as part of the educational process in progress
3. seeking for changes on the environmental conditions of the course and the corresponding and expected status of the student.

The search for the new information contained in the profile of the student, the results of the course so far, and the most recent information from the above filtering process is enriched in order to constitute elements of a sample space to be analyzed by SCI in depth.

Part of the analysis of the SCI will deal with events such as A1 = {the student has successfully cover the material in ordinary time}, A2 = {the student failed to meet successfully the learning material} and A3 = {the student cover the learning material successfully in smaller than expected time}.

7   Use of Events Probabilities

Based on the recording performance of the student, SCI can assess the probability that a student belongs to the event A1, A2 or A3, and based on the level of probabil-ity, to help the educator undertake an action either assisting the student or accelerating the tempo of learning. Generally, the SCI acts as an expert system which can answer questions such as: what is the new material that should be given to "good" students? is it necessary for a student to repeat the last one or some previous educational material? etc. Whatever the event, E can be described by the characteristic registered on the Units (U1, U2, U3). An estimated probability could be attributed helping educators and students to better interprete the materialization of this event.

It is obvious that the complexity of the educational process will encounter situations that require multiple types of events to be described. A schematic example is shown in Figure 3 in which various types of events are shown as squares. With Boolean operations we may describe many other events, like E1, E2, and E3 shown as circles (Fig. 4).

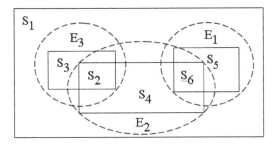

Fig. 3. A schematic example

The SCI aims to achieve two main objectives: (1) shaping the experience of the history data that have been recorded using data mining methods, and (2) to assess the likelihood of future events estimation methods such as the method of maximum entropy [9], in order to facilitate trainers to improve the efficiency of the learning process.

8   Methodological and Algorithmic Aspects

As mentioned in the introduction, the methodology we shall follow combined techniques from official statistics, where knowledge has been accumulated on struc-turing metadata (codification of information, classification of units, etc.), multidimensional data analysis (data mining techniques clustering- and classification, decision trees for learners and courses profiling, etc.) and parameter estimation in probability modeling. The central idea is to identify "learning events" allowing the classification of any potential learning situations observed.

Having identify and describe all the possible events, the SCI allows us to develop models for the probabilistic distribution of the events. The determinations of probabilities are estimated by applying well established methods, like maximum entropy methods. Based on these probabilities, the SCI will assist the educator in finding in real time, unjustified deviations from the prescribed learning trajectory and it will calculate the probability of the behavior of the student to be maintained in the future. This way, the trainer will have to decide whether or not she/he should intervene correctively, answering the questions: who, how, when and which way to 'encourage' a desired behavior or promptly and effectively "remove" an undesirable behavior.

Of course, based on the accumulated in time experience, from past attitudes of students and the relevant decisions of the trainers, the SCI will be able to suggest some best practices on appropriate educational material, successful educational paths and correct learning speeds (tempo of the education).

9   Conclusions and Future Work

The development and implementation of the SCI will contribute to:

- the better organization and monitoring of the educational process
- the prediction and prevention of adverse events,
- the fast and effective educators' intervention when a student's behavior is outside these limits,
- the correct recording, structuring and continuous enrichment of elements that describe the educational material, students, educators and all those who take part in the learning process,
- provide assistance and, in some cases, automation of addressing the difficulties encountered during the learning process.

Finally, it should be noted that at the moment the development and the testing of the subsystems of the SCI are taking place.

References

1. Hillman, D., Willis, D., Gunardena, C.: Learner-interface interaction in distance education: An extension of contemporary models and strategies for practitioners. American J. of Distance Education 8(2) (1994)
2. Benson, L., Elliot, D., Grant, M., Holschuh, D., Kim, B., Kim, H., et al.: Usability andinstructional design heuristics for e-Learning evaluation. In: Barker, P., Rebelsky, S. (eds.) Proceedings of World Conference on Educational Multimedia, Hypermedia and Telecommunications 2002, pp. 1615–1621 (2002); Presented at the World Conferenceon Educational Multimedia, Hypermedia and Telecommunications (EDMEDIA)Chesapeake. AACE, VA
3. Clark, R.: Six principles of effective e-Learning: What works and why. The e-Learning Developer's Journal, 1–10 (2002)
4. Dede, C.: The evolution of distance education: Emerging technologies anddistributed learning. The American Journal of Distance Education 10(2), 4–36 (1996)
5. Вагнер В. Н. Методика преподавания русского языка англоговорящим и франкоговорящим на основе межъязыкового сопоставительного анализа: Фонетика. Графика. Словообразование. Структуры предложений, порядок слов: Гуманит. изд. центр ВЛАДОС (2001)
6. Гарцов, А. Д. Электронная лингводидактика. Инновации языкового образования – Германия, Саарбрюкен: LAP (2010)
7. Балыхина, Т. М. От методики к этнометодике / Т. М. Балыхина, Ч. Юйтзян. М.:РУДН (2010)
8. Οικονόμου Κ.: Δοκίμιον περί της πληρεστάτης συγγενείας της σλαβωνορωσικής γλώσσης προς την ελληνικήν – Πετρούπολη (1828)
9. Wu, N.: The Maximum Entropy Method. Springer Series In Information Science. Springer, Berlin (1997)

Classification Models for Alzheimer's Disease Detection

Christos-Nikolaos Anagnostopoulos[1], Ioannis Giannoukos[1], Christian Spenger[2],
Andrew Simmons[3], Patrizia Mecocci[4], Hikka Soininen[5], Iwona Kłoszewska[6],
Bruno Vellas[7], Simon Lovestone[3], and Magda Tsolaki[8]

[1] Cultural Technology and Communication Dpt. University of the Aegean,
81100, Mytilene, Greece
[2] Department of Clinical Science, Intervention and Technology, 14183 Stockholm, Sweden
[3] King's College London, UK
[4] Section of Gerontology and Geriatrics, Department of Clinical and Experimental Medicine,
University of Perugia, Perugia, Italy
[5] Department of Neurology, University of Eastern Finland and
Kuopio University Hospital, 21 Kuopio, Finland
[6] Department of Old Age Psychiatry & Psychotic Disorders,
Medical University of Lodz, 92-17 216 Lodz, Poland
[7] Department of Internal and Geriatrics Medicine, Hôpitaux de Toulouse, Toulouse, France
[8] Third Department of Neurology, Aristotle University of Thessaloniki, Thessaloniki, Greece

Abstract. This paper presents a classification fusion for Alzheimer's Disease
(AD) and Mild Cognitive Impairment (MCI) classification based on dataset
acquired basically from an automated structural MRI image processing pipeline.
The dataset includes eighty-one regional cortical volume and cortical thickness
features produced by the automated pipeline, along with two demographic
measurements and three manual volume measurements of the hippocampus. This
high-dimensional pattern classification problem is tested in a large database that
contains clinical tests from six medical centers in Europe. The assessment of the
results has shown that with a careful selection of combined classifiers, subject
classification in three classes (Normal Controls, patients with MCI or with AD) is
fairly accurate and can be used as an assistive tool to clinical examinations.

Keywords: Alheimer's Disease detection, Mild Cognitive impairment,
classification fusion.

1 Introduction

Alzheimer's disease (AD) is a neurodegenerative disorder and the most common
cause of dementia. Mild Cognitive Impairment (MCI) is an early stage of AD where
patients present cognitive impairments not expected for their age and education,
without, however, significant interference in their daily activities. Definitive diagnosis
can only be made with histopathological confirmation of amyloid plaques and
neurofibrillary tangles, usually at autopsy (post-mortem). Early diagnosis of AD,
especially in the MCI stage, is of great importance since treatment may be most

L. Iliadis, H. Papadopoulos, and C. Jayne (Eds.): EANN 2013, Part II, CCIS 384, pp. 193–202, 2013.
© Springer-Verlag Berlin Heidelberg 2013

efficacious if introduced as early as possible. Usually, to provide an ante-mortem MCI or AD diagnosis, the NINCDS–ADRDA criterion [1],[2] is used.

Methods of brain imaging (BI) have been used to rule out alternative causes of dementia, being consistent with the NINCDS-ADRDA diagnostic criterion. The BI techniques include Magnetic Resonance Imaging (MRI), Positron Emission Tomography (PET), Single Photon Emission Computed Tomography (SPECT) or electroencephalography (EEG). These techniques provide features of the patients' brain, which can then be forwarded to classifiers to distinguish patients diagnosed with AD or MCI from Normal Controls (NC).

Several techniques and methods based on machine learning have been proposed in the literature for detecting and diagnosing Alzheimer's disease (AD) and Mild Cognitive Impairment (MCI). These techniques however have not been assessed extensively. The majority of the papers in the literature classify the data into 2 classes, (NC vs. AD or NC vs. MCI) [3],[4],[5],[6], while the three class classification problem is rarely assessed with very small samples [7],[8].

In this paper, we present a combined classification schema for classification of three classes, (NC, MCI, AD) using a large database of 358 records with high-dimensional feature vectors (86 features for every record). The features were extracted mainly by utilizing an automated structural MRI image processing pipeline [9],[10], along with two demographic measurements and three manual volume measurement of the hippocampus (e.g. see Image 1 below).

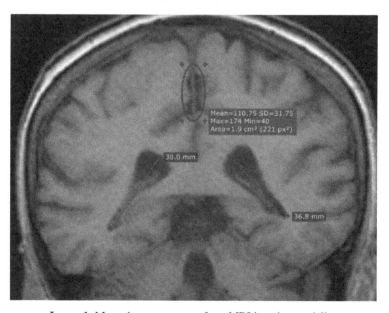

Image 1. Manual measurements from MRI imaging modality

The paper is structured as follows. Section 2 provides a short description of the dataset and the materials/methods of this work. Section 3 presents the experimental results of this study and Section 4 concludes this paper.

2   Materials and Methods

2.1   Dataset Acquisition

A total of 122 MCI subjects (62 females), 123 AD patients (76 females), and 113 age-matched healthy controls (64 females) were selected from the AddNeuroMed study [9],[10], which is a prospective, longitudinal multicenter study to discover biomarkers for AD. Data were collected from 6 medical centers across Europe: University of Kuopio, Finland; University of Perugia, Italy; Aristotle University of Thessaloniki, Greece; King's College London, UK; Medical University of Lodz, Poland; and University of Toulouse, France. All the MCI subjects had successfully undergone magnetic resonance imaging at baseline and cognitive tests evaluated at baseline. In addition, clinical evaluation and cognitive tests were repeated 1 year later.

This study was approved by ethical review boards in each participating country. The inclusion criteria for AD, MCI and healthy controls are shown in Table 1.

Table 1. Inclusion criteria for AD and MCI respectively

Inclusion criteria for AD:	Inclusion criteria for MCI:
1. National Institute of Neurological and Communicative Disease and Stroke and Alzheimer's disease (NINCDS / ADRDA) [1]	1. Subject aged 65 years or above 2. Memory complaint by patient, family or physician 3. Normal activities of daily living 4. Mini Mental State Examination (MMSE) score range between 24 and 30 [14] 5. Score equal to or less than 5 in Geriatric Depression Scale 6. Memory score of 0.5 or 1 in Clinical Dementia Rating Scale (CDR)[15], absence of dementia according to NINCDS/ADRDA criteria for AD [1] and following Petersen's criteria [16].

Data acquisition took place using six different 1.5T Magnetic Resonance systems (4 General Electric, 1 Siemens and 1 Picker). At each site a quadrature birdcage coil was used for radio frequency transmission and reception. Data acquisition was designed to be compatible with the Alzheimer Disease Neuroimaging Initiative (ADNI) [11]. The imaging protocol included high resolution sagittal 3-dimensional T1-weighted Magnetization Prepared RApid Gradient Echo (MPRAGE) volume (voxel size 1.1 x 1.1 x 1.2 mm3) and axial proton density/T2-weighted fast spin echo images. Full brain and skull coverage was required for both latter datasets and detailed quality control was carried out on all Magnetic Resonance images [12],[13]. All images received a clinical read by an on-site radiologist in order to exclude any subjects with non-AD related pathologies.

Table 2. The 86 measurements of the feature vector

Two demographic features:
Gender (Male=1, Female=0), Age (years)
Forty-six (46) Normalized modules (in mm3):
Brain volume, third ventricle, fourth ventricle, fifth ventricle, brain volume, brainstem, corpus callosum central, corpus callosum midanterior, corpus callosum midposterior, corpus callosum posterior, cerebrospinal fluid, left accumbens, left amygdala, left caudate, left cerebellum cortex, left cerebellum white matter, left cerebral cortex, left cerebral white matter, left hippocampus, left inferior lateral ventricle, left lateral ventricle, left pallidum, left putamen, left thalamus, left ventral DC, left choroid plexus, left vessel (total), optic chiasm, right accumbens, right amygdala, right caudate, right cerebellum cortex, right cerebellum white matter, right cerebral cortex, right cerebral white matter, right hippocampus, right inferior lateral ventricle, right lateral ventricle, right pallidum, right putamen, right thalamus, right ventral DC, right vessel (total), white matter hypointensities, non-white matter hypointensities
Thirty-five (35) Cortical thickness measuremens (in mm):
Banks of superior temporal sulcus, insula, caudal middle frontal gyrus, corpus callosum, cuneus cortex, entorhinal cortex, frontal pole, fusiform gyrus, inferior parietal cortex, inferior temporal gyrus, isthmus of cingulated cortex, lateral occipital cortex, lateral orbitofrontal cortex, lingual gyrus, medial orbitalfrontal cortex, middle temporal gyrus, paracentral sulcus, parahippocampal gyrus, frontal operculum, orbital operculum, triangular part of inferior frontal gyrus, pericalcarine cortex, postcentral gyrus, posterior cingulate cortex, precentral gyrus, precuneus cortex, rostral middle frontal gyrus, superior frontal gyrus, superior parietal gyrus, superior temporal gyrus, supramarginal gyrus, temporal pole, transverse temporal cortex, unknown cortex thickness
Three (3) normalized manual hippocampal measurements:
Hippocampus left hemisphere, hippocampus right hemisphere, hippocampus total

2.2   Feature Set

For feature acquisition an automated MRI image processing pipeline is used [17],[18],[19] producing regional cortical thickness and regional volume features, incorporating the next steps: (a) non-brain tissue removal [20], (b) transformation, segmentation of the subcortical white matter and deep gray matter volumetric structures [18],[19], (c) intensity normalization, (d) tessellation of the gray-white matter boundary, automated topology correction [21], (e) surface deformation following intensity gradient calculation. Surface inflation [22] then follows in order to register to a spherical atlas which utilizes individual cortical folding patterns to match cortical geometry across subjects and parcellation of the cerebral cortex into units based on gyral and sulcal structure [23].

As a result forty-six (46) regional cortical volume features that were normalized by the subjects' intracranial volume, along with thirty-five (35) cortical thickness features measured in millimeters (mm) were generated. Those 81 features are presented in Table 2 along with two (2) demographic measurements (gender, age) and three (3) manual volume measurement of the hippocampus (normalized by the subjects' intracranial volume as well). The three manually acquired hippocampal volume measurements were measured with the HERMES Multimodality software package [24] by a neuroradiologist who was unaware of clinical information.

2.3   Combination of Machine Learning Methods

In order to categorize subjects into age-matched NC, patients with MCI and AD, the performance of seven machine learning techniques were combined in this paper, selected based on their popularity and frequency in related literature. These techniques include Feed-Forward Neural Networks (FFNN), Support Vector Machines (SVMs), Probabilistic Ensemble Simplified Fuzzy ARTMAP (PESFAM), Probabilistic Neural Networks (PNNs) and the k-nearest neighbour (KNN) algorithm.

Each technique may produce different results depending on their initial parameters and their different architectures. Therefore, their outputs tend to be uncorrelated. This phenomenon can be used to create a single stronger classifier by fusing their outputs.

Three fusion methods were evaluated using the seven classifiers, along with the performance of each single classifier method, namely majority voting, gentle adaptive boosting and the combined classifier.

2.3.1   Majority Voting

Different ways are proposed in the literature for the combination of multiple classifiers into a single stronger one [25]. The simplest method is to assume that each classifier has an equal weight to classify an input. Let a counting function a_{ij} defined as shown in equation (1). Then, a majority voting combining technique can be formulated using equation (2).

$$a_{ij} = \begin{cases} 1 & correct \quad class \\ 0 & wrong \quad class \end{cases} \quad (1) \quad \text{and} \quad O_j = \arg \max_k (\sum_i a_{ij}) \quad (2)$$

where O_j is the classifying decision for the jth input and k (k>2) is the number of the classes of the classification problem.

2.3.2   Gentle Adaptive Boosting

To compare also in a qualitative manner to findings in related literature, the experimental results of majority voting are also shown along with a variation of adaptive boosting, called Gentle Adaptive Boosting [26]. The latter was found in the literature to be more accurate than the original Adaptive Boosting technique and other boosting methods.

2.3.3   Combined Classifier

Due to differences in the classifier architectures, training epochs or initialization parameters, we use a trained combining classifier, in order to "learn" the level of success of the ensemble classifier training and generalization abilities on the specific dataset. We define that the combining classifier estimates the ensemble function F_{ens} which merges the outputs a_{ij} as depicted in equation (3).

$$O_j = F_{ens}(a_{1j}, a_{2j}, a_{3j}, ..., a_{nj}) \quad (3)$$

To ensure that over-training has not occurred, we chose to use a feed-forward neural network as the combining classifier, with a small number of neurons (three neurons) in its hidden layer.

3   Experimental Results

Table 3 presents the performance of each individual classifier for the two-class (NCvsMCI, NCvsAD and MCIvsAD) and the three-class problems. The results are similar and no technique outperforms the others. As one may expect, better discrimination is presented when NCvsAD is examined, while as far as the MCI versus AD classification problem is concerned, the classification were less accurate. It is important to note that there is a significant reduction in performance when the three-class problem is assessed. Obviously, the classification of patients with MCI is harder due to the higher inter-class MRI-extracted feature variation with the other two classes.

Table 4 highlights the experimental results of the proposed ensemble with the combining classifier. The proposed ensemble achieved 77.1% accuracy in classifying patients in the three classes, compared to 73.2% and 65.1% for the majority voting ensemble and gentle boosting classifier. Also, the proposed ensemble outperformed the other techniques as far as the 2-class problem is concerned.

Finally, Table 5 presents the confusion matrix of the proposed ensemble results for the three-class problem with an overall accuracy of 77.1% (276/358 correctly classified instances). As one may observe, the ensemble classification error occurs mostly in distinguishing the MCI class from the other two. As discussed previously, MCI may be a transitional stage between normal aging and AD, as it is found that people with MCI probably will characterized as patients with AD as the time passes.

Therefore, this paper proposes a machine learning ensemble with a combined classification schema (an ANN) that reports similar results in the problem of two-class classification, but quite better performance in the harder to solve three-class classification problem. It is also important to note, that: (1) our research is performed in a significantly larger dataset compared to those found in the literature and (2) the ensemble output moderates the limitations in performance of each separate classifier. All classifiers were developed in Matlab in a Core 2 Duo at 2.83GHz PC with 3GB RAM. The experimental results were cross-validated by the leave-one-out method which gives a lower limit for the probability of correct classification.

The training session using Leave-One-Out validation takes about 5 hours to complete, while classifying a single input with the trained ensemble requires less than 1 second. The important parameters of the individual classifiers, as well as those of the combining classifier are shown in Table 6.

Table 3. Performance of individual machine learning techniques

		2 classes		3 Classes
	NCvsMCI (%)	NCvsAD (%)	MCIvsAD (%)	(%)
FFNN-LM	61.7	81.4	67.0	53.4
FFNN-BR	65.1	90.7	67.4	59.0
KNN	68.5	84.8	70.6	58.1
PESFAM	61.3	86.0	66.5	55.3
PNN	68.5	86.0	69.0	58.4
SVM - 1vsall	67.2*	89.0*	67.8*	56.4
SVM - 1vs1				58.4

Table 4. Experimental results of the proposed ensemble with (i) a combined classifier, (ii) majority voting classifier and (iii) gentle boosting classifier

	Two classes			Three Classes
	NCvsMCI (%)	NCvsAD (%)	MCIvsAD (%)	(%)
Ensemble with combining classifier	72.8	93.2	73.1	77.1
Majority voting ensemble	71.1	89.4	71.2	73.2
Gentle boosting classifier	64.7	84.2	68.6	65.1

Table 5. Confusion matrix of the proposed machine learning ensemble with a combining classifier for the three-class problem

	NC	MCI	AD
NC	106	7	0
MCI	24	75	23
AD	5	13	105

Table 6. The basic parameters of the 7 classifiers and the combining classifier

SVMs: Gaussian kernel, $C=2^{10}$, $\gamma=2^{-7}$, least squares method to separate hyper-planes

PNN: 5 hidden units, radial basis functions' spread value $\sigma = 0.1$

FFNN-LM: 23 hidden units, sigmoid, training rate = 0.86, training time = 950 (Mean Squared Error = 0.001)

FFNN–BR: 14 hidden units, sigmoid, training rate = 0.58, training time = 1500 (Mean Squared Error = 0.001)

KNN: Euclidean distance metric, nearest neighbors=3, majority rule

PESFAM: vigilance=0. 5, alpha= 0.001, beta= 0.8, epochs=50, error=0.01

Combining classifier: 3 hidden units, sigmoid function, learning ratio=0.74, epochs = 850 (Mean Squared Error = 0.001)

4   Conclusions

In this paper, we presented a combined classification schema for the three-class classification problem (NC, MCI and AD simultaneously) using a large heterogeneous database of 358 records with high-dimensional feature vector. The features were extracted mainly by utilizing an automated structural MRI image processing pipeline, along with 2 demographic measurements and three manual volume measurement of the hippocampus.

This procedure could be of great interest for the Alzheimer's disease researchers, as we: (i) evaluate for the first time the three-class classification problem assessing numerous Artificial Intelligence techniques and (ii) assess the results in a large and heterogeneous collection of measurements from 6 countries in Europe. The dataset includes 3 manual measurements, 46 automated volume and 35 cortical thickness features and 2 demographic measures for 358 individual subjects (patients and healthy people). The experimental results depict that the performance of proposed method is slightly lower for the 2-class problems, but it yields better results in classifying data into three classes simultaneously, compared in a qualitative manner with related literature.

We expect that the goal of machine learning based automated AD, MCI and NC is quite promising for ante-mortem diagnosis. The method that is described in this paper has clearly the potential in achieving more accurate dementia diagnosis and therefore can be used as an assistive tool to clinical examinations for this purpose.

References

1. McKhann, G., Drachman, D., Folstein, M., Katzman, R., Price, D., Stadlan, E.M.: Clinical diagnosis of alzheimer's disease: report of the NINCDS-ADRDA work group under the auspices of department of health and human services task force on alzheimer's disease. Neurology 34, 939–944 (1984)
2. Dubois, B., Feldman, H.H., Jacova, C., DeKosky, S.T., Barberger-Gateau, P., Cummings, J., Delacourte, A., Galasko, D., Gauthier, S., Jicha, G., Meguro, K., O'Brien, J., Pasquier, F., Robert, P., Rossor, M., Salloway, S., Stern, Y., Visser, P.J., Scheltens, P.: Research criteria for the diagnosis of alzheimer's disease: revising the NINCDS-ADRDA criteria. Lancet. Neurol. 6, 734–746 (2007)
3. Lehmann, A., Koenig, T., Jelic, V., Prichep, L., John, R.E., Wahlund, L.O., Dodge, Y., Dierks, T.: Application and comparison of classification algorithms for recognition of alzheimer's disease in electrical brain activity (EEG). J. Neurosci. Meth. 161, 342–350 (2007)
4. Chaves, R., Ramírez, J., Górriz, J.M., López, M., Salas-Gonzalez, D., Alvarez, I., Segovia, F.: SVM-based computer-aided diagnosis of the alzheimer's disease using t-test NMSE feature selection with feature correlation weighting. Neurosci. Lett. 461, 293–297 (2009)
5. López, M.M., Ramírez, J., Górriz, J.M., Alvarez, I., Salas-Gonzalez, D., Segovia, F., Chaves, R.: SVM-based CAD system for early detection of the alzheimer's disease using kernel PCA and LDA. Neurosci. Lett. 464, 233–238 (2009)

6. Vemuri, P., Gunter, J.L., Senjem, M.L., Whitwell, J.L., Kantarci, K., Knopman, D.S., Boeve, B.F., Petersen, R.C., Jack, C.R.: Alzheimer's disease diagnosis in individual subjects using structural MR images: validation studies. NeuroImage 39, 1186–1197 (2008)

7. Colliot, O., Chételat, G., Chupin, M., Desgranges, B., Magnin, B., Benali, H., Dubois, B., Garnero, L., Eustache, F., Lehéricy, S.: Discrimination between alzheimer disease, mild cognitive impairment, and normal aging by using automated segmentation of the hippocampus. Radiology 248, 194–201 (2008)

8. Tripoliti, E.E., Fotiadis, D.I., Argyropoulou, M.: An automated supervised method for the diagnosis of alzheimer's disease based on fMRI data using weighted voting schemes. In: Proc. of IEEE International Workshop on Imaging Systems and Techniques, pp. 340–345. IEEE, Chania (2008)

9. Liu, Y., Paajanen, T., Zhang, Y., Westman, E., Wahlund, L.O., Simmons, A., Tunnard, C., Sobow, T., Mecocci, P., Tsolaki, M., Vellas, B., Muehlboeck, S., Evans, A., Spenger, C., Lovestone, S., Soininen, H.: Combination analysis of neuropsychological tests and structural MRI measures in differentiating AD, MCI and control groups - The AddNeuroMed study. Neurobiol. Aging 32, 1198–1206 (2011)

10. Lovestone, S., Francis, P., Strandgaard, K.: Biomarkers for disease modification trials - the innovative medicines initiative and AddNeuroMed. J. Nutr. Health Aging 11, 359–361 (2007)

11. Jack Jr, C.R., Bernstein, M.A., Fox, N.C., Thompson, P., Alexander, G., Harvey, D., Borowski, B., Britson, P.J., Whitwell, J.L., Ward, C., Dale, A.M., Felmlee, J.P., Gunter, J.L., Hil, D.L., Killiany, R., Schuff, N., Fox-Bosetti, S., Lin, C., Studholme, C., DeCarli, C.S., Krueger, G., Ward, H.A., Metzger, G.J., Scott, K.T., Mallozzi, R., Blezek, D., Levy, J., Debbins, J.P., Fleisher, A.S., Albert, M., Green, R., Bartzokis, G., Glover, G., Mugler, J., Weiner, M.W.: The Alzheimer's disease neuroimaging initiative (ADNI): MRI methods. J. Magn. Reson. Im. 27, 685–691 (2008)

12. Simmons, A., Westman, E., Muehlboeck, S., Mecocci, P., Vellas, B., Tsolaki, M., Kloszewska, I., Wahlund, L.O., Soininen, H., Lovestone, S., Evans, A., Spenger, C.: MRI measures of alzheimer's disease and the AddNeuroMed study. Ann. NY. Acad. Sci. 1180, 47–55 (2009)

13. Simmons, A., Westman, E., Muehlboeck, S., Mecocci, P., Vellas, B., Tsolaki, M., Kloszewska, I., Wahlund, L.O., Soininen, H., Lovestone, S., Evans, A., Spenger, C.: The AddNeuroMed framework for multi-centre MRI assessment of longitudinal changes in alzheimer's disease: experience from the first 24 months. Int. J. Ger. Psych. 26, 75–82 (2011)

14. Folstein, M.F., Folstein, S.E., McHugh, P.R.: Mini-mental state. A practical method for grading the cognitive state of patients for the clinician. J. Psychiat. Res. 12, 189–198 (1975)

15. Hughes, C.P., Berg, L., Danziger, W.L., Coben, L.A., Martin, R.L.: A new clinical scale for the staging of dementia. Brit. J. Psychiat. 140, 566–572 (1982)

16. Petersen, R.C., Smith, G.E., Waring, S.C., Ivnik, R.J., Tangalos, E.G., Kokmen, E.: Mild cognitive impairment: clinical characterization and outcome. Arch. Neurol-Cigago 56(6), 303–308 (1999)

17. Fischl, B., Dale, A.M.: Measuring the thickness of the human cerebral cortex from magnetic resonance images. Proc. of the National Academy of Sciences 97, 11050–11055 (2000)

18. Fischl, B., Salat, D.H., Busa, E., Albert, M., Dieterich, M., Haselgrove, C., van der Kouwe, A., Killiany, R., Kennedy, D., Klaveness, S., Montillo, A., Makris, N., Rosen, B., Dale, A.M.: Whole brain segmentation: automated labeling of neuroanatomical structures in the human brain. Neuron 33, 341–355 (2002)
19. Fischl, B., Salat, D.H., van der Kouwe, A.J., Makris, N., Segonne, F., Quinn, B.T., Dale, A.M.: Sequence-independent segmentation of magnetic resonance images. Neuroimage 23, S69–S84 (2004)
20. Segonne, F., Dale, A.M., Busa, E., Glessner, M., Salat, D., Hahn, H.K., Fischl, B.: A hybrid approach to the skull stripping problem in MRI. Neuroimage 22, 1060–1075 (2004)
21. Fischl, B., Liu, A., Dale, A.M.: Automated manifold surgery: constructing geometrically accurate and topologically correct models of the human cerebral cortex. IEEE T. Med. Imaging 20, 70–80 (2001)
22. Fischl, B., Sereno, M.I., Dale, A.M.: Cortical surface-based analysis. II: Inflation, flattening, and a surface-based coordinate system. II: Inflation, flattening, and a surface-based coordinate system. Neuroimage 9, 195–207 (1999)
23. Fischl, B., Sereno, M.I., Tootell, R.B., Dale, A.M.: High-resolution intersubject averaging and a coordinate system for the cortical surface. Hum. Brain Mapp. 8, 272–284 (1999)
24. Hermes medical solutions (2012), http://www.hermesmedical.com (accessed March 12, 2013)
25. Kittler, J., Hatef, M., Duin, R.P.W., Matas, J.: On combining classifiers. IEEE T. Pattern Anal. 20, 226–239 (1998)
26. Friedman, J., Hastie, T., Tibshirani, R.: Additive logistic regression: a statistical view of boosting. Ann. Stat. 38, 337–374 (2000)

Combined Classification of Risk Factors
for Appendicitis Prediction in Childhood

Theodoros Iliou[1], Christos-Nikolaos Anagnostopoulos[1], Ioannis M. Stephanakis[2],
and George Anastassopoulos[3]

[1] Cultural Technology and Communication Department, University of the Aegean,
81100 Lesvos, Greece
{th.iliou,canag}@ct.aegean.gr
[2] Hellenic Telecommunication Organization S.A. (OTE),
99 Kifissias Avenue, GR-151 24, Athens, Greece
stephan@ote.gr
[3] Medical Informatics Lab. Medical School, Democritus University of Thrace
68100 Alexandroupolis, Greece
anasta@med.duth.gr

Abstract. Abdominal pain is a common symptom associated with transient disorders or serious disease. Diagnosing the cause of abdominal pain can be difficult, because many diseases can cause this symptom. One of the most common conditions associated with acute abdominal pain is acute appendicitis. Diagnosis is based on patient history and physical examination. The present study is based on a data set consisting of 516 children's medical records. Each record consists of 15 factors that are used in the routine clinical practice for the assessment of the acute appendicitis. The importance of these factors is examined in this paper, with the use of many Artificial Intelligence and classification methods. As a result, only 5 factors of the initial 15 factors can be used, in order to have equal or even better diagnosis.

Keywords: Classification, Risk Factors, Computational Intelligence, Abdominal Pain, Appendicitis, Peritonitis.

1 Introduction

Abdominal pain is pain that is felt in the abdomen and is commonly categorized as acute or chronic based on the duration or occurrence. One of the most common conditions associated with acute abdominal pain is acute appendicitis. It is classified as a medical emergency and many cases require removal of the inflamed appendix. Untreated, mortality is high, mainly because of the risk of rupture leading to infection and inflammation of the intestinal lining and eventual sepsis, clinically known as peritonitis which can lead to circulatory shock. The incidence of Acute Appendicitis (AA) is 4 cases per 1000 children. However appendicitis despite pediatric surgeons' best efforts remains the most commonly misdiagnosed surgical condition. Although

L. Iliadis, H. Papadopoulos, and C. Jayne (Eds.): EANN 2013, Part II, CCIS 384, pp. 203–211, 2013.
© Springer-Verlag Berlin Heidelberg 2013

diagnosis and treatment have improved, appendicitis continues to cause significant morbidity and still remains, although rarely, a cause of death [1].

Diagnosis is a medical term denoting an attempt of physician to accurately estimate how a patient's disease will progress, and whether there is chance of recovery, based on an objective set of factors that represent that situation. The diagnosis of abdominal pain is based on the signs and symptoms noted and the physical examination performed by the physician. Following this, additional tests (laboratory tests such as blood test, urine test, and stool tests) may be advised to identify or rule out the presence of some of the common causes of abdominal pain. Finally, the diagnosis of a disease is the outcome of combination of clinical and laboratorial examinations through medical techniques. Physicians use specific medical protocols in order to diagnose diseases, as well as, to observe the response of a patient to a treatment [2, 3]. As diagnosis, there are four stages of appendicitis, including acute focal appendicitis, acute supurative appendicitis, gangrenous appendicitis and perforated appendicitis. These distinctions are vague, and only the clinically relevant distinction of perforated (gangrenous appendicitis includes into this entity as dead intestine functionally acts as a perforation) versus non-perforated appendicitis (acute focal and supurative appendicitis) should be made. Table 1 highlights the classes of the problem, as well as the number of distinctive cases as identified by clinicians.

Several reports have described clinical scoring systems incorporating specific elements of the history, physical examination, and laboratory studies decision rules can predict which children are at risk for appendicitis [4 – 6]. To date, all efforts to find clinical features or laboratory tests, either alone or in combination, that are able to diagnose appendicitis with 100% sensitivity or specificity have proven futile. Also, in bibliography there are some research works [7 - 13] dealing with the abdominal pain prognosis that are based on Artificial Neural Networks.

The present study is based on data set that is obtained from the Pediatric Surgery Clinical Information System of the University Hospital of Alexandroupolis, Greece. It consisted of 516 children's medical records. In the routine clinical practice the appendicitis diagnosis is based on 15 clinical and laboratorial factors. These factors are: Sex, Age, Religion, Demographic data, Duration of Pain, Vomitus, Diarrhea, Anorexia, Tenderness, Rebound, Leucocytosis, Neutrophilia, Urinalysis, Temperature and Constipation. Some of above 516 cases had different stages of appendicitis and, therefore, underwent operative treatment.

Table 1. Classes of the problem and corresponding cases

Diagnosis	Abbreviation	Cases
Focal Appendicitis	FA	34
No Findings	NF	15
Observation	Obs	186
Phlegmonous or Supurative Appendicitis	PSA	29
Discharge	Dis	236
Peritonitis	Per	8
Gangrenous Appendicitis	GA	8

2   Materials and Methods

2.1   Feature Selection

In order to select the most important features and optimise the classification time, a subset evaluator was used. Subset evaluators take a subset of features and return a number which measure a quality of the subset and guides the further search. For the selection of the method, the WEKA data mining tool was used [14]. WEKA offers many feature selection and feature ranking methods, where each method is a combination of feature search and evaluator of currently selected features. Several combinations have been tested in order to assess the feature selection combination that gives the optimum performance for our problem. The feature evaluator and search method (offered in WEKA) that presented the best performance in the data set were (i) Correlation-based Feature Selection Sub Set Evaluator and (ii) BestFirst search method.

The Correlation-based Feature Selection Sub Set Evaluator (CfsSUbsetEval) assesses the predictive ability of each feature individually and the degree of redundancy among them. It prefers sets of features that are highly correlated with the class but are not correlated with other features. An option iteratively adds attributes that have the highest correlation with the class, provided that the set does not already contain an attribute whose correlation with the attribute in question is even higher. On the other hand, Best First search method searches the space of attribute subsets using the greedy hill-climbing approach and backtracking. Setting the number of consecutive non-improving nodes allowed controls the level of backtracking done. Best first may start with the empty set of attributes and search forward, or start with the full set of attributes and search backward, or start at any point and search in both directions (by considering all possible single attribute additions and deletions at a given point).

The combination of the above mentioned methods proposed 5 from the total of 15 features that formed originally the feature set. These features are: (i) duration of pain, (ii) tenderness, (iii) leucocytosis, (iv) neutrophilia and (v) urinalysis.

2.2   Diversity-Dependence of Classifiers

Classification was performed using WEKA. Numerous classification schemas have been evaluated in the dataset as shown in Table 2, in order to select the most suitable set for creating a combined classifier. The classifiers were selected as follows.

Let $F^* = \{F_1, F_2, ..., F_N\}$ be set of potential classifiers to be combined. According to literature, finding a subset $F \subseteq F^*$ of independent classifiers is one central aim of classifier fusion methods. The notion of dependence between classifiers can be perceived as lack of independence but there are various ways of further interpretation associated with diversity, orthogonality, complementarity, etc. It has been recognized that quantifying and studying the dependencies is an important issue

Table 2. List of classifiers that were used to assess the dataset

Trees	Description	Classification Result
ADTree	Alternating decision trees	81.39 %
Id3	Basic divide-and-conquer decision tree algorithm	71.245%
J48 C4.5	Decision tree learner (implements C4.5)	82.36%
LMT	Logistic model trees	79.65 %
M5P	M5' model tree learner	71.38%
NBTree	Decision tree with Naïve Bayes classifiers at the leaves	81.78 %
RF	Random forests	81.39 %
RandomTree	A tree that considers a given no. of random features at each node	78.87 %
REPtree	Fast tree learner that uses reduced-error pruning	81.00 %
Rules		
ConjunctiveRule	Simple conjunctive rule learner	50.77 %
JRip	RIPPER algorithm for fast, effective rule induction	80.00 %
M5Rules	Obtain rules from model trees built using M5	71.45
Nnge	Nearest-neighbor method of generating rules using non-nested generalized exemplars	77.90 %
OneR	1R classifier	68.99 %
Part	Rules from partial decision trees built using J4.8	81.39 %
Prism	Simple covering algorithm for rules	72.75
Ridor	Ripple-down rule learner	78.29 %
Lazy learners		
IB1	Basic nearest-neighbor instance-based learner	81.39 %
K-NN	k-nearest-neighbor classifier	80.42%
KStar (K)*	Nearest neighbor with generalized distance function	80.4 %
LBR	Lazy Bayesian Rules classifier	72.46
LWL	General algorithm for locally weighted learning	75.58 %
Other classifiers		
MLP	Backpropagation neural network	81.39%
RBFNetwork	Implements a radial basis function network	78.48 %
SMO	Sequential minimal optimization algorithm for support vector classification	79.65 %
Bagging	Bags a classifier, works for regression too	82.6 %

in combining classifiers. It is also recognized that a negative correlation should be pursued when designing classifier ensembles. It is to be noted however, that experimental results support the intuition that negative dependence is beneficial although not straightforwardly related to the accuracy [15-21]. In this paper, the set F* includes the 26 classification methods that appear in Table 2.

Numerous measures of dependence and diversity have been proposed. In this paper, we have chosen to incorporate a well-known framework that dictates pairwise measurements calculated for each pair of classifiers in F and then averaged [REFS].

For these experiments, the Yule's Q statistic is derived as the equivalent of the correlation coefficient for binary (correct/incorrect) valued measurements.

Suppose that $Z = \{z_1, z_2, ..., z_N\}$ is a labeled data set, coming from the classification problem in question. For each classifier Fi an N-dimensional output vector $y_i = \{y_{1,i}, y_{2,i}, ..., y_{N,i}\}^T$ of correct classification is created, such that yj,i = 1, if Fi recognizes correctly zj, and 0, otherwise. Yule [22] suggested that the Q statistic can be used as a measure of association. The Yule's Q statistic measure calculates the diversity/dependency between the two classifiers using equation (1).

$$Q_{i,k} = \frac{N^{11}N^{00} - N^{01}N^{10}}{N^{11}N^{00} + N^{01}N^{10}}$$ (1)

Where, Nxy is the number of elements zj of Z, for which yj,i = x and yj,k = y as exactly shown in Table 3. Positive Q values show positive dependency, negative values show negative dependency and zero shows no dependency.

Table 3. A 2x2 table of the relationship between a pair of classifiers

	F_k correct (1)	F_k wrong (0)
F_i correct (1)	N^{11}	N^{10}
F_i wrong (0)	N^{01}	N^{00}

The values of the Q statistic for all 26 classifiers (in pairs) revealed several interesting things about their performance in the dataset. First, some classification methods are highly correlated (i.e., they tend to misclassify the same cases), while some of them are not. As already mentioned, negative or small correlation should be pursued when designing classifier ensembles. In our case, Q<0 or Q near zero was not possible. Therefore, we have chosen to keep all classifiers that present correlation values less than 0.5 (Q<0.5) and recognition performance more than 80% to the proposed ensemble. Using this methodology, three classifiers (K*, JRip and Bagging in conjuction with REPtree as base classifier) in WEKA were found to fulfill the above criteria, which were used to form the ensemble classifier. A brief theoretical background of the selected classifiers follows.

KStar (K*) is a lazy learning method that, as all lazy methods, stores the training instances and do not real work until the classification process. Lazy learning is a learning method in which generalization beyond the training data is delayed until a query is made to the system. Lazy classifiers are most useful for large datasets with few attributes. K* is an instance-based classifier, that is the class of a test instance is based upon the class of those training instances similar to it, as determined by some similarity function. It differs from other instance-based learners in that it uses an entropy-based distance function [23].

JRip [24] implements repeated incremental pruning to produce error reduction method (RIPPER), including heuristic global optimization of the rule set. Initial rule set for each class is generated using IREP. The Minimum Description Length (MDL) based stopping condition is used. Once a rule set has been produced for each class, each rule is reconsidered and two variants are produced.

Bagging bags a classifier to reduce variance. This implementation works for both classification and regression, depending on the base learner. In the case of classification, predictions are generated by averaging probability estimates, not by voting. One parameter is the size of the bags as a percentage of the training set. Another is whether to calculate the out-of-bag error, which gives the average error of the ensemble members [25].

3   Results

3.1   Individual Classifiers

Each technique produced different results depending on their initial parameters and their different architectures. Therefore, their outputs tend to be uncorrelated. This phenomenon can be used to create a single stronger classifier by combining (fuse) their outputs. Tables 4, 5 and 6 present the confusion matrices of the 3 selected individual classifiers.

Table 4. Confusion matrix for K* (correct instances 415/516, performance: 80.4%)

	FA	NF	Obs	PSA	Dis	Per	GA
FA	16	6	3	9	0	0	0
NF	5	17	1	1	21	0	0
Obs	7	0	157	7	15	0	0
PSA	2	1	0	24	0	1	1
Dis	0	9	7	0	190	0	0
Per	0	0	0	1	0	6	1
GA	0	0	0	1	0	2	5

Table 5. Confusion matrix for JRIP (correct instances 413/516, performance: 80.0%)

	FA	NF	Obs	PSA	Dis	Per	GA
FA	26	3	4	1	0	0	0
NF	12	15	0	5	13	0	0
Obs	0	0	174	0	12	0	0
PSA	3	0	0	14	0	8	4
Dis	0	0	19	7	180	0	0
Per	0	0	0	0	0	1	7
GA	0	0	0	1	0	4	3

Table 6. Confusion matrix for Bagging (correct instances 426/516, performance: 82.6%)

	FA	NF	Obs	PSA	Dis	Per	GA
FA	25	1	2	6	0	0	0
NF	6	5	1	1	32	0	0
Obs	1	0	173	0	12	0	0
PSA	9	0	0	18	0	1	1
Dis	0	0	8	0	198	0	0
Per	0	0	0	0	0	4	4
GA	0	0	0	1	0	4	3

3.2   Ensemble of Three Classifiers

Different ways are proposed in the literature for the combination of multiple classifiers into a single stronger one [19-21]. The simplest method is to assume that each classifier has an equal weight to classify an input. Let a counting function a_{ij} defined as shown in equation (2). Then, a majority voting combining technique can be formulated using equation (3).

$$a_{ij} = \begin{cases} 1 & correct & class \\ 0 & wrong & class \end{cases} \tag{2}$$

$$O_j = \arg \max_{k} (\sum_{i} a_{ij}) \tag{3}$$

where O_j is the classifying decision for the jth input and k (k>2) is the number of the classes of the classification problem. As a result, a stronger classifier was created by combining the outputs of the three individual classifiers. The results after classification fusion, are illustrated in Table 7.

Table 7. Confusion matrix for the combined classifier (correct instances 453/516, performance: 87.8%)

	FA	NF	Obs	PSA	Dis	Per	GA
FA	28	1	1	4	0	0	0
NF	7	19	2	2	15	0	0
Obs	0	0	176	0	10	0	0
PSA	0	0	0	24	0	3	2
Dis	0	5	4	2	195	0	0
Per	0	0	0	1	0	6	1
GA	0	0	0	0	0	3	5

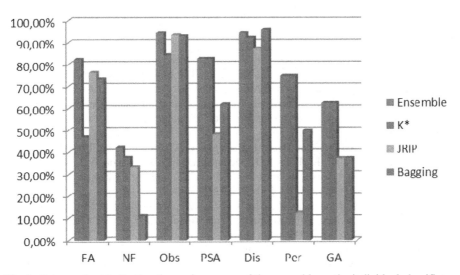

Fig. 1. Column chart indicating the performances of the ensemble vs the individual classifiers. Only in "Discharge" class, Bagging was better than the proposed ensemble.

4   Conclusions - Discussion

Diagnosing the cause of abdominal pain in children is difficult, because many diseases can cause this symptom. Abdominal pain is a symptom associated either with transient disorders or serious disease. In the latter case, accurate diagnosis of abdominal pain is essential for seasonable and appropriate treatment, and avoidance of serious complications that may require urgent intervention.

The prediction of appendicitis is a major issue in medicine. In this paper, we have assessed twenty-six individual machine learning methods for generating predictive models for the 516 children's medical records obtained from the Pediatric Surgery Clinical Information System of the University Hospital of Alexandroupolis, Greece. All methods were validated using 10-fold cross validation method and the classification parameters that literature proposes as optimal. After the assessment of each individual classifier, we proceeded to the creation of a classification ensemble in order to increase the performance. Specifically, we chose to keep all classifiers that present correlation values less than 0.5 (Q<0.5) and recognition performance more than 80%. The experimental results reveal that the overall performance of the ensemble classifier with a majority voting combining technique reached 87.8% (vs 82.6% in Bagging). In addition, the values in the main diagonal have increased or stayed unchanged, with an exception in the "Discharge" class where Bagging was slightly more efficient (see Fig. 1). In any case, a very interesting and useful result is the insight into the tradeoff between overall performance and diversity of the classifiers. Our contribution is that in literature has not been used combined classification for appendicitis prediction or Yule's Q statistic for classifier selection. The major finding in our work is that with Yule's Q statistic for classifiers selection and in conjuction with ensembled classification we achieved high percentage of correct classification in appendicitis prediction. To this end, our team intends to exploit this finding and plan a future research to further explore this tradeoff.

References

1. Addiss, D., Shaffer, N., Fowler, B., Tauxe, R.: The Epidemiology of Appendicitis and Appendectomy in the United States. Am. J. Epidemiol. 132, 910–925 (1990)
2. Głuszek, S., Kozieł, D.: Prevalence and progression of acute pancreatitis in the świętokrzyskie voivodeship population. Pol Przegl Chir. 84(12), 618–625 (2012), doi:10.2478/v10035-012-0102-4
3. Bachur, R.G., Dayan, P.S., Bajaj, L., Macias, C.G., Mittal, M.K., Stevenson, M.D., Dudley, N.C., Sinclair, K., Bennett, J., Monuteaux, M.C., Kharbanda, A.B.: The effect of abdominal pain duration on the accuracy of diagnostic imaging for pediatric appendicitis. Ann. Emerg. Med. 60(5), 582.e3–590.e3 (2012)
4. Blazadonakis, M., Moustakis, V., Charissis, G.: Deep Assessment of Machine Learning Techniques Using Patient Treatment in Acute Abdominal Pain in Children. Artificial Intelligence in Medicine 8, 527–542 (1996)
5. Hamada, T., Yasunaga, H., Nakai, Y., Isayama, H., Horiguchi, H., Fushimi, K., Koike, K.: Japanese severity score for acute pancreatitis well predicts in-hospital mortality: a nationwide survey of 17,901 cases. J. Gastroenterol. (February 19, 2013)

6. Øhrn, A., Komorowski, J.: Diagnosing Acute Appendicitis with Very Simple Classification Rules. In: Żytkow, J.M., Rauch, J. (eds.) PKDD 1999. LNCS (LNAI), vol. 1704, pp. 462–467. Springer, Heidelberg (1999)
7. Papadopoulos, H., Gammerman, A., Vovk, V.: Reliable Diagnosis of Acute Abdominal Pain with Conformal Prediction. Engineering Intelligent Systems 17(2-3), 127–137 (2009)
8. Papadopoulos, H., Gammerman, A., Vovk, V.: Confidence Predictions for the Diagnosis of Acute Abdominal Pain. In: Iliadis, L., Vlahavas, I., Bramer, M. (eds.) Artificial Intelligence Applications & Innovations III. IFIP, vol. 296, pp. 175–184. Springer, Heidelberg (2009)
9. Mantzaris, D., Anastassopoulos, G., Iliadis, L., Adamopoulos, A.: A hybrid multi-objective genetic algorithm for evaluation of essential sets of medical diagnostic factors. Engineering Intelligent Systems 17(2-3), 99–104 (2009)
10. Keogan, M., Lo, J., Freed, K., Raptopoulos, V., Blake, S., Kamel, I., Weisinger, K., Rosen, M., Nelson, R.: Outcome Analysis of Patients with Acute Pancreatitis by Using an Artificial Neural Network. Academic Radiology 9(4), 410–419 (2002)
11. Mantzaris, D., Anastassopoulos, G., Adamopoulos, A., Gardikis, S.: A non-Symbolic Implementation of Abdominal Pain Estimation in Childhood. Information Sciences 178(20), 3860–3866 (2008)
12. Son, C.S., Jang, B.K., Seo, S.T., Kim, M.S., Kim, Y.N.: A hybrid decision support model to discover informative knowledge in diagnosing acute appendicitis. BMC Medical Informatics and Decision Making 12, 17 (2012), doi:10.1186/1472-6947-12-17.
13. Anastasopoulos, G., Iliadis, L.: Intelligent hybrid modeling towards the prognosis of abdominal pain. International Journal of Hybrid Intelligent Systems 6(4), 245–255 (2009)
14. Waikato Environment for Knowledge Analysis, http://www.cs.waikato.ac.nz/ml/weka/downloading.html
15. Kuncheva, L.I., Whitaker, C.J., Shipp, C.A.: Limits on the Majority Vote Accuracy in Classifier Fusion. Pattern Anal. Appl. 6, 22–31 (2003)
16. Lam, L.: Classifier combinations: implementations and theoretical issues. In: Kittler, J., Roli, F. (eds.) MCS 2000. LNCS, vol. 1857, pp. 77–86. Springer, Heidelberg (2000)
17. Bonissone, P.P., Eklund, N.H., Goebel, K.: Using an ensemble of classifiers to audit a production classifier. In: Oza, N.C., Polikar, R., Kittler, J., Roli, F. (eds.) MCS 2005. LNCS, vol. 3541, pp. 376–386. Springer, Heidelberg (2005)
18. Kuncheva, L.: Switching Between Selection and Fusion in Combining Classifiers: An Experiment. IEEE T. Syst. Man Cy. B 32(2), 146–156 (2002)
19. Rosen, B.E.: Ensemble learning using decorrelated neural networks. Connect. Sci. 8(3/4), 373–383 (1996)
20. Roli, F., Giacinto, G., Vernazza, G.: Methods for Designing Multiple Classifier Systems. In: Kittler, J., Roli, F. (eds.) MCS 2001. LNCS, vol. 2096, pp. 78–87. Springer, Heidelberg (2001)
21. Kuncheva, L.: Is independence good for combining classifiers? In: Proc. of 15th International Conference on Pattern Recognition, Barcelona, Spain, vol. 2, pp. 168–171 (2000)
22. Yule, G.U.: On the association of attributes in statistics. Phil. Trans. A 194, 257–319 (1900)
23. Cleary, J.G., Trigg, L.E.: K*: An Instance-based Learner Using an Entropic Distance Measure. In: 12th International Conference on Machine Learning, pp. 108–114 (1995)
24. Cohen, W.W.: Fast Effective Rule Induction. In: Twelfth International Conference on Machine Learning, pp. 115–123 (1995)
25. Breiman, L.: Stacked regression. Machine Learning 24(1), 49–64 (1996)

Analysis of DNA Barcode Sequences Using Neural Gas and Spectral Representation

Antonino Fiannaca, Massimo La Rosa, Riccardo Rizzo, and Alfonso Urso

ICAR-CNR, National Research Council of Italy,
viale delle Scienze Ed.11, 90128 Palermo, Italy
{fiannaca,larosa,ricrizzo,urso}@pa.icar.cnr.it

Abstract. In this paper we present an application of the neural gas network to the classification of the DNA barcode sequences. The proposed method is based on the identification of distinctive words, extracted from the spectral representation of DNA sequences. In particular we calculated the "signatures" that are a characteristic of the DNA sequence at different taxonomic levels. In order to demonstrate the efficacy of the proposed method, we tested it over 10 real barcode datasets belonging to different animalia species, provided by on-line resource Barcode of Life Database (BOLD).

Keywords: DNA barcode, DNA k-mer, alignment-free, spectral representation, neural gas.

1 Introduction

One of the most interesting challenge in bioinformatics is the identification of living species by means of analysis of their DNA sequences. More recently, in order to both improve and accelerate this process, scientists investigate on fragments (markers) of DNA sequences extracted by standard gene regions, that contain enough information for the assignment of the proper taxa. In the last years, the "cytochrome c oxidase subunit 1" (COI) gene, located into mitochondrial DNA of the animal kingdom, has proved to be a good marker for DNA sequences; for this reason, it has been used as a barcode sequence for identification and taxonomic rank assignment of animals [6]. In the last years, alignment-free methods have been proposed for the analysis of short sequences aimed to taxonomic classification. These approaches, in fact, are able to overcome the drawbacks of classical methods as highlighted in [14]

In this paper, we investigate a new alignment-free method for DNA barcoding based on both spectral representation and neural network unsupervised clustering. The spectral representation introduces the advantage of using fixed-length sub-sequences, called DNA k-mers (DNA 'words' composed by k nucleotides); whereas the unsupervised clustering allows us to extract group of sequences sharing common DNA words. The main goal of this work is to investigate how much the characteristics of different species are related to their spectral distribution, even considering "noisy" sequences, i.e. sequences with "wrong" or undefined

L. Iliadis, H. Papadopoulos, and C. Jayne (Eds.): EANN 2013, Part II, CCIS 384, pp. 212–221, 2013.

nucleotides. In order to accomplish this goal, we do not take into account the whole spectrum of a DNA sequence, but only a small number of words with very high frequency. The smallest set of k-mers able to assign a DNA sequence to its proper taxonomic category is defined as the *signature* of the DNA sequence. Taking into account only a few very high frequency words allows us to obtain a robust syste for barcode sequence classification with respect to both sequence length and noise in sequences.

2   Background

Alignment-free analysis for classification of DNA barcode sequences is an open challenge in bioinformatics. Several authors proposed different solutions based on machine learning and/or statistical methods.

For instance, in order to perform a taxonomic analysis, authors in [9] use two compression-based methods implementing a non-computable Universal Similarity Metric (USM) class of distances. This parameter-free approach builds phylogenetic trees comparable with those obtained with standard evolutionary distances. This interesting approach is sensitive to noisy experimental data, because noisy sequences provide a low compression ratio. Better results can be obtained by means of models based on spectral representation. As introduced in the previous Section, these models are characterized by DNA k-mers (words) of fixed-length. More in detail, during spectral analysis, all the information related to parameters such as nucleotides position are discarded, and only distribution of words is taken into account. In [3] authors emonstrate how the analysis of DNA words distribution in the whole genome can be representative for many living organism. In particular, authors investigate on the modality of DNA k-mers distribution in the whole curve corresponding to sequence spectrum.

Regarding our work, the most interesting analysis was carried out by [8]. In fact, in addition to standard spectral representation, the authors introduce a mismatch kernel method that increases barcode sequence spectrum density adding inexact sequence matching. In other words, the new spectrum contains both words of k length obtained with sliding window extraction over the DNA sequence, and all the k-mers that differ in at most m bases to each extracted word. This edited version introduces information about similarity among words (for less than m bases), that can be read as a sort of site mutation and/or noise balance technique. The authors empirically demonstrate the effectiveness of this mismatch representation for DNA barcode sequences. In order to take advantage of using [8] representation, it is important to choice the right number of k and m, because results strictly depend on these two parameters. A few years ago, authors in [2] presented a study about DNA fragments encoded in different oligonucleotide frequencies. As result of this research, authors noticed that an increase of nucleotide frequency order may deteriorate the assignment of DNA sequences to the correct category. Conversely classification tests with tri, tetra and penta-nucleotide frequencies, can reach an adequate degree of correctness.

Finally, since the spectrum representation of a DNA sequence defines the occurrence of a fixed number of features, we can consider the distance between two sequences as the euclidean distance between their respective feature vectors. Taking advantage of this possibility, we can identify some groups of sequences that share common properties (principal directions or discriminant features), adopting a clustering technique. For this purpose, we use an unsupervised neural network algorithm for prototype-based clustering. As stated by different authors, such as [13,5], those machine learning techniques applied to bioinformatics can offer very intuitive and robust results. In particular, we adopt the well know Neural Gas algorithm (NG) by [11] that, with respect to other unsupervised techniques, does not suffer from the problem of local minima or topological restrictions [4].

3   Classification Pipeline

In this section we explain the workflow of the proposed classification approach. The goal of the proposed method is to define a small set of words, called High Frequency Words (HFW) useful to classify the DNA sequence. First of all we introduce how to choose a DNA spectrum that is representative for a class of sequences. Then, we introduce the evaluation criteria and comparative techniques used in order to obtain the number of high representative words, necessary to identify the taxa of a test sequence.

In order to investigate the representative fragments of a DNA barcode sequence, it is necessary to identify a new sequence model that preserve the major characteristics of a group of sequences. To obtain a consistent fixed-length representation for all DNA sequences we choose to consider words with $k = 5$ bases and $m = 1$ mismatch, for our experiments [8,2]. Therefore, we translate each sequence (seq) into its spectral representation (according to [8]), obtaining a 5-mers occurrence vector $\bar{s}_{seq} = [s_1, s_2, \ldots, s_N]$, where s_i is the frequency of the i-th word in sequence seq and $N = 4^k = 4^5 = 1024$ is the number of words.

As stated in Section 2, we adopt the Neural Gas algorithm to cluster sequences that share common properties [10] and NG parameters have been tuned through a series of experiments (not shown in this paper), in order to obtain the best clustering results. After the network is trained, we obtain an unsupervised clustering of input spectra representing DNA sequences. The last step of the procedure is to find the "reference spectra" for each different category of sequences; these reference spectra are represented by n centroids $C = \{\bar{c}_i | i = 1, 2, \ldots, n\}$, where $\bar{c}_i \in \Re^{1024}$. In order to find a set of n centroids for fixed n, we adopt the centroid neighbourhood graph algorithm implemented by Leisch [10]. This technique reports how well the corresponding clusters are separated by means of a weighted graph, where edges contain information about cluster separation. Since this technique is computed off the fitted partition, it can be easily combined with the NG algorithm.

Since we are working with pre-labelled sequences, we calculate the Overall Accuracy [1], $OA(n)$, on varying n, in order to find a small number of centroids

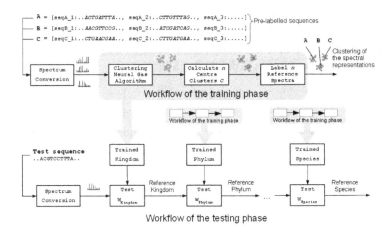

Fig. 1. Pipelines of both training (top) and test (bottom) methods

that effectively distinguish different groups of spectra, and we take as number of reference spectra (n_s) the smallest n that maximizes $OA(n)$, as defined below:

$$n_s = \underset{n}{\operatorname{argmin}}\left\{ \max\{OA(n)\} \right\}. \tag{1}$$

For each n, $OA(n)$ value ranges between 0 (if all elements are misclassified) and 1 (if all elements are well classified). Moreover, this measure is used for defining the training error on varying n, $e_t(n)$, as reported in Eq.2. In other words, n_s can be defined as the smallest integer that minimize $e_t(n)$.

$$e_{t_{min}} = e_t(n_s) = 1 - \max\left\{ OA(n_s) \right\} \tag{2}$$

The reference spectra identified are responsible for representing a group of sequences belonging to a specific taxonomy rank (such as phylum, class, order, and so on).

The top part of Figure 1 shows the workflow of the training method. In this figure we suppose to have a dataset composed by some sequences, belonging to three different species, called respectively A, B and C. Since during training phase all the sequences are labelled, we can assign the right label to each reference spectrum, at the end of this process. Now it is necessary to use this method at each taxonomic level in order to obtain a set of reference spectra for each taxa. For this reason, the learning process is repeated for all taxonomic levels and, at the end of training phase, we will have the set of reference spectra for each taxonomic rank level.

On the basis of concepts related to genomic DNA k-mer spectra, developed by different authors [3,8], we notice that all these spectra have an irregular distribution of words, with a few peaks representing words with high occurrence. For these reasons, we take into account only those words with the highest frequencies.

In order to establish how many words are needed to identify the correct taxonomy of a DNA sequence, we search the smallest number of words needed to assign the spectrum representation of a new test sequence, $\overline{s}_{seq_t}$, to its own reference centroid, $\overline{c}_{ref}$, where the ref index is defined as follows:

$$ref = \operatorname*{argmin}_i \left\{ dist\left(\overline{c}_i, \overline{s}_{seq_t}\right) \right\} \tag{3}$$

At this point, we take into account only w High Frequency Words (HFW), i.e. the w greatest components (highest peaks in frequency histograms) of $\overline{s}_{seq_t}$ in all centroids C. For each reference centroid, $\overline{c}_i = [c_{i_1}, c_{i_2}, \ldots, c_{i_N}]$, the vector components are sorted in decreasing order: $\overline{c^*}_i = \left[c_{i_{\lambda_1}}, c_{i_{\lambda_2}}, \ldots, c_{i_{\lambda_N}}\right]$ so that $c_{i_{\lambda_1}} \geq c_{i_{\lambda_2}} \geq \ldots \geq c_{i_{\lambda_N}}$. In the same way, we obtain an ordered vector $\overline{s^*}_{seq}$. At this point, we consider only the first w components $c_{i_{\lambda_l}}$, where $l = 1, 2, \ldots, w$. Words corresponding to $c_{i_{\lambda_l}}$ components, define a fingerprint for the sequence.

Finally, in order to measure the dissimilarity between words of test sequence and centroids set, we calculate the Jaccard distance [7], J_δ (considering only the w most frequent words) between test sequence and centroid set. Values of this distance are between 1 (if two sets have no common words) and 0 (if two sets have the same words). According to Jaccard distance definition, we can define the validation error e_v at varying of w, as follows.

$$e_v(w) = \min \left\{ J_\delta \bigg|_{HFW=w} \left(\overline{s^*}_{seq}, \overline{c^*}_i\right) \right\} \tag{4}$$

In order to obtain a low percentage of error, we will calculate $e_v(w)$ to determine how many words can discriminate a test sequence among different taxonomic ranks.

The bottom part of Figure 1 shows the workflow of the testing method. In particular, after a test DNA sequence is translated into its own spectral representation ($\overline{s}_{seq_t}$), we test fingerprint for Kingdom taxonomic ($w_{Kingdom}$) level comparing its w highest peaks with as much highest peaks of all the reference spectra belonging to kingdom taxonomy. After the reference kingdom spectrum is found, the same comparison technique will be done at phylum taxonomy level, and so on. This way, our technique introduces an alignment-free method for taxonomy identification.

4   Experimental Results

This Section contains information about experimental data used for testing the effectiveness of the proposed technique. Obtained results are showed and discussed, in order to explain our contribution to DNA barcode sequence analysis.

4.1   Barcode Datasets

We tested the proposed technique using real datasets, provided by the on-line resource Barcode Of Life Database (BOLD) [12]. Since BOLD collects a lot of

Table 1. Ten experimental datasets. Each one of these datasets comes from a different BOLD project.

#	Code	Title	Specimens	Familia	Genera	Species
1	**AHNFE**	*Ahnfeltiales of Canada*	133	1	1	3
2	**ANOD**	*Revision of Malagasy Anochetus and Odontomachus*	447	1	2	8
3	**BACA**	*Churchill Ant Barcode Accumulation Curves*	438	1	4	7
4	**BBST**	*Bumblebees of Subterraneobombus*	164	1	1	10
5	**EBFSF**	*Billfish and Swordfish COI Identification*	286	2	6	10
6	**FUCUB**	*Canadian Fucus*	111	1	1	3
7	**LHSMI**	*Blackflies of the New World_2009*	103	1	1	11
8	**LTERH**	*Lumbricus terrestris/herculeus*	219	1	1	5
9	**PMF**	*Passerines from the Malvinas - Falkland Islands*	102	6	7	8
10	**RFPB**	*Pemphigus - sugarbeet root aphid*	209	1	1	3

barcode datasets in a less than 15 separate projects representative for different phyla in animalia kingdom, we consider 10 datasets, one for a different project. Those datasets differ each other also for the number of species, the number of barcode sequences per species (specimens), the sequence length and the sequence quality, expressed in terms of number of sequences with undefined nucleotides. In order to develop a quantitative analysis in a consistent environment, we filtered all those sequences that are classified in BOLD as *non-barcode compliant*. Moreover, since typical sequence length of COI barcode gene is about 650 bp [6], longer sequences contain a part of information content related to other genes, whereas shorter sequences could not contain all the gene information. To avoid problems during training, we removed all sequences that are shorter than 500 bp and longer than 800 bp. In addition, we also removed all the barcode sequences that are unique exemplars of a single species, because if a specific species has only a specimen, no test algorithm will be able to identify those species with a cross validation technique. The complete list of the barcode datasets of our experiments is reported in Table 1. All datasets contain a total of 2212 sequences parted into 68 species. Moreover 4 datasets (ANOD, BACA, EBFSF and PMF) contain more than a genus and two of them (EBFSF and PMF) are parted into more than one family. Since for each dataset all sequences belong to the same order, this field is not reported into the table. Obviously, also the specimens/species ratio is not constant for each dataset, in fact a consistent difference is observed at intra species level for several datasets. These unbalanced distributions of specimens over respective species, could represent a problem both for training and for testing phases. In fact, since the training method is based on an

Fig. 2. Training error for BBST dataset, varying number of reference spectra (centre clusters). The plot line shows the number of reference spectra, n_s, for which $e_t(n_s)$ is the lowest.

unsupervised clustering algorithm, it tends to generate groups of sequences that are equally spread out, bringing about any sort of distortions. With respect to testing phase, any cross-validation technique could miss the trained model, due to high variability of the input distribution. As we will demonstrate later, the proposed technique appears robust with respect to specimens distribution.

4.2   Validation of the Proposed Technique

Validation of the proposed technique is divided into two parts: firstly we demonstrate the efficacy of the method used for finding reference spectra, then we test the whole training method using a cross-validation process.

As regards the training method, we introduced in Section 3 the training error $e_t(n)$, as defined in Eq.2. Since in this phase we aim at obtaining the lowest $e_t(n)$, we calculate this error against the n value from the number of different classes to a reasonable large value. For instance, in Figure 2, we show $e_t(n)$ for the 4th dataset, BBST. More in detail, this chart reports the number of reference spectra versus the error due to sequences assigned to a wrong reference spectrum. The x-axis goes from 10 (the number of species contained into BBST dataset) to 25 (that we consider a number of cluster centres large enough for representing 10 species). According to the plot line, we can assert that n_s (as defined in Eq. 1) is equal to 17. For this value $e_t(17) = 0.012$, because the proposed training method misclassifies only 2 sequences over 164. A summary table about training error at species taxonomy level for analysed datasets is reported in Table 2. The first four columns contain some technical data about the datasets; the fifth column reports the number of reference spectra, n_s, observed during training error analysis. The last column gives the percentage of the smallest training error for each dataset. This table contains those (six) datasets that are classified only at species taxonomic level. Since we adopt a hierarchical analysis, it follows that four datasets (2nd,3rd,5th,9th rows in Table 1) are split into different genera, before making the species classification analysis. For this reason it is not possible to have a unique n_s for each dataset. Table 2 shows how the number of reference spectra is uncorrelated with both the dimension of datasets and the number of

Table 2. Training error at species taxonomy level

#	Code	Specimens	Species	n_s	$e_t(n_s)\%$
1	AHNFE	133	3	4	0
4	BBST	164	10	17	1.2
6	FUCUB	111	3	3	0
7	LHSMI	103	11	17	0.9
8	LTERH	219	5	9	0.7
10	RFPB	209	3	3	0

different species, just because n_s strictly depends on the particular similarity among DNA barcode sequences into the datasets. In any case, this table gives information about the performance of proposed method, with respect to both unsupervised clustering technique and centroids selection method. In fact, with all experimental data, at each taxonomic level, we found at least a value of n_s, that allows to reach a low degree of e_t. At this point, it is possible to complete the learning phase, as defined in Section 3.

In order to estimate the performance of the classification method defined in Section 3, we adopt two well-know statistical methods, belonging to cross-validation techniques: the "ten-fold cross-validation" and the "leave-one-out cross-validation". More in detail, we select the validation technique according to the distribution of sequences over different taxonomic groups. In particular, we adopt the "ten-fold cross-validation" at phylum taxonomic level, because each dataset consists of more than 100 sequences; whereas we use the "leave-one-out cross-validation" for all the other taxa, because some familia, as well as some genera or species, are populated by very few sequences. For instance, the *Anochetus* genus of the ANOD dataset is composed of 290 sequences and it contains only 3 sequences belonging to the *Anochetus pattersoni* species; in this case, the "ten-fold cross-validation" could not include those sequences for the training phase and produce no reference spectra for that species. Results of cross-validation tests are reported in Figure 3. This figure contains four charts, one for each analysed taxa, i.e. phylum (Figure 3(a)), familia (Figure 3(b)), genus (Figure 3(c)) and species (Figure 3(d)). Each chart reports the number of high frequency words (w), from 10 to 100, versus the validation error, $e_v(w)$, as defined in Eq.4. The chart 3(a) (phylum classification) reports the validation error calculated during cross-validation, whereas the other charts show both the mean errors (the main plot) and the standard deviations (error bars). More in detail, 3(a) is calculated over 10 datasets; 3(b) is calculated over those 2 datasets that contains more than one familia; 3(c) is calculated over 6 groups of sequences (2 datasets contain more than one genus and two datasets are split in 4 sub-datasets for genus classification, according to the proposed hierarchical analysis); 3(c) is calculated over 17 groups of sequences (6 datasets contain more than one species and the other ones are split in 11 sub-datasets for species classification, according to the proposed hierarchical analysis). As regarding chart 3(a), if we take into account only 25 HFW, it follows that $e_v(25) = 0.021$; this way, we can

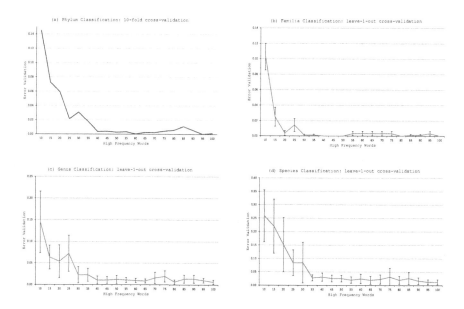

Fig. 3. Validation error for respectively Phylum, Familia, Genus and Species classification. All the charts are defined as number of high frequency words versus validation error.

classify a barcode sequence in the correct phylum at 97.8%. According to charts 3(b) and 3(c), in order to classify a sequence into the correct family and/or genus with both a low percentage of e_v and a low standard deviation, we need at most 30 - 35 HFW. Finally, in order to obtain a correct classification of a barcode sequence with an approximation of 97.3% (with a standard deviation equal to 0.022), the proposed hierarchiacal method needs 35 HFW. Obviously, we are not surprised if it is necessary a growing number of high frequency words, when taxonomic analysis became deeper (from phylum to species). In fact, we need more information to distinguish similar sequences. Of course the number of datasets used for training the system at familia taxonomic level is too small to have any statistical value, but obtained plot appears quite consistent with other results obtained at different taxa.

5   Conclusion and Future Works

In this paper we proposed an alignment-free method for DNA barcoding, based on the spectral representation of DNA sequences. We used the Neural Gas unsupervised algorithm for prototype-based clustering to identify some groups of sequences sharing common properties. We tested the method for the hierarchical classification of barcode sequences, over 10 datasets from BOLD on-line resource, at different taxonomic levels (i.e. phylum, familia, genus and species). The obtained results are quite encouraging because for all the datasets the taxonomic classification was good. Although the proposed technique was tested with short sequences (DNA barcode), it could be applied also to longer DNA sequences, where advantage of a fixed-length representation is more relevant. In the future, we are planning to provide a web service containing a large number of trained species belonging to different BOLD projects; this way, an user can quickly obtain the taxonomic classification of a DNA sequence.

References

1. Alberg, A.J., Park, J.W., Hager, B.W., Brock, M.V., Diener-West, M.: The use of "overall accuracy" to evaluate the validity of screening or diagnostic tests. Journal of General Internal Medicine 5(pt. 1), 460–465 (2004)
2. Chan, C.K.K., Hsu, A.L., Tang, S.L., Halgamuge, S.K.: A Method for Evaluating Quality of Clustering DNA Fragments Encoded in Different Nucleotide Frequencies. In: Proceeding of FBIT 2007, pp. 60–63. IEEE (2007)
3. Chor, B., Horn, D., Goldman, N., Levy, Y., Massingham, T.: Genomic DNA k-mer spectra: models and modalities. Genome Biology 10(10), R108 (2009)
4. Cottrell, M., Hammer, B., Hasenfuss, A., Villmann, T.: Batch and median neural gas. Neural Networks 19(6-7), 762–771
5. Fiannaca, A., Di Fatta, G., Rizzo, R., Urso, A., Gaglio, S.: Simulated annealing technique for fast learning of SOM networks. Neural Computing and Applications (2011)
6. Hebert, P.D.N., Ratnasingham, S., DeWaard, J.R.: Barcoding animal life: cytochrome c oxidase subunit 1 divergences among closely related species. Proceedings of the Royal Society. Series B, Biological sciences 270(suppl.), S96–S99 (2003)
7. Jaccard, P.: Nouvelles recherches sur la distribution florale. Bul. Soc. Vaudoise Sci. Nat. 44, 223–270 (1908)
8. Kuksa, P., Pavlovic, V.: Efficient alignment-free DNA barcode analytics. BMC Bioinformatics 10(suppl. 14), S9 (2009)
9. La Rosa, M., Fiannaca, A., Rizzo, R., Urso, A.: Alignment-free Analysis of Barcode Sequences by means of Compression-Based Methods. BMC Bioinformatics 14 (suppl. 7), S4 (2013)
10. Leisch, F.: A toolbox for -centroids cluster analysis. Computational Statistics & Data Analysis 51(2), 526–544 (2006)
11. Martinetz, T.M., Berkovich, S.G., Schulten, K.J.: "Neural-gas" network for vector quantization and its application to time-series prediction. IEEE Transactions on Neural Networks 4(4), 558–569 (1993)
12. Ratnasingham, S., Hebert, P.D.N.: bold: The Barcode of Life Data System. Molecular Ecology Notes 7(3), 355–364 (2007), http://www.barcodinglife.org
13. Seiffert, U., Hammer, B., Kaski, S., Villmann, T.: Neural networks and machine learning in bioinformatics-theory and applications. In: European Symposium on Artificial Neural Networks, pp. 521–532 (2006)
14. Vinga, S., Almeida, J.: Alignment-free sequence comparison-a review. Bioinformatics 19(4), 513–523 (2003)

A Genetic Algorithm for Pancreatic Cancer Diagnosis

Charalampos Moschopoulos[1,2], Dusan Popovic[1,2], Alejandro Sifrim[1,2],
Grigorios Beligiannis[3], Bart De Moor[1,2], and Yves Moreau[1,2]

[1] Department of Electrical Engineering-ESAT, SCD-SISTA, KU Leuven, Kasteelpark
Arenberg 10, Box 2446, 3001, Leuven, Belgium
[2] iMinds Future Health Department, KU Leuven, Kasteelpark Arenberg 10, Box 2446, 3001,
Leuven, Belgium
[3] Department of Business Administration of Food and Agricultural Enterprises,
University of Western Greece, G. Seferi 2, GR-30100 Agrinio, Greece
{Charalampos.Moschopoulos,Dusan.Popovic,Alejandro.Sifrim,
Bart.DeMoor,Yves.Moreau}@esat.kuleuven.be, gbeligia@uwg.gr

Abstract. Pancreatic cancer is one of the leading causes of cancer-related death
in the industrialized countries and it has the least favorable prognosis among
various cancer types. In this study we aim to facilitate early detection of the
pancreatic cancer by finding minimal set of genetic biomarkers that can be used
for establishing diagnosis. We propose a genetic algorithm and we test it on
gene expression data of 36 pancreatic ductal adenocarcinoma tumors and
matching normal pancreatic tissue samples. Our results show that a minimum
group of genes are able to constitute a high reliability pancreatic cancer
predictor.

Keywords: genetic algorithm, support vector machines, pancreatic cancer,
biomarkers, pancreatic ductal adenocarcinoma, microarrays.

1 Introduction

In present, pancreatic cancer is considered as one of the most lethal of common
cancer types [1]. At this moment, pancreatic cancer holds the eighth most common
cause of cancer related deaths with survival rate of less than 5%, five years after the
diagnosis. Its lethality is largely due to the fact that is diagnosed at a later stage,
which significantly decreases the chance of patient's survival. The most common type
of pancreatic cancer, accounting for 95% of these tumors, is adenocarcinoma or
PDAC. An additional problem is that PDAC has an extremely poor prognosis, as it
seems that pancreas emits few clues to signal the carcinogenic process.

During the last years, several research teams have tried to detect molecular markers
that facilitate early detection of the disease so that appropriate treatment could be
applied timely [2-3]. There are several cases where such a set of biomarkers has been
proposed, but none of them has been shown to be robust enough to constitute a
diagnosis classifier [4]. Additionally, given the biomarker discovery problem, it is
rather difficult to extract knowledge from high throughput data as it suffers from
curse of dimensionality and high level of noise [5].

L. Iliadis, H. Papadopoulos, and C. Jayne (Eds.): EANN 2013, Part II, CCIS 384, pp. 222–230, 2013.
© Springer-Verlag Berlin Heidelberg 2013

In this contribution we apply a genetic algorithm on an publically available gene expression dataset (from GEO database [6] by [7]) and we try to obtain the minimum set of biomarkers that can be used for detection of pancreatic cancer. Due to the non-deterministic nature of the genetic algorithms, we performed extensive experiments and we show that the each time selection of biomarkers produces a classifier with a relatively robust performance. We also formed a list out of the most "popular" genes presented in the final results of each run and show the biological relevance of them for pancreatic cancer. Finally, we provide numerous performance metrics such as F-measure and accuracy. For these we obtain high values for almost all runs proving the efficiency of the proposed method.

The remaining of the paper is organized as follows: section 2 introduces the proposed genetic algorithm relative to chromosome encoding, initialization procedure, fitness function and genetic operators. Also, more information about the dataset used in our experiments is given. Section 3 presents our results and gives a overview of the biological significance of the biomarkers found. Finally, in the last section, we present our conclusions and directions for future work.

2   Methods

2.1   Microarray Gene Expression Data for Pancreatic Cancer

The dataset used in our experiments contains pairs of normal and tumor tissue samples which were obtained at the time of surgery from resected pancreas of 36 pancreatic cancer patients [8]. All the patients were suffering from PDAC. Gene expressions were obtained using Affymetrix U133 plus 2.0 whole genome microarrays. Also, 6 control samples (3 normal and 3 tumor) were present in dataset, which were used to test the quality of the rest obtained samples. In total, this dataset includes 19898 genes and 78 genechip hybridizations have been performed. This dataset is freely available at GEO database where the reader can find more information (http://www.ncbi.nlm.nih.gov/geo/query/acc.cgi?acc=GSE15471).

2.2   The Proposed Genetic Algorithm (GA)

The efficiency of GAs has been proved by successful applications in many different scientific fields, including bioinformatics [9, 10], where they surpassed other algorithmic strategies for optimization and search. The GAs are stochastic algorithms that simulate the process of natural evolution. Based on this model, all GAs use three simple operators, which allows them to evolve and reach near optimal solutions [11, 12]. These are the Selection, Crossover and Mutation operators.

Initially, we randomly divided the expression dataset to 70% training (53 samples) and 30% test (25 samples) data. For the chromosome encoding we chose a one dimensional binary array representation where each position corresponds to biological gene (Figure 1) to be included as a biomarker. We chose this representation because it gave us the opportunity to perform experiments with different mutation and crossover operators. In the end, as there were no significant changes in the performance of the GA, we chose to use the single point crossover and uniform mutation.

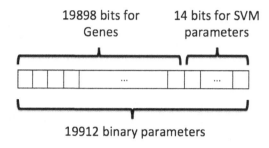

Fig. 1. Chromosome Encoding

As a selection operator we chose to use a stochastic universal [13]. In the fitness function, which is the heart of each GA, we tried to achieve balance between the F-measure of a classifier and the size of the corresponding solution, penalizing solutions that contain many genes. Following is the formal expression of the fitness function used to evaluate each chromosome:

$$fitness\_function = 1 + F\_measure - \frac{size\_of\_chromosome}{mean\_size\_of\_initial\_population}$$

where the *size_of_chromosome* is the number of genes that are used as biomarkers in the chromosome and *mean_size_of_initial_population* is the mean number of genes that are used as biomarkers in the initial population of the GA, across all chromosomes. This particular form of fitness function pushes GA to select solutions that contain a small number of genes and are as accurate as possible. We experimented with different weights in order to boost one metric over the other but we obtained similar results, just with slower convergence. The F-measure metric has been preferred as it is geometric average of the precision and recall, both being of great interest when designing diagnostics tests. The efficiency of this strategy is clearly shown in the next section were our experimental results are presented in more detail. To classify samples given biomarkers selected in each chromosome of the population, we used a Support Vector Machine (SVM) with Gaussian Radial Basis Function (RBF) kernel function [14]. This also adds two parameters in the system that has to be tuned: the sigma and the penalty parameter c. These two parameters were represented, as the additional chromosome for each individual solution, constituted of 7 bits per each parameter, which were also tuned as the by-product of GA evolution.

The structure and operation of the proposed algorithm is presented in Figure 2. In the initialization procedure, a random population was created, where the 10% of the bits that represents the gene panel and 50% of the bits that represents the SVM parameters were set to 1. This way, an initial generated chromosome contains around 200 genes in average. We reassured that the individuals that constitute the first generation were spread in the whole search space of possible solutions in order to avoid local optima solutions. In the next step, each chromosome is evaluated by the fitness function. Then, if the maximum number of generations is not reached, the three operators (selection, crossover and mutation) were applied consequently to create the new population. When the maximum number of generations is reached, the GA outputs the best solution generated throughout its execution. In all the performed experiments, the probability of was 0.7, while that of mutation was equal to 10^{-4} per bit, for all bits and all chromosomes. This means that in average two bits per chromosome were changing during the mutation. The size of the population has been set to 500 and the maximal number of generations to 500 (the number of generations was used as termination criterion).

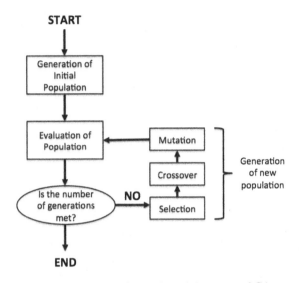

Fig. 2. Structure and operation of the proposed GA

3   Results and Discussion

As the GAs are essentially stochastic, each run produced slightly different list of biomarkers. However, after performing 100 runs of the algorithm using the same operators and parameters, all the generated results had similar performance, achieving high values for performance metrics. Additionally, the number of genes that were selected in each run was approximately stable (around 16 with a deviation of 2). These values of the GA's parameters were selected by trial-and-error method after performing additional experiments.

The convergence of our algorithm and the reduction of the number of genes taking part in the best chromosome of each generation were similar in all our experiments, following the pattern presented in Figure 3. It is clear that the performance of the algorithm stabilizes approximately after 200 generations, and it remains literally unchanged approximately after the 300th generation. After the training process, we use the best chromosome produced to classify the test data using a SVM classifier. Note that during the training process, the parameters of a SVM the also optimized.

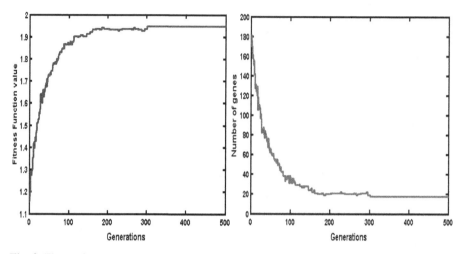

Fig. 3. Fitness function value and size of best chromosome during GA evolution. Most of our experiments followed the same pattern.

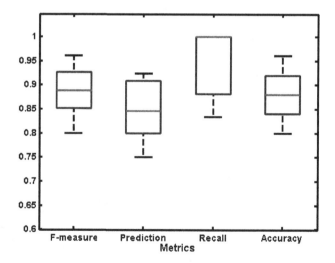

Fig. 4. Box plot presenting the variance of the evaluation metrics

In order to evaluate the classification results, we used well-established metrics for classification such as F-measure, Prediction, Recall and Accuracy [15]. As it can be seen in Figure 4, the GA algorithm achieves on average 88% for F-measure and Accuracy metrics.

Even the list of biomarkers generated by the GA varies for every execution of the algorithm, one could expect that the genes that are somehow functionally related to pancreatic cancer should be present most of the times in results. To examine this, we counted genes that were present in solutions with corresponding fitness values higher than the average in the last generation of the GA. We repeated this procedure for every run of the algorithm. The 20 most "popular" genes are presented in Figure 5.

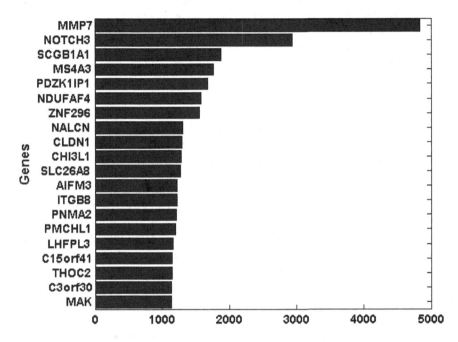

Fig. 5. The first 20 most popular genes in 100 different GAs runs

We assessed the functional characteristics of the resulting set of genes using the Ingenuity Pathway Analysis suite. Out of the top 15 genes 7 genes were found to be associated with cancer (MMP7, NOTCH3, PDZK1IP1, NDUFAF4, CLDN1, NALCN, PNMA2). The top ranking gene, MMP7, has previously been reported to be overexpressed in pancreatic ductal adenocarcinomas and not in normal pancreatic tissue [16-17] and is believed to apply apoptotic pressure to epithelial cells. The second best ranking gene, NOTCH3, is part of the Notch signaling pathway, which is well studied in carcinogenesis of many different types of cancer [18-19]. NALCN, together with other genes involved in axon guidance, has recently been shown by Blankin et al. to show a significant higher number of aberrations in pancreatic cancer compared to control [20]. The remaining cancer-related genes (PDZK1IP1, NDUFAF4, CLDN1, PNMA2) were associated with non-pancreatic cancers or involved in apoptosis and/or cell proliferation pathways.

4   Conclusions and Future Work

In this manuscript a GA has been proposed, developed and applied on PDAC data to classify tissue samples. Our benchmark shows that the method results in robust classifiers for pancreatic cancer. In addition, the algorithm provides a list of biomarkers that play the most important role in this lethal disease as a by-product. However, due to stochastic nature of the GA, we decided to statistically measure the importance of the genes by measuring their appearances on "good" solutions generated by the GA in all our experiments, which is also suggested procedure if one should use the method for biomarker mining. Our results were verified by the biological significance of these genes in specific biological functions in human organism.

Our future plans include the optimization of the classification process of tissue samples by using the generated "popular" gene list as biomarkers and trying to optimize different kind of classifiers through a new genetic algorithm. Moreover, to further examine robustness of the method we are going to apply the resulting classifier on additional gene expression datasets on this type of cancer.

Acknowledgments. The authors would like to acknowledge support from:
Research Council KU Leuven: GOA/10/09 MaNet, KUL PFV/10/016 SymBioSys, START 1, OT 09/052 Biomarker, several PhD/postdoc & fellow grants. Industrial Research fund (IOF): IOF/HB/10/039 Logic Insulin, IOF: HB/12/022 Endometriosis. Flemish Government: FWO: PhD/postdoc grants, projects: G.0871.12N (Neural circuits), research community MLDM, IWT: PhD Grants; TBM-Logic Insulin, TBM Haplotyping, TBM Rectal Cancer, Hercules Stichting: Hercules III PacBio RS, iMinds: SBO 2013; Art&D Instance, IMEC: phd grant. Federal Government: FOD: Cancer Plan 2012-2015 KPC-29-023 (prostate). COST: Action BM1104: Mass Spectrometry Imaging, Action BM1006: NGS Data analysis network.

References

1. Jemal, A., Siegel, R., Xu, J., Ward, E.: Cancer statistics. CA Cancer J. Clin. 60(5), 277–300 (2010)
2. Brandt, R., Grutzmann, R., Bauer, A., Jesnowski, R., Ringel, J., Lohr, M., Pilarsky, C., Hoheisel, J.D.: DNA microarray analysis of pancreatic malignancies. Pancreatology 4(6), 587–597 (2004)
3. Bauer, A.S., Keller, A., Costello, E., Greenhalf, W., Bier, M., Borries, A., Beier, M., Neoptolemos, J., Buchler, M., Werner, J., Giese, N., Hoheisel, J.D.: Diagnosis of pancreatic ductal adenocarcinoma and chronic pancreatitis by measurement of microRNA abundance in blood and tissue. PLoS One 7(4), e34151 (2012)
4. Bussom, S., Saif, M.W.: Methods and rationale for the early detection of pancreatic cancer. In: Highlights from the 2010 ASCO Gastrointestinal Cancers Symposium, Orlando, FL, USA, January 22-24, vol. 11(2), pp. 128–130. JOP (2010)

5. Phan, J.H., Moffitt, R.A., Stokes, T.H., Liu, J., Young, A.N., Nie, S., Wang, M.D.: Convergence of biomarkers, bioinformatics and nanotechnology for individualized cancer treatment. Trends Biotechnol. 27(6), 350–358 (2009)
6. Edgar, R., Domrachev, M., Lash, A.E.: Gene Expression Omnibus: NCBI gene expression and hybridization array data repository. Nucleic Acids Res. 30(1), 207–210 (2002)
7. Badea, L., Herlea, V., Dima, S.O., Dumitrascu, T., Popescu, I.: Combined gene expression analysis of whole-tissue and microdissected pancreatic ductal adenocarcinoma identifies genes specifically overexpressed in tumor epithelia. Hepatogastroenterology 55(88), 2016–2027 (2008)
8. Valentini, G., Tagliaferri, R., Masulli, F.: Computational intelligence and machine learning in bioinformatics. Artif. Intell. Med. 45(2-3), 91–96 (2009)
9. Bandyopadhyay, S., Pal, S.K.: Classification and Learning Using Genetic Algorithms: Applications in Bioinformatics and Web Intelligence. Natural Computing Series, vol. 311. Springer (2007)
10. Bäck, T., Fogel, D.B., Michalewicz, Z., Beck, T.: Evolutionary Computation 1: Basic Algorithms and Operators. Institute of Physics Publishing, Bristol (2000)
11. Bäck, T., Fogel, D.B., Michalewicz, Z., Beck, T.: Evolutionary Computation 2: Advanced Algorithms and Operators. Institute of Physics Publishing, Bristol (2000)
12. Fishman, G.S.: Monte Carlo: Concepts, Algorithms, and Applications. Springer, New York (1995)
13. Baker, J.E.: Reducing Bias and Inefficiency in the Selection Algorithm. In: 2nd International Conference on Genetic Algorithms and their Application, Cambridge, Massachusetts, USA, pp. 14–21 (1987)
14. Parker, C.: An Analysis of Performance Measures for Binary Classifiers. In: IEEE 11th International Conference on Data Mining (ICDM), Corvallis, OR, USA, pp. 517–526 (2011)
15. Xu, K., Cui, J., Olman, V., Yang, Q., Puett, D., Xu, Y.: A comparative analysis of gene-expression data of multiple cancer types. PLoS One 5(10), e13696 (2010)
16. Crawford, H.C., Scoggins, C.R., Washington, M.K., Matrisian, L.M., Leach, S.D.: Matrix metalloproteinase-7 is expressed by pancreatic cancer precursors and regulates acinar-to-ductal metaplasia in exocrine pancreas. J. Clin. Invest. 109(11), 1437–1444 (2002)
17. Tan, X., Egami, H., Abe, M., Nozawa, F., Hirota, M., Ogawa, M.: Involvement of MMP-7 in invasion of pancreatic cancer cells through activation of the EGFR mediated MEK-ERK signal transduction pathway. J. Clin. Pathol. 58(12), 1242–1248 (2005)
18. Doucas, H., Mann, C.D., Sutton, C.D., Garcea, G., Neal, C.P., Berry, D.P., Manson, M.M.: Expression of nuclear Notch3 in pancreatic adenocarcinomas is associated with adverse clinical features, and correlates with the expression of STAT3 and phosphorylated Akt. J. Surg. Oncol. 97(1), 63–68 (2008)
19. Vo, K., Amarasinghe, B., Washington, K., Gonzalez, A., Berlin, J., Dang, T.P.: Targeting notch pathway enhances rapamycin antitumor activity in pancreas cancers through PTEN phosphorylation. Mol. Cancer 10, 138 (2011)

20. Biankin, A.V., Waddell, N., Kassahn, K.S., Gingras, M.C., Muthuswamy, L.B., Johns, A.L., Miller, D.K., Wilson, P.J., Patch, A.M., Wu, J., Chang, D.K., Cowley, M.J., Gardiner, B.B., Song, S., Harliwong, I., Idrisoglu, S., Nourse, C., Nourbakhsh, E., Manning, S., Wani, S., Gongora, M., Pajic, M., Scarlett, C.J., Gill, A.J., Pinho, A.V., Rooman, I., Anderson, M., Holmes, O., Leonard, C., Taylor, D., Wood, S., Xu, Q., Nones, K., Fink, J.L., Christ, A., Bruxner, T., Cloonan, N., Kolle, G., Newell, F., Pinese, M., Mead, R.S., Humphris, J.L., Kaplan, W., Jones, M.D., Colvin, E.K., Nagrial, A.M., Humphrey, E.S., Chou, A., Chin, V.T., Chantrill, L.A., Mawson, A., Samra, J.S., Kench, J.G., Lovell, J.A., Daly, R.J., Merrett, N.D., Toon, C., Epari, K., Nguyen, N.Q., Barbour, A., Zeps, N., Kakkar, N., Zhao, F., Wu, Y.Q., Wang, M., Muzny, D.M., Fisher, W.E., Brunicardi, F.C., Hodges, S.E., Reid, J.G., Drummond, J., Chang, K., Han, Y., Lewis, L.R., Dinh, H., Buhay, C.J., Beck, T., Timms, L., Sam, M., Begley, K., Brown, A., Pai, D., Panchal, A., Buchner, N., De Borja, R., Denroche, R.E., Yung, C.K., Serra, S., Onetto, N., Mukhopadhyay, D., Tsao, M.S., Shaw, P.A., Petersen, G.M., Gallinger, S., Hruban, R.H., Maitra, A., Iacobuzio-Donahue, C.A., Schulick, R.D., Wolfgang, C.L., Morgan, R.A., Lawlor, R.T., Capelli, P., Corbo, V., Scardoni, M., Tortora, G., Tempero, M.A., Mann, K.M., Jenkins, N.A., Perez-Mancera, P.A., Adams, D.J., Largaespada, D.A., Wessels, L.F., Rust, A.G., Stein, L.D., Tuveson, D.A., Copeland, N.G., Musgrove, E.A., Scarpa, A., Eshleman, J.R., Hudson, T.J., Sutherland, R.L., Wheeler, D.A., Pearson, J.V., McPherson, J.D., Gibbs, R.A., Grimmond, S.M.: Pancreatic cancer genomes reveal aberrations in axon guidance pathway genes. Nature 491(7424), 399–405 (2012)

Enhanced Weighted Restricted Neighborhood Search Clustering: A Novel Algorithm for Detecting Human Protein Complexes from Weighted Protein-Protein Interaction Graphs

Christos Dimitrakopoulos[1], Konstantinos Theofilatos[1], Andreas Pegkas[1],
Spiros Likothanassis[1], and Seferina Mavroudi[1,2]

[1] Department of Computer Engineering and Informatics, University of Patras, Greece
[2] Department of Social Work, School of Sciences of Health and Care,
Technological Educational Institute of Patras, Greece
{dimitrakop,theofilk,pegkas,likothan,mavroudi}@ceid.upatras.gr

Abstract. Proteins and their interactions have been proven to play a central role in many cellular processes. Although there are many experimental techniques for protein-protein interaction prediction, only a few exist for predicting protein complexes. For the sake of this, researchers have emphasized lately in the computational prediction of protein complexes from Protein-Protein Interaction (PPI) data. The two major limitations of the current advances in the prediction of protein complexes are that most of the algorithms do not take into consideration the participation of a protein to many protein complexes and that they cannot handle weighted PPI graphs. In the present paper, we altered the original Restricted Neighborhood Search Clustering (RNSC) algorithm to overcome the above limitations. The Enhanced Weighted Restricted Neighborhood Search Clustering (EWRNSC) permits the participation of a protein to many protein complexes by modifying the moves of the original RNSC. In addition, EWRNSC can accept and process weighted PPI graphs as inputs by altering the cost functions of the original RNSC cost clustering schemes. When experimented using atasets from Human, the proposed algorithm proved to outperform the original RNSC and the MCL algorithms which are two of the most broadly used methods in the field of protein complexes prediction.

Keywords: Protein Complexes Prediction, Human, Cost-based Clustering, Clustering Protein-Protein Interaction Networks, Weighted Protein-Protein Interaction Networks.

1 Introduction

Proteins are nowadays considered to be the most important participants in molecular interactions. Specifically, they play a significant role in almost all the cellular functions such as regulatory signals transmission in the cell and they catalyze a huge number of chemical reactions. Except for functioning alone, proteins are also combined to each other in functional modules called protein complexes. The prediction of the protein complexes is crucial for understanding the cellular

L. Iliadis, H. Papadopoulos, and C. Jayne (Eds.): EANN 2013, Part II, CCIS 384, pp. 231–240, 2013.
© Springer-Verlag Berlin Heidelberg 2013

mechanisms and for predicting the function of uncharacterized proteins. The experimental prediction of protein complexes is mainly limited to Tandem Affinity Purification (TAP) [1] which provide erroneous data and demand high cost without being time-efficient.

Because of the above fact, researchers have emphasized lately in the computational prediction of protein complexes from Protein-Protein Interaction (PPI) data. To achieve this goal, several clustering methods have been applied to the protein interactome graph in order to detect highly connected subgraphs [2,3,4]. These algorithms rely on very different approaches. Each of them requires specifying several parameters, some of which may drastically affect the results.

The Restricted Neighborhood Search Clustering (RNSC) [17], is a cost-based local heuristic search algorithm that explores the solution space to minimize a cost function, calculated according to the numbers of intra-cluster and inter-cluster edges. Starting from an initial random solution, RNSC iteratively moves a vertex from one cluster to another if this move reduces the general cost. When a (user-specified) number of moves has been reached without decreasing the cost function, the program ends up. The algorithm is analytically described in section 2.2.

In the present paper, we propose a fully unsupervised clustering algorithm, EWRNSC, which is an enhancement of the original Restricted Neighborhood Search Clustering (RNSC) algorithm. The original RNSC algorithm was altered so that a) it permits the participation of a protein to many protein complexes b) the initial estimation of the clusters is allocated using an analytical estimation method and c) takes advantage of the information which lies within the weights of weighted PPI graphs. The participation of a protein to many protein complexes (in terms of clusters) was achieved by altering the moves of the original RNSC. Two new operators were added together with the move of a node from a cluster to another random cluster. The algorithm chooses each move with a given probability. The process of weighted PPI graphs as inputs was achieved by altering the cost functions of the original RNSC cost clustering schemes (section 2.3).

When experimented using public available protein complex datasets from Human, the proposed algorithm proved to outperform the original one. The proposed method was tested on one weighted PPI graph from Human using two evaluation datasets (section 2.1).

2   Materials and Methods

2.1   Protein-Protein Interactions Datasets

In the present paper a Human PPI dataset was used to build the PPI graph which is used as input for the protein complex prediction methods. The examined dataset is weighted where the value of a confidence score is assigned to each protein pair.

The PPI dataset for the Human organism consists of the protein interactions included in the HPRD database [6]. These protein interactions were filtered using the method proposed in [14] and a confidence score was assigned to each protein pair using the same methodology. In this way we achieved to incorporate sequential, functional and structural information on the extracted PPI graph. The extracted PPI graph consists of 7450 proteins and 21.475 interactions.

2.2   Enhanced Weighted Restricted Neighborhood Clustering Algorithm

The Enhanced Weighted RNSC is a novel algorithm for detecting protein complexes based on the original RNSC algorithm with two major improvements aiming at the increase of its efficiency as well as its flexibility. The first improvement is the modification of the metrics so that the process of weighted PPI graphs is enabled. The second is the modification of the algorithm so that the moving process enables the possibility of generating overlapping clusters.

As in the original RNSC, the Enhanced Weighted RNSC uses two cost functions for evaluating solutions, the naive cost function and the scaled cost function. The naive cost function is simple in its computation and is used as a preprocess solution tool. For computational matters, it processes the network and finds an initial approximative solution by ignoring the weights of the network. The cost function that determines the final solution is the scaled cost function which is computationally more pretentious as it takes into consideration the weights of the network.

Naive cost function: For each node α_v is computed, which is the sum of the nodes' "bad connections", naming the sum of the weights of the node's edges with nodes that belong to different clusters plus the sum of the weights of the nodes that belong to the same cluster with the subjective node and they are not connected to it (equation 2.1). Since the node might participate in more than one cluster we take the average of a_v over all clusters.

$$C_n(G,C) = \frac{1}{2} \sum_{u \in V} \frac{\sum_{u \in Cu} a_v}{|C_u|} \tag{2.1}$$

where V is the set of the nodes of graph G and C_u is the set of all the clusters that node u belongs.

Scaled cost function. Let us define:
- $w_{v,u}$ the weight that connects the nodes v and u
- C_v the cluster where the node v belongs

Then we define for each node v:

$$\gamma_v = \sum_{u \notin C_v} w_{v,u} + \sum_{u \in C_v} (1 - w_{v,u}) \tag{2.2}$$

Moreover, for each node v, if N(v) is the number of nodes connecting to it, then we define:

$$\beta_v = |N(v) \cup C_v| \tag{2.3}$$

which means that β_v is equal to the number of nodes that belong to either the "neighbors" of v or to the same cluster with v. If n is the total number of the graph's nodes, then the scaled cost function of EWRNSC is defined as:

$$C_s(G,C) = \frac{n-1}{3} \sum_{v \in V} \sum_{v \in C_u} \frac{\frac{\gamma_v}{\beta_v}}{|C_u|} \tag{2.4}$$

As in the original RNSC, the ultimate goal of the algorithm is the minimization of the cost functions where ideally the nodes of a cluster connect to each other (all to all) whereas they do not connect to any other nodes (nodes of other clusters). After the execution of the EWRNSC, we filtered out the clusters with size equal to one protein.

EWRNSC has also many parameters which need to be tuned. The number of the initial random clusters of the algorithm was chosen based on the number of the benchmark dataset's clusters. Other parameters like tabu list length, diversification parameters and stopping tolerance were chosen empirically based on the size of the input PPI graph as described in [4]. In specific, the parameters are the number of different experiments that we ran the algorithm (10), the length of the list with the forbidden moves (50), the frequency of the random diversification moves (50), the number of nodes shuffled when diversification is performed (10), and the number of moves without improvement in cost for the naive and the scale schemes (10).

Our intention is to transform RNSC algorithm to a new form that takes into consideration the participation of the proteins to more than one clusters. For the sake of this transformation, we alter the moving procedures conducted during the naive and the scaled cost scheme. In the original RNSC one node randomly moves from one cluster to another. In the Enhanced Weighted RNSC one node has the possibility to perform one of three following operators:

- **Operator 1**: The node is <u>moved</u> from its cluster i to a random cluster j ~ i with probability Pr_mov.
- **Operator 2**: The node is <u>copied</u> from its cluster i to a random cluster j ~ i with probability Pr_cop.
- **Operator 3**: The node is <u>deleted</u> from its cluster i with probability Pr_del.

In general, the probabilities Pr_cop and Pr_del must be lower that the Pr_mov probability so that we do not fall into high cluster overlapping. Various experimentations have been tested for different values of the three probabilities and they are presented in Section 3. By using the above operators one node is allowed to move to another cluster and at the same time to remain to its current cluster (operator 2). Operator 3 is a prerequisite operator generated by the existence of operator 2, because its absent would result to the creation of extremely large -extremely overlapping clusters. In Operators 1 and 2 the node is moved or copied to another cluster or to a singleton cluster with equal probability. In Operator 3 the node is deleted with equal probability from one of the clusters that it belongs to. As in the original RNSC, we consider moving one node only to the clusters of its neighbors (or to a singleton cluster). A move to a cluster that contains none of its neighbors never does as well, in terms of either the naive cost function or the scaled cost function, as moving the vertex to an empty cluster.

The initial estimation of the clusters is allocated using an analytical estimation method. We consider the datasets for the protein complexes of Yeast as the most valuable, due to the fact that the existing knowledge about protein complexes of the yeast organism is in satisfactory levels (compared to the Human dataset). For the Yeast organism, there are three well studied protein complex datasets. The first is the BT_409 dataset [8] which contains of 409 protein complexes. The second is the Aloy dataset [8] which contains 101 protein complexes derived using structure based protein matching with known structures and screened with the electron microscopy method. The third dataset is the Pu dataset [9] which contains 408 protein complexes. The three datasets contain in total 811 complexes (without duplications). The most known current available PPI dataset for the Yeast (Saccharomyces cerevisiae) was published by [7] and contains information about 5195 proteins. We use the proportion of proteins and complexes for Yeast as a reference (5195 proteins and 811 complexes) to compute the expectation complexes for Human. In this way, we set the number of 1163 initial clusters.

The EWRNSC algorithm was built in Matlab R2010b. The algorithm's pseudocode follows:

```
Input:
    PPI network (an undirected weighted graph) G(V,E)
    Number of experiments: Ne=10
    Tabu Length = 50
    Diversification frequency = 50
    Diversification length = 10
    T_n = 10
    T_s = 10
Output:
    The predicted protein clusters: Clusters

Algorithm:
    Initialize clustering based on analytical method
    for 1 to Ne
        Call Naive_Scheme
        Call Scaled_Scheme
    end
    Store the best Clustering of the Experiments: Clusters

Routine Naive_Scheme:
    Until Best cost has improved in the last T_n moves:
        Choose a Node not in Tabu List.
        Make a Move with Probability 0.8 that decreases the total Naive Cost.
        Make a Copy with Probability 0.1 that decreases the total Naive Cost.
        Make a Delete with Probability 0.1 that decreases the total Naive Cost.
        (destroy or create clusters in the process)
        Update Tabu List
    end
    Store the best Naive Clustering

Routine Scaled_Scheme:
    Until Best cost has improved in the last T_s moves:
        Choose a Node not in Tabu List.
        Make a Move with Probability 0.8 that decreases the total Scaled Cost.
        Make a Copy with Probability 0.1 that decreases the total Scaled Cost.
        Make a Delete with Probability 0.1 that decreases the total Scaled Cost.
        (destroy or create clusters in the process)
        Update Tabu List
    end
    Store the best Scaled Clustering
```

2.3   Evaluation Metrics

Sensitivity, positive predictive value (*PPV*) and geometric accuracy are classically used to measure the correspondence between the result of a clustering on a set of reference complexes. Considering the annotated complexes as a reference classification, *sensitivity* is defined as the fraction of proteins of complex i which are found in cluster j. To characterize the general sensitivity of a clustering result, the *clustering-wise sensitivity is computed* as the weighted average of Sn_{co_i} over all complexes.

$$Sn = \frac{\sum_{i=1}^{n} N_i Sn_{co_i}}{\sum_{i=1}^{n} N_i} \tag{2.5}$$

Defined on one cluster, the positive predictive value is the proportion of members of cluster j which belong to complex i, relative to the total number of members of this cluster assigned to all complexes. To characterize the general PPV of a clustering result as a whole, we compute a *clustering-wise* PPV as the weighted average of the individual PPV_{cl_j} of all clusters.

$$PPV = \frac{\sum_{j=1}^{m} T_{.j} PPV_{cl_j}}{\sum_{j=1}^{m} T_{.j}} \tag{2.6}$$

The *geometric accuracy* (*Acc*) indicates the tradeoff between sensitivity and predictive value. It is obtained by computing the geometrical mean of the Sn and the *PPV*.

$$Acc = \sqrt{Sn * PPV} \tag{2.7}$$

The advantage of taking the geometric rather than arithmetic mean is that it yields a low score when either the Sn or the PPV metric is low and as a result it balances better the tradeoff between the two metrics. The sensitivity and PPV individually give a false idea of quality in the trivial cases where all proteins are assigned to a single cluster ($Sn = 1 \Rightarrow Acc > 0.5$) or where, on the contrary, each protein is assigned to a single-element cluster ($PPV = 1 \Rightarrow Acc > 0.5$). To avoid these erroneous interpretations we also have used the Separation metric [11] which takes into consideration the fact that clustering predictions with fewer known complexes should be regarded as the ones with the higher quality.

2.4   Evaluation Datasets

We built two different protein complex datasets by filtering out the protein complexes which are published in CORUM [12]. The first human dataset (443 protein complexes) was created by filtering out all protein complexes which include a protein that is not present in interactions annotated in the HPRD database. The second human evaluation dataset (1097 protein complexes) consists of protein complexes in CORUM when filtering out all complexes with more than half of their proteins not included in the HPRD PPIs. Protein complexes with one protein are considered as poor interconnecting components and are discarded.

3   Experimental Results and Discussion

EWRSNC focuses on local searching and as a result we used diversification moves and multiple experiments to aid it in escaping from local optimum solutions. In specific, we ran 10 different experiments and calculated the mean values for its evaluation metrics. For the probabilities of the operator moves we chose 0.8 for operator 1, 0.1 for operator 2 and 0.1 for operator 3. 111 proteins were found to participate in different clusters. The performance of EWRNSC was compared with the performance of the original RNSC as well as with the performance of MCL [18] which is one of the most well-established and frequently used methods for the prediction of protein complexes. The results for these two algorithms were calculated using the Superclusteroid Tool [13] which uses the optimized values for their parameters as described in [11].

In Figures 3.1 and 3.2 we present the evaluation metrics of the aforementioned algorithms when evaluating their outcome with the two Human dataset. EWRNSC exceeds almost all the classical metrics of sensitivity, PPV and geometric accuracy compared to both MCL and RNSC (except for the PPV case in the CORUM Complexes with more than 50% in HPRD benchmark dataset) maintaining a small improvement in the separation metric. In specific, the improvement for the geometric accuracy metric varies between 6.8% and 19,1% for the original RNSC and between 4,8% and 11,4% for the MCL algorithm. The minor improvement in the separation metric can be attributed to the absence of overlapping clusters in Human evaluation datasets.

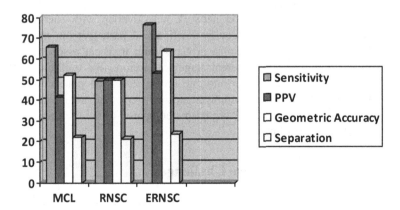

Fig. 3.1. Comparative Results for the Human organism (CORUM Complexes with only proteins included in HPRD)

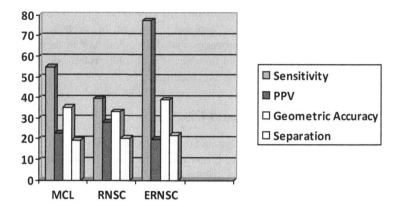

Fig. 3.2. Comparative Results for the Human organism (CORUM Complexes with more than 50% proteins included in HPRD)

For the reasons described in section 2.4, we consider the three classical measurements (Sensitivity, PPV, geometric accuracy) as biased. Separation is the only metric that takes into consideration the participation of one protein to more than one protein complexes (by avoiding duplications during computation) and as a result we consider as highly important the EWRNSC's trend on better separation results.

We can mainly attribute the improvements of the EWRNSC algorithm to its handling of the edges' weights. All methods for predicting Human protein-protein interactions are known to yield a nonnegligible amount of noise (false positives) and to miss a fraction of existing interactions (false negatives) [11]. Therefore, the protein interaction data available for clustering are very noisy. The framework of the EWRNSC that incorporates weighted PPI graphs definitely faces that problem directly by evaluating the confidence of each PPI.

Clusters of a protein interaction network may overlap with each other. Most proteins have more than one molecular function and participate in more than one biological process. For example, some proteins form transient associations and are part of several complexes at different stages. Therefore, the traditional clustering approaches of putting each protein into one single cluster do not suit this problem well. Hence, the EWRNSC algorithm creates clusters that are closer to the real interaction models that exist in the organisms compared to the clusters produced by the MCL and RNSC algorithms.

4   Conclusion and Future Challenges

In the post-genomic era, an important issue is to analyze biological systems at the network level, in order to understand the topological organization of protein interaction networks, identify protein complexes and functional modules, discover functions of uncharacterized proteins, and obtain more exact networks. To achieve this aim, a series of clustering approaches have been proposed.

In the context of the present paper, we proposed an unsupervised clustering algorithm, EWRNSC, which is an enhancement of the original Restricted Neighborhood Search Clustering (RNSC) algorithm which was altered in order to permit the participation of a protein to many protein complexes and to handle the input of weighted PPI graphs. The participation of a protein to many protein complexes (in terms of clusters) was achieved by altering the moves of the original RNSC. To handle weighted graphs, we have affected RNSC's cost functions. Two new operators were added together with the move of a node from a cluster to another random cluster. Moreover, the initial estimation of the clusters is allocated using an analytical estimation method.

As a future work we intend to implement EWRNSC method to more available datasets of PPI networks and known protein complexes from different organisms. It is a matter of robustness to prove the superiority of the method to more than one datasets. Moreover, our goal is to also compare our method with other promising algorithms [7, 15, 16] except for the well-established ones such as MCL and RNSC.

Enhanced Weighted RNSC algorithm has a large number of parameters that are chosen empirically [5] as there is little a priori knowledge for them (cluster number, tabu list length and diversification length). As a future direction, the effective tuning of the parameters within the heuristic procedure of the algorithm would give an improved clustering solution.

Current clustering approaches mainly focus on detecting clusters in static protein interaction networks. However, both the protein-protein interactions and protein complexes are dynamically organized when implementing special functions. Dynamic modules generally correspond to the sequential ordering of molecular events in cellular systems. The way to explore dynamic modules from static protein interaction networks is a very difficult task and should definitely be addressed by the EWRNSC algorithm in a future direction.

Acknowledgments. This research has been co-financed by the European Union (European Social Fund - ESF) and Greek national funds through the Operational Program "Education and Lifelong Learning" of the National Strategic Reference Framework (NSRF) - Research Funding Program: Heracleitus II. Investing in knowledge society through the European Social Fund.

References

1. Theofilatos, K.A., Dimitrakopoulos, C.M., Tsakalidis, A.K., Likothanassis, S.D., Papadimitriou, S.T., Mavroudi, S.P.: Computational Approaches for the Prediction of Protein-Protein Interactions: A Survey. Current Bioinformatics 6(4), 398–414 (2011)
2. Spirin, V., Mirny, L.A.: Protein complexes and functional modules in molecular networks. Proc. Natl. Acad. Sci. USA 100(21), 12123–12128 (2003)
3. Bader, G.D., Hogue, C.W.V.: An automated method for finding molecular complexes in large protein interaction networks. BMC Bioinformatics 4, 2 (2003)
4. Tamas, N., Haiyuan, Y., Alberto, P.: Detecting overlapping protein complexes in protein-protein interaction nteworks. Nature Methods 9, 471–472 (2012)

5. King, A.D.: Graph clustering with restricted neighbourhood search. Master's thesis, University of Toronto (2004)
6. Keshava Prasad, T.S., Goel, R., Kandasamy, K., Keerthikumar, S., Kumar, S., Mathivanan, S., Telikicherla, D., Raju, R., Shafreen, B., Venugopal, A., Balakrishnan, L., Marimuthu, A., Banerjee, S., Somanathan, D.S., Sebastian, A., Rani, S., Ray, S., Harrys Kishore, C.J., Kanth, S., Ahmed, M., Kashyap, M.K., Mohmood, R., Ramachandra, Y.L., Krishna, V., Rahiman, B.A., Mohan, S., Ranganathan, P., Ramabadran, S., Chaerkady, R., Pandey, A.: Human Protein Reference Database—2009 update. Nucleic Acids Research 37, D767- D772 (2009)
7. Theofilatos, K., Pavlopoulou, N., Papasavvas, C., Likothanassis, S., Dimitrakopoulos, C., Georgopoulos, E., Moschopoulos, C., Mavroudi, S.: Evolutionary Enhanced Markov Clustering - EEMC: A Novel Unsupervised Methodology for Predicting Protein Complexes From Weighted Protein-Protein Interaction Graphs. In: Artificial Intelligence in Medicine (submitted)
8. Friedel, C., Krumsiek, J., Zimmer, R.: Bootstrapping the Interactome: Unsupervised Identification of Protein Complexes in Yeast. Journal of Computational Biology 16(8), 971–987 (2009)
9. Aloy, P., Bottcher, B., Ceulemans, H., Leutwein, C., Mellwig, C., Fischer, S., Gavin, A.C., Bork, P., Superti-Furga, G., Serrano, L., Russell, R.B.: Structure-based assembly of protein complexes in yeast. Science 303, 2026–2029 (2004)
10. Pu, S., Vlasblom, J., Emili, A., et al.: Identifying functional modules in the physical interactome of Saccharomyces cerevisiae. Proteomics 7, 944–960 (2007)
11. Brohee, S., van Helden, J.: Evaluation of clustering algorithms for protein-protein interaction networks. BMC Bioinformatics 7, 488 (2006)
12. Ruepp, A., Waegele, B., Lechner, M., Brauner, B., Dunger-Kaltenbach, I., Fogo, G., Frishman, G., Montrone, C., Mewes, H.W.: CORUM: the comprehensive resource of mammalian protein complexes—2009. Nucleic Acids Research 38(database issue), D497–D501 (2009)
13. Ropodi, A., Sakkos, N., Moschopoulos, C., Magklaras, G., Kossida, S.: Superclusteroid: a Web tool dedicated to data processing of protein-protein interaction networks. EMBnet Journal 17(2), 10–15 (2011)
14. Theofilatos, K., Dimitrakopoulos, C., Kleftogiannis, D., Moschopoulos, C., Papadimitriou, S., Likothanassis, S., Mavroudi, S.: HINT-KB: The human interactome knowledge base. In: Iliadis, L., Maglogiannis, I., Papadopoulos, H., Karatzas, K., Sioutas, S. (eds.) Artificial Intelligence Applications and Innovations, Part II. IFIP AICT, vol. 382, pp. 612–621. Springer, Heidelberg (2012)
15. Wang, X., Zhengzhi, W., Jun, Y.: HKC: An Algorithm to Predict Protein Complexes in Protein-Protein Interaction Networks. Journal of Biomedicine and Biotechnology, Article ID 480294, 14 pages (2011), doi:10.1155/2011/480294
16. Wu, M., Li, X., Kwoh, C.K., Ng, S.K.: A core-attachment based method to detect protein complexes in PPI networks. BMC Bioinformatics 10, 169 (2009), doi:10.1186/1471-2105-10-169
17. King, A.D., Przulj, N., Jurisica, I.: Protein complex prediction via cost-based clustering. Bioinformatics 20(17), 3013–3020 (2004)
18. Van Dongen, S.: Graph clustering by flow simulation. In: PhD thesis Centers for mathematics and computer science (CWI), University of Utrecht (2000)

A Hybrid Approach to Feature Ranking
for Microarray Data Classification

Dusan Popovic, Alejandro Sifrim, Charalampos Moschopoulos,
Yves Moreau, and Bart De Moor

ESAT-SCD / iMinds-KU Leuven Future Health Department, KU Leuven,
Kasteelpark Arenberg 10, Box 2446, 3001, Leuven, Belgium
{Dusan.Popovic,Alejandro.Sifrim,Charalampos.Moschopoulos,
Yves.Moreau,Bart.DeMoor}@esat.kuleuven.be

Abstract. We present a novel approach to multivariate feature ranking in context of microarray data classification that employs a simple genetic algorithm in conjunction with Random forest feature importance measures. We demonstrate performance of the algorithm by comparing it against three popular feature ranking and selection methods on a colon cancer recurrence prediction problem. In addition, we investigate biological relevance of the selected features, finding functional associations of corresponding genes with cancer.

Keywords: genetic algorithms, Random forest, feature ranking, feature selection, gene prioritization, microarrays, classification, cancer.

1   Background

High-throughput technologies, such are mass-spectrometry, microarrays and next-generation sequencing, recently empowered biomedical researchers with capabilities to study biological phenomena at the molecular level. Consequently, these technological advances facilitated development of targeted therapies and non-invasive diagnostic tests for certain diseases [1]. However, ever-increasing utilization of these techniques also gave rise to plethora of new problems that seriously challenge traditional views on data analysis. The data sets resulting from high-throughput experiments are often characterized by a large number of highly correlated features, severe signal/noise ratios and a small number of biologically very heterogeneous samples. The described problems are especially prominent when analyzing diseases that display complex patterns of molecular changes, such as cancer. These issues eventually promoted extensive utilization of machine learning algorithms in the field. In this work we present a method that aids identification of relevant predictor variables in this context by combining optimization capabilities of genetic algorithms with robust Random forest feature importance estimation.

Genetic algorithms [2],[3] (GA) are class of search and optimization meta-heuristics inspired by the process of natural selection. They represent a potential solution to the problem at hand as an individual (also called chromosome) that is defined

L. Iliadis, H. Papadopoulos, and C. Jayne (Eds.): EANN 2013, Part II, CCIS 384, pp. 241–248, 2013.

over several, usually binary, variables, called genes. A set of individuals constitutes a population, from which the fittest individuals are selected to be combined and sometimes otherwise altered in order to produce a next generation of solutions. The fitness value reflects the desired quantitative aspect(s) of an individual, and is obtained trough application of a user-defined fitness function. This is essentially an iterative stochastic process that terminates when the optimization objective is achieved or when certain stopping criterion is met. Applications of the genetic algorithms in bioinformatics include amongst others: multiple sequence alignment [4], RNA structure prediction [5] and biomarker discovery [6].

The Random forest [7] (RF) is popular classification method that has been applied to numerous scientific fields so far. It essentially operates by constructing an ensemble of fully-grown decision trees built on different bootstraps extracted from the data, with additional randomness injected in algorithm by selecting splitting variables from random subsets of all possible candidate splits. Random forests are especially suitable for dealing with problems characterized by high dimensionality and severe correlation between predictors. They also produces internally estimated feature importance scores (for discussion on different methods together with their possible biases see [8]) that take into account interactions between features, which is a utility of great interest in many applications. The later capability is often exploited for multivariate feature selection, as well as in explanatory analyses ("opening a black-box"). These two advantages are the main reason behind growing popularity of the method in bioinformatics, where it has been used for analysis of microarray data [9] and DNA sequencing [10] , among other applications.

2   Materials and Methods

2.1   A Hybrid Approach to Feature Ranking Problem

The proposed algorithm is depicted in Figure 1. It is essentially wrapper approach to feature selection/ranking [11]. We represent subset of biological genes to be used for subsequent classification (chromosome in GA) as a vector of binary values. The initial generation is created randomly, with probability of 1% for each feature to be selected in single chromosome. Prior to a GA generation run we take class-balanced bootstrap [12] from the original data, after which subsets are created for every chromosome given their particular selections of features. These subsets are then used for training Random forest classifiers, whose out-of-bag accuracy values are then fed back to the genetic algorithm as a fitness of corresponding chromosomes. We apply bootstrapping aiming to mitigate the GA potential for over-fitting, which is an especially severe problem when the later is used in wrapper-based feature selection context [13]. Also, this procedure should promote "longevity" of solutions that are robust against small perturbations in data, and thus generalize well. Settings of other parameters of GA and RF are provided in Table 1.

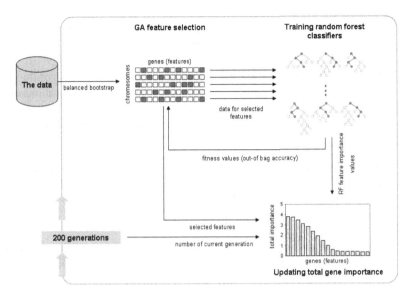

Fig. 1. Workflow of the feature ranking using hybrid approach

When GA converges to the general region of maximal fitness; longevity and classification performance of particular genes are continuously rewarded until the end of algorithm execution. In particular, the importance of a biological gene is increased by corresponding RF-FI value during every generation and for every chromosome where the gene is present. This principle essentially mimics usage of gene conservation scores for prediction [14]. Formally, if s_{ijk} is a binary function that indicates if feature j ($j=1..f$, here f stands for number of biological genes) is selected in chromosome i ($i = 1..c$, here $c=100$) during generation k ($k = 0..g$, here $g=200$), r_{ijk} is the random forest feature importance measure (note that $r_{ijk}=0$ if $s_{ijk}=0$) and **1** is an unit vector of size $1xf$; vector of importance across all of the genes and for a single generation becomes :

$$GI_k = \mathbf{1} \left(\begin{bmatrix} s_{11k} & s_{12k} & \cdot & \cdot & s_{1fk} \\ s_{21k} & s_{22k} & \cdot & \cdot & s_{2fk} \\ \cdot & \cdot & & & \cdot \\ s_{c1k} & s_{c2k} & \cdot & \cdot & s_{cfk} \end{bmatrix} \circ \begin{bmatrix} r_{11k} & r_{12k} & \cdot & \cdot & r_{1fk} \\ r_{21k} & r_{22k} & \cdot & \cdot & r_{2fk} \\ \cdot & \cdot & & & \cdot \\ r_{c1k} & r_{c2k} & \cdot & \cdot & r_{cfk} \end{bmatrix} \right) \tag{1}$$

That is, values of s_{ijk} multiplied by r_{ijk} are summed across chromosomes for each gene, resulting in a gene importance vector for given generation. Accordingly, the final gene importance vector is:

$$GI = \sum_{k=100}^{g} GI_{k} ,$$ (2)

where k starts from 100, which is the approximate moment when the algorithm starts exploring the general area of a solution (half of the total number of generations).

Finally, to assess utility of our method for aiding microarray classification we compare it against three additional feature selection methods that are often used in microarray analyses:

- *Wilcoxon rank-sum test:* Features are ranked by values of the test obtained by comparing a vector of single variable values corresponding to positive cases and that corresponding to negatives. Final subset of features can be than selected according to predefined p-value or cut-off.
- *t-test:* Similarly to former.
- *Random forest feature importance:* The features are ranked by difference between out-of-bag mean square error obtained by single trees when values of a feature are shuffled to that achieved on unaffected data. To produce global estimate these values are averaged over the entire ensemble and divided by standard deviation. As before, the cut-off for selecting feature subset can be chosen arbitrary or by statistical modeling.

Table 1. The parameter setting of the genetic algorithm and random forest classifiers that are used within fitness function

	Parameter	Value
Genetic algorithm	Type of selection	Stochastic universal sampling
	Sigma scaling	On
	Sigma scaling coefficient	1
	Size of the population	100
	Number of generation	200
	Type of crossover	Uniform
	Probability of crossover	0.7
	Type of mutation	Simple (flipping a value of single bit)
	Probability of mutation per bit	0.5/number of genes
RF	Feature importance estimation method	Permutation accuracy importance
	Number of trees in ensemble	10
	Number of variables randomly selected for a split	Square root of total number of variables
	Minimal number of observations in a leaf	1

Once we have variables ordered by all four methods, we train classifier on the whole training set, selecting the first five, ten and fifty top-ranked features from each list. After this, resulting classifiers (12 in total) are tested against independent test set. The classifier of choice is a Random forest, with the number of trees set to one hundred.

2.2   Data Sets

We demonstrate our method using three publicly available microarray data sets of colon cancer samples that have been generated on Affymetrix HG U133 2.0 Plus platform. For all three, we consider cancer recurrence as an outcome of interest. Two data sets [15] have been merged to be used for training, while the third set [16] was used for testing. Further details on data can be found in Table 2.

Table 2. Details on the used data sets

Data set GEO accession no.	GSE17536	GSE17537	GSE5206
Author	Smith (MMC) [15]	Smith (VMC) [15]	Aronow [16]
Preprocessing method	MAS5	MAS5	RMA
No. positive outcomes	36	20	16
No. negative outcomes	109	35	58
Role in the study	a part of train-ing set	a part of train-ing set	the test set

3   Results and Discussion

The ROC curves obtained on the test set using four different feature selection methods and three different numbers of selected features are depicted in Figure 2. The corresponding AUC (area under the curve) values can be found in Table 3. It is immediately apparent from these that, in this particular setup, our method outperforms the other three regardless to the number of features selected. Furthermore, most of AUC gain obtained with the hybrid approach seems to be concentrated in regions of low false positive rates. This is often of great importance in the real-world applications, due to the high costs associated with confirmatory functional experiments.

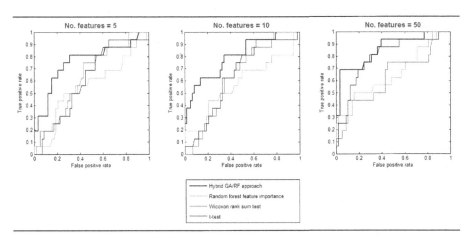

Fig. 2. ROC curves obtained on the test set using four different feature ranking methods and three different numbers of selected features

Additionally, these results are comparable or better than that reported in literature for the same classification problem. For example, the genetic signature reported in Wang et al. [17] achieves AUC of 0.74 on the data set that has been used in the study, while further refinement of the same method [18] reaches an AUC of 0.66 on an external validation set. Also, a study by Lin et al. [19] reports AUCs of 0.73 and 0.80 obtained on the two data sets using genetic signatures augmented with the clinical data.

Table 3. Values of area under the ROC curve for classifiers using different number of top-ranked variables as suggested by four feature ranking methods. The best values per number of selected features are indicated in bold.

Method	Number of features		
	5	10	50
Hybrid GA/RF approach ·	0.7575	0.7985	0.8572
Random forest feature importance	0.5571	0.5765	0.6546
Wilcoxon rank sum test	0.6342	0.6325	0.6352
T-test	0.6121	0.6433	0.7936

To investigate a possible functional relation of resulting highly ranked genes to colorectal cancer, we also performed a functional analysis using the Ingenuity Pathway Analysis suite. Out of the twenty top ranked genes, eight (ALDH1A3, DNAJA3, FAM65C, HOXA7, MCM8, TM4SF1, PXDN, SEC31A) have been reported to be cancer-related. ALDH1A3, DNAJA3 and HOXA7 play a role in apoptopis, an important hallmark in oncogenesis. However, TM4SF1 is the only gene to be reported specifically for colorectal cancer recurrence. Interestingly a gene of unknown function, TSPAN11, was found which belongs to the same protein family of tetraspanins as TM4SF1. We blasted the nucleotide sequence of TSPAN11 to find sequence paralogs. One of the top ranking hits was CD151 which has been shown to show differential expression in colorectal cancer [20]. However we can't exclude the possibility that the oligoprobe on the microarray platform shows aspecific hybridization in relation to CD151 and TSPAN11 due to their high sequence similarity.

4   Conclusions and the Future Work

We propose and demonstrate a novel method for feature ranking that combines genetic algorithm-facilitated search and Random forest feature importance measures. We tested it against three feature ranking algorithms in context of microarray-based colon cancer classification, achieving superior results in terms of area under the ROC curves. Furthermore, we observe functional association of several genes from top of our prioritized list with cancer, indicating that the method might be usable in wider context of biomarker discovery research.

However, as these genes have been judged predictive thought "guilt-by-association" rather than by proving their causality given the disease, further analyses are needed to establish the utility of the method beyond feature ranking. In the future, we also plan to test this hybrid approach in different high-throughput setups and for various biological classification problems. In addition, we will further investigate the idea of utilizing artificial "conservation scores" in optimization by genetic algorithm in general, perhaps for guiding the search process via a fitness function.

Acknowledgements. The authors would like to acknowledge support from:

- Research Council KU Leuven: GOA/10/09 MaNet, KUL PFV/10/016 SymBioSys, START 1, OT 09/052 Biomarker, several PhD/postdoc & fellow grants
- Industrial Research fund (IOF): IOF/HB/10/039 Logic Insulin, IOF: HB/12/022 Endometriosis
- Flemish Government:
  o FWO: PhD/postdoc grants, projects: G.0871.12N (Neural circuits), research community MLDM
  o IWT: PhD Grants; TBM-Logic Insulin, TBM Haplotyping, TBM Rectal Cancer
  o Hercules Stichting: Hercules III PacBio RS
  o iMinds: SBO 2013; Art&D Instance
  o IMEC: phd grant
- Federal Government: FOD: Cancer Plan 2012-2015 KPC-29-023 (prostate)
- COST: Action BM1104: Mass Spectrometry Imaging, Action BM1006: NGS Data analysis network

The scientific responsibility is assumed by its authors.

References

1. Glas, A.M., Floore, A., Delahaye, L.J., Witteveen, A.T., Pover, R.C., Bakx, N., Lahti-Domenici, J.S., Bruinsma, T.J., Warmoes, M.O., Bernards, R., Wessels, L.F., Van't Veer, L.J.: Converting a breast cancer microarray signature into a high-throughput diagnostic test. BMC Genomics 7, 278 (2006)
2. Fraser, A.: Simulation of genetic systems by automatic digital computers. I. Introduction. Aust. J. Biol. Sci. 10, 484–491 (1957)
3. Holland, J.H.: Adaptation in natural and artificial systems: an introductory analysis with applications to biology, control, and artificial intelligence. University of Michigan Press (1975)
4. Gondro, C., Kinghorn, B.P.: A simple genetic algorithm for multiple sequence alignment. Genetics and Molecular Research 6(4), 964–982 (2007) PMID 18058716
5. Van Batenburg, F.H., Gultyaev, A.P., Pleij, C.W.: An APL-programmed genetic algorithm for the prediction of RNA secondary structure. Journal of Theoretical Biology 174(3), 269–280 (1995) PMID 7545258, doi:10.1006/jtbi.1995.0098

6. Popovic, D., Sifrim, A., Pavlopoulos, G.A., Moreau, Y., De Moor, B.: A simple genetic algorithm for biomarker mining. In: Shibuya, T., Kashima, H., Sese, J., Ahmad, S. (eds.) PRIB 2012. LNCS, vol. 7632, pp. 222–232. Springer, Heidelberg (2012)

7. Breiman, L.: Random Forests. Machine Learning 45(1), 5–32 (2001)

8. Strobl, C., Boulesteix, A.L., Zeileis, A., Hothorn, T.: Bias in random forest variable importance measures: illustrations, sources and a solution. BMC Bioinformatics 8, 25 (2007)

9. Huang, X., Pan, W., Grindle, S., Han, X., Chen, Y., Park, S.J., Miller, L.W., Hall, J.: A comparative study of discriminating human heart failure etiology using gene expression profiles. BMC Bioinformatics 6, 205 (2005)

10. Bureau, A., Dupuis, J., Falls, K., Lunetta, K.L., Hayward, B., et al.: Identifying SNPs predictive of phenotype using random forests. Genetic Epidemiology 28, 171–182 (2005)

11. Saeys, Y., et al.: A review of feature selection techniques in bioinformatics. Bioinformatics 23, 2507–2517 (2007)

12. Efron, B., Tibshirani, R.: An Introduction to the Bootstrap. Chapman & Hall/CRC, Boca Raton (1993)

13. Loughrey, J., Cunningham, P.: Overfitting in wrapper-based feature subset se lection: the harder you try the worse it gets. In: Proceedings of International Conference on Innovative Techniques and Applications of Artificial Intelligence, vol. 33, p. 43 (2004)

14. Loots, G.G., Locksley, R.M., Blankespoor, C.M., Wang, Z.E., Miller, W., Rubin, E.M., Frazer, K.A.: Identification of a coordinate regulator of interleukins 4, 13, and 5 by cross-species sequence comparisons. Science 288, 136–140 (2000)

15. Smith, J.J., Deane, N.G., Wu, F., Merchant, N.B., et al.: Experimentally derived me tastasis gene expression profile predicts recurrence and death in patients with colon cancer. Gastroenterology 138(3), 958–968 (2010)

16. Kaiser, S., Park, Y.K., Franklin, J.L., Halberg, R.B., et al.: Transcriptional recapitula tion and subversion of embryonic colon development by mouse colon tumor models and human colon cancer. Genome Biol. 8(7), R131 (2007)

17. Wang, Y., Jatkoe, T., Zhang, Y., Mutch, M.G., Talantov, D., Jiang, J., McLeod, H.L., Atkins, D.: Gene expression profiles and molecular markers to predict recur rence of Dukes' B colon cancer. J. Clin. Oncol. 22, 1564–1571 (2004)

18. Jiang, Y., Casey, G., Lavery, I.C., Zhang, Y., Talantov, D., Martin-McGreevy, M., Skacel, M., Manilich, E., Mazumder, A., Atkins, D., Delaney, C.P., Wang, Y.: Development of a clinically feasible molecular assay to predict recurrence of stage II colon cancer. J. Mol. Diagn. 10, 346–354 (2008)

19. Lin, Y.H., Friederichs, J., Black, M.A., Mages, J., Rosenberg, R., Guilford, P.J., Phillips, V., Thompson-Fawcett, M., Kasabov, N., Toro, T., Merrie, A.E., van Rij, A., Yoon, H.S., McCall, J.L., Siewert, J.R., Holzmann, B., Reeve, A.E.: Multiple gene expression classi fiers from different array platforms predict poor prognosis of colorectal cancer. Clin. Cancer. Res. 13, 498–507 (2007)

20. Lin, P.C., Lin, S.C., Lee, C.T., Lin, Y.J., Lee, J.C.: Dynamic change of tetraspanin CD151 membrane protein expression in colorectal cancer patients. Cancer Invest. 29(8), 542–547 (2011)

Derivation of Cancer Related Biomarkers from DNA Methylation Data from an Epidemiological Cohort

Ioannis Valavanis, Emmanouil G. Sifakis, Panagiotis Georgiadis,
Soterios Kyrtopoulos, and Aristotelis A. Chatziioannou[*]

Institute of Biology, Medicinal Chemistry & Biotechnology, National Hellenic Research
Foundation, Athens, Greece
{ivalavan,sifakise,panosg,skyrt,achatzi}@eie.gr

Abstract. DNA methylation profiling methods exploit microarray technologies and provide a wealth of high-volume data. This data solicits generic, analytical pipelines for the meaningful systems-level analysis and interpretation. In the current study, an intelligent framework is applied, encompassing epidemiological.DNA methylation data produced from the Illumina's Infinium Human Methylation 450K Bead Chip platform, in an effort to correlate interesting methylation patterns with cancer predisposition and in particular breast cancer and B-cell lymphoma. Specifically, feature selection and classification are exploited in order to select the most reliable predictive cancer biomarkers, and assess their classification power for discriminating healthy versus cancer related classes. The selected features, which could represent predictive biomarkers for the two cancer types, attained high classification accuracies when imported to a series of classifiers. The results support the expediency of the methodology regarding its application in epidemiological studies.

Keywords: epigenomic analysis, Classification, Cancer, DNA methylation profiling, feature selection, microarrays.

1 Introduction

Epigenetic methylation events comprise heritable modifications that regulate gene expression without altering the DNA sequence itself and can serve as regulatory mechanisms for a wide range of biological processes [1]. Methylation patterns have been correlated with certain cancers. Specifically hypermethylation of CpG islands, typically a sequence of 300-3,000 base pairs in length within or near to approximately 40% of promoters [2], have been found to switch off tumor suppression genes. DNA methylation, the first epigenetic alteration to be observed in cancer cells, may be altered by a number of factors like aging of tissues, nutrition, and environment [3-4].

Recently, the rapid progress in microarray technologies, which has enabled the interrogation of a larger number of DNA/RNA transcripts more efficiently and less costly, has opened new avenues for epigenetic monitoring since the use of cell lines

---

[*] Corresponding author.

L. Iliadis, H. Papadopoulos, and C. Jayne (Eds.): EANN 2013, Part II, CCIS 384, pp. 249–256, 2013.
© Springer-Verlag Berlin Heidelberg 2013

which were used for testing only for non global epigenetic effects [5]. Specifically, two broad microarray-based assay categories have been developed to measure DNA methylation, i.e. enrichment-based microarrays, and bisulfite sequencing microarrays, with the latter mainly adopted by Illumina [6]. One of the newest microarray platforms is the Illumina's Infinium Human Methylation 450K Bead Chip, which can detect CpG methylation changes in more than 480,000 cytosines distributed over the whole genome [7].

Cancer is the first group of diseases associated with DNA methylation and considered for DNA-methylation-targeted therapeutics [8]. Several epigenetic effects in DNA methylation and in proteins involved in DNA methylation occur in cancer: hypermethylation of tumor suppressor genes, aberrant expression of DNMT1 and other DNMTs, hypomethylation of unique genes and repetitive sequences [9]. Mostly, silencing of tumor suppressor genes by DNA methylation provides a powerful molecular mechanism by which DNA methylation can trigger cancer and such genes have been targeted by studies aiming to identify DNA methylation biomarkers of cancer. However, hypomethylation is equally important, since critical genes for cancer growth and metastasis are hypomethylated in cancer [9].

From an artificial intelligence perspective, the identification of biomarkers can be viewed as a feature selection task that usually precedes the classification task, which discriminates among various physiological states (i.e. disease stages) [10,11]. The derivation of a small feature set that best explains the difference between the biological states, is aiming to yield robust, well performing classifiers. Feature selection represents in general, a prerequisite for the setup of reliable classification models in the area of bioinformatics, given the usually high dimensionality of the feature spaces observed in microarray analyses [12]. While feature selection and classification methods have been comprehensively explored in the context of gene expression data, little work has been done on how to perform feature selection or classification in the context of epigenetic data. Given the importance of epigenomics in cancer and other complex genetic diseases, it is critical to identify the appropriate statistical methods to be used in this novel context. So far, related studies have focused on the derivation of biomarkers using non genome-wide epigenetic data (cell lines) or the 27K DNA Methylation Array by Illumina [13-15].

In the current study, we employ a data mining framework towards the analysis of genome-wide epigenetic data that have been produced by Illumina's Infinium Human Methylation 450K Bead Chip. We aim to examine retrospectively the manifestation of two cancer types (breast cancer and B-cell lymphoma) through DNA methylation measurements that are fed to a workflow consisting of a feature selection and a three class (control plus two cancer types) classification module. It is the first time to the authors' knowledge that an artificial intelligence based pipeline is applied to this extended version of Ilumina Bead Chip Arrays. Data come from an Italian epidemiological cohort and consist of 261 samples organized in control class, breast cancer class and B-cell lympoma class. The available samples have been carefully split into two independent datasets: i) training set used for feature selection based on an evolutionary algorithm, training various popular classifiers and their evaluation through resampling and ii) testing set which consists of samples that have never been seen by classifiers and is used for blind testing. Methylation data has been previously pre-processed by a composite statistical framework presented by authors in [16].

This pre-process considers i) the correction of methylation signal using a novel intensity-based correction method and appropriate quality controls and ii) a statistical pre-selection of candidate CpG sites to be used for our data mining purposes here (See Section 2). Data is analyzed here through Rapidminer, a freely available open-source data mining platform that integrates fully the machine learning WEKA library, and permits easily data mining algorithms integration, process and usage of data and metadata [17,18]. Preliminary results reported here show that subset of features, corresponding to CpG sites, delivered by the feature selection could represent predictive biomarkers for the two cancer types studied here. Furthermore, encouraging classification performance measurements could be obtained by the series of classifiers used here.

2   Dataset and Pre-processing

High-throughput DNA methylation analysis based on Illumina's Infinium technology was first introduced with the Infinium Human Methylation 27K BeadChip The dataset studied here contains methylation data extracted using the more recent chip of Illumina, i.e. the Infinium Human Methylation 450K BeadChip, that includes 485,577 probes (482,421 CpG sites, 3,091 non-CpG sites and 65 random SNPs). The available Italian cancer dataset encompassed 261 samples which correspond to 131 controls, 48 breast cancer cases (BCCA) and 82 B-cell lymphoma cases (LYCA). Methylation of each CpG site is measured based on two channel intensities I_{Meth} and $I_{Un-Meth}$, available for all probes and samples, and specifically $M=\log_2(I_{Meth}/I_{Un-Meth})$ is utilized as DNA methylation measurement.

As a usual practice when pre-processing microarray data, M-values have been corrected. This has been done here using a novel intensity based method previously presented by authors in [16] which additional uses quality controls embedded in the methylation arrays. Furthermore, two statistical components have been used to pre-select a wide subset of candidate biomarkers prior to applying the data mining framework presented here. These components correspond to i) a scaled coefficient variation (Scaled CV) measurement and ii) robust p-value measurements corrected by bootstrap (see [16] for a detailed description). Scaled CV represents a robust measure of the real inter-class variability observed for a probe in the whole sample pool (controls∪cases), when compared to that observed among quality control samples, which measures solely the technical variation. The bootstrap corrected p-value measurements originate from a typical paired t-test for extracting statistically significant differentially methylated probes (Controls vs BCCA or Controls vs LYCA). The classical statistical test is followed, however, by a bootstrap p-value correction that immunizes statistical findings against the effect of multiple hypothesis.

Prior to applying the framework presented here that deals with data within a three-class (Controls, BCCA, LYCA), M-value correction and the two statistical components described in [16] were applied for the two separate experiments (Controls vs BCCA or Controls vs LYCA). For each of the experiments, the top 1% criterion in paired t-test bootstrap corrected p-values and a Scaled CV>1 criterion were combined in order to derive a finally selected robust subset of CpG sites that are both significantly differentially methylated and immunized against technical variation. Unifying the two selected robust subsets of CpG sites yielded a total of 5716 CpG sites (~1.2% of the genome-wide methylation array) that comprise the features used in this study. The total of 261 samples organized in Controls[1], BCCA and LYCA were split in two distinct sets: i) training set (50 Controls, 35 BCCA, 46 LYCA) used for feature selection, training various classifiers and their evaluation through resampling and ii) testing set (81 Controls, 13 BCCA, 36 LYCA) used for blind testing of classifiers. The sizes of training and testing sets and their internal classes are estimated so as to meet the need of a balanced occurrence of classes in the training set adjusted by the abundance percentages in the originally available dataset.

3   Methods

In order to select the most important features (CpG sites) corresponding to candidate biomarkers and test their relevance in terms of accuracy they can provide when fed to popular classifiers, an appropriate workflow was built using the Rapidminer platform [18]. An evolutionary feature selection algorithm using the training set and an embedded k-nn classifier was initially applied. Then various classification modules where the selected features were imported, i.e. k-nn classifiers, a decision tree and a feed-forward artificial neural network of one hidden layer were constructed and evaluated using the training set and leave-one-out resamplng. Then all classifiers were tested using the totally unknown testing set. In the following, the feature selection algorithm and the classifiers used are described.

Evolutionary Feature Selection: This feature selection method uses a Genetic Algorithm (GA) [19] to select the most robust features that feed an internally used, 10-nn weighted classifier [20]. GA mimics the mechanisms of natural evolution and applies biologically inspired operators on a randomly chosen initial population of candidate solutions, in order to optimize a fitness criterion. Here, an initial population of 500 random chromosomes is created during the first step of the GA. Each chromosome, representing a candidate solution for the feature selection task, is a binary mask of N binary digits (0 or 1), equal in number to the dimensionality of the complete features set, that reflect the use (or not) of the corresponding feature. Each chromosome is restricted to be a binary mask corresponding to a solution that included 50 to 150 features so as to keep the number of the finally selected features low. The genetic operators of selection, crossover and mutation are then applied to the initial generation of chromosomes. The selection operator selects, through a roulette wheel scheme, the chromosomes to participate in the operators of crossover and mutation, based on the fitness value of each chromosome previously calculated, i.e. the total accuracy (number of samples correctly classified) that the corresponding feature yields using the 12-nn weighted classifier on a 3-fold cross validation basis

---

[1] Control samples have actually originated from two different control samples subsets available for the experiments Controls vs BCCA and Controls vs LYCA. The original control samples have been thus unified in the wider control samples subset appropriate for the three-class problem studied here.

(Fig. 1). The crossover operator mates random pairs of the selected chromosomes with probability $P_c=0.5$ based on a uniform crossover operator, while bits within a chromosome are mutated (switched from 0 to 1 or vice-versa) with probability $P_m=1/N$ when the mutation operator is applied. The whole procedure is repeated until the stopping criterion of attaining a maximum number (50) of generations is reached and the best performing chromosome in the last generation is the one that represents the finally chosen feature subset. The k-nn classifier was used internally in the evolutionary algorithm due to the rather low computational cost it raises, compared to other alternatives e.g. artificial neural network, and the need for executing and evaluating the classifier for a large number of rounds within feature selection algorithm.

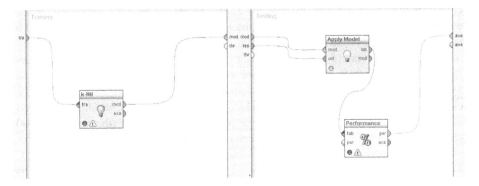

Fig. 1. The embedded training and evaluation process for the 12-nn classifier with weights within the frame of the evolutionary feature selection algorithm in Rapidminer

Classification: Following the evolutionary feature selection, classification is performed to construct classifiers fed by the selected features and measure their performance, thus validating the relevance of the selected features. Here, three nearest neighbor classifiers (k-NN, k=1,6,12) with weights, a classification tree (the gini index was used a split criterion), and a feed-forward artificial neural network (ANN) of one hidden layer were used. All classification algorithms except ANN are described in more detail in [16]. The ANN used here was trained using the back-propagation algorithm for 1000 epochs with a learning rate equal to 0.3 and momentum equal to 0.2 that were found best choices on a trial and error basis. The hidden layer used sigmoid activation function and contained ((num of features+num of classes)/2 + 1) nodes. Classifiers' performance in terms of total accuracy (number of samples correctly classified), and class sensitivity (number of true positives in a class that were correctly classified in this class) was measured using training set and leave-one-out resampling. The same classification algorithms were evaluated, utilizing the totally unknown testing set: classifiers were constructed once using the training set and applied to the samples belonging to the testing set.

4   Results and Discussion

Evolutionary feature selection provided 142 features, corresponding to equal in number CpG sites. These features achieved moderate or high accuracies and class sensitivity measurements both during the selection process (64% to 89%) (Table 1, Column 2) and additionally when evaluated by the constructed classifiers by the use of the leave-one-out strategy using the training set (58% to 87%) (Table 1, Columns 3-7). The most well performing classifiers seem to be ANN and 12-nn, with the latter being expected since it is comprises the classifier embedded into the feature selection process.

It is also encouraging that the different learning algorithms tested here, were all found to perform consistently adequately well (more than 50% and far more than 33% corresponding to a random classifier for the current three-class problem), when evaluated on a resampling basis using the training set. It should be noted that two more learning algorithms were added to the k-nn weighted classifier, which was used internally by the feature selection modules, when selected features and classifiers were evaluated for their generalization ability. Thus, any doubts for a bias introduced by the use of an embedded classifier into the feature selection module could be eliminated. The evolutionary feature selection method itself seems that performed well since the algorithm could screen the whole solution space (the setting of 500 chromosomes in the initial population each corresponding to 50-150 features ensures

Table 1. Accuracy and class sensitivity measurements (%) obtained by i) the embedded 12-nn classifier within evolutionary feature selection fed by the 142 finally selected features (3-fold cross validation using training set) and ii) all applied classifiers fed by the 142 finally selected features (leave-one-out resampling using training set).

	Evolutionary Feature Selection (embedded 12-nn)	1-nn	6-nn	12-nn	Tree	ANN
Total Accuracy	79.39	65.65	67.94	**87.02**	59.54	**87.02**
Controls Sensitivity	64.00	62.00	58.00	**84.00**	48.00	**84.00**
BCCA Sensitivity	88.57	80.00	80.00	**88.57**	74.29	**88.57**
LYCA Sensitivity	89.31	58.70	69.57	**89.13**	60.87	**89.13**

Table 2. Accuracy and class sensitivity measurements (%) obtained by all applied classifiers fed by the 142 finally selected features when trained by training set and tested to the totally unknown testing

	1-nn	6-nn	12-nn	Tree	ANN
Total Accuracy	47.69	47.69	45.38	32.13	**60.00**
Controls Sensitivity	54.32	55.56	54.32	40.74	**69.14**
BCCA Sensitivity	53.85	46.15	38.46	23.08	**61.54**
LYCA Sensitivity	30.56	30.56	27.78	16.67	**38.89**

that all features out of 5716 could be handled as can candidate features for the final solution).

Regarding the application of the classifiers constructed by the training set and evaluated with the independent testing set, performance is lower a finding conforming with the expectations (Table 2). However, many of the accuracy or class sensitivity measurements are greater than 50% and most importantly the ANN classifier (the best one among the five classifiers) seems to attain high performance measurements (Table 2, Column 6). Probably its superiority is due to its high capability of capturing non-linear effects within the input set. Its greater drawback is the sensitivity of lyphoma class, which samples are mislabeled as controls or breast cancer cases. Future work will include the use of other classifiers (e.g. SVM) and sophisticated algorithms for combination of various classifiers, as well as effective treatment of class imbalance problems.

Preliminary results presented here show that CpG sites selected by the workflow presented can comprise candidate methylation biomarkers for the two cancer types studied here. Since they will go under biological validations, CpGs sites or corresponding gene names are not reported here. In future, the utilization of functional information based on CpG sites extracted here, gene enrichment analysis tools and controlled biological vocabularies (Gene Ontology Terms, KEGG Pathways) could promote the understanding of epigenetic effects towards cancer (breast and B-cell lymphoma) development.

Acknowledgements. This work has been partially co-funded by the EU/FP7/EnviroGenomarkers project and the Entrepreneurial Program "Competitiveness and Entrepreneurship" Action COOPERATION entitled "PIK3CA Oncogenic Mutations in Breast and Colon Cancers: Development of Targeted Anticancer Drugs and Diagnostics" [POM].

References

1. Bird, A.: DNA methylation patterns and epigenetic memory. Genes & Development 16, 6–21 (2002)
2. Fatemi, M., et al.: Footprinting of mammalian promoters: use of a CpG DNA methyltransferase revealing nucleosome positions at a single molecule level. Nucleic Acids Res. 33, e176 (2005)
3. Esteller, M., Herman, J.G.: Cancer as an epigenetic disease: DNA methylation and chromatin alterations in human tumours. J. Pathol. 196, 1–7 (2002); Bascands, J.L., Schanstra, J.P.: Obstructive nephropathy: Insights from genetically engineered animals. Kidney Int. 68, 925–937 (2005).
4. Johnson, X., Jacobsen, S.: Interplay between two epigenetic marks. DNA methylation and histone H3 lysine 9 methylation. Curr. Biol. 12, 1360–1367 (2002)
5. Steensel, V., Henikoff, S.: Epigenomic profiling using microarrays. BioTechniques 35, 346–350 (2003)
6. Siegmund, K.D.: Statistical approaches for the analysis of DNA methylation microarray data. Hum. Genet. 129, 585–595 (2011)

7. Sandoval, J., et al.: Validation of a DNA methylation microarray for 450,000 CpG sites in the human genome. Epigenetics 6, 692–702 (2011)
8. Szyf, M.: DNA methylation properties: consequences for pharmacology. Trends Pharmacol. Sci. 15, 233–238 (1994)
9. Szyf, M.: DNA methylation signatures for breast cancer classification and prognosis. Genome Medicine 4, 26 (2012)
10. Abeel, T., Helleputte, T., Van de Peer, Y., Dupont, P., Saeys, Y.: Robust biomarker identification for cancer diagnosis with ensemble feature selection methods. Bioinformatics 1(26) (2010)
11. Valavanis, I., Maglogiannis, I., Chatziioannou, A.: Intelligent Utilization of Biomarkers for the Recognition of Obstructive Nephropathy. Intelligent Decision Technologies Journal 7(1), 11–22 (2013)
12. Saeys, Y., Inza, I., Larrañaga, P.: A review of feature selection techniques in bioinformatics. Bioinformatics 23(19), 2507–2517 (2007)
13. Zhuang, J., Widschwendter, M., Teschendorff, A.E.: A comparison of feature selection and classification methods in DNA methylation studies using the Illumina Infinium platform. BMC Bioinformatics 13, 59 (2012)
14. Carmen, J., et al.: DNA Methylation Array Analysis Identifies Profiles of Blood-Derived DNA Methylation Associated With Bladder Cancer. Journal of Clinical Oncology 29(9), 1133–1139 (2011)
15. Marchevsky, A.M., Tsou, J.A., Laird-Offringa, I.A.: Classification of individual lung cancer cell lines based on DNA methylation markers: use of linear discriminant analysis and artificial neural networks. J. Mol. Diagn. 6(1), 28–36 (2004)
16. Valavanis, I., Sifakis, E., Georgiadis, P., Kyrtopoulos, S., Chatziioannou, A.: Analysis of DNA methylation epidemiological data through a generic composite statistical framework. In: BIBE 2012 Proceedings, pp. 632–637. IEEE Computer Society (2012)
17. Mierswa, I., Wurst, M., Klinkenberg, R., Scholz, M., Euler, T.: YALE: Rapid Prototyping for Complex Data Mining Tasks. In: Proceedings of the 12th ACM SIGKDD International Conference on Knowledge Discovery and Data Mining, KDD 2006 (2006)
18. http://rapid-i.com/
19. Goldberg, D.: Genetic algorithms in search, optimization and machine learning. Addison-Wesley Publishing Company (1989)
20. Wu, Y., Ianakiev, K., Govindaraju, V.: Improved k-nearest neighbor classification. Pattern Recognition 35, 2311–2318 (2002)

A Particle Swarm Optimization (PSO) Model for Scheduling Nonlinear Multimedia Services in Multicommodity Fat-Tree Cloud Networks

Ioannis M. Stephanakis[1], Ioannis P. Chochliouros[2], George Caridakis[3], and Stefanos Kollias[3]

[1] Hellenic Telecommunication Organization S.A. (OTE),
99 Kifissias Avenue, GR-151 24, Athens, Greece
stephan@ote.gr
[2] Research Programs Section, Hellenic Telecommunications Organization (OTE),
99 Kifissias Avenue, GR-151 24, Athens, Greece
ichochliouros@oteresearch.gr
[3] National Technical University of Athens, Department of Electrical Engineering,
GR-157 73, Zographou, Greece
gcari@image.ece.ntua.gr, stefanos@cs.ntua.gr

Abstract. Cloud computing delivers computing services over virtualized networks to many end-users. Virtualized networks are characterized by such attributes as on-demand self-service, broad network access, resource pooling, rapid and elastic resource provisioning and metered services at various qualities. Cloud networks provide data as well as multimedia and video services. They are classified into private cloud networks, public cloud networks and hybrid cloud networks. Linear video services include broadcasting and in-stream video that may be viewed in a video player whereas non-linear video services include a combination of in-stream video with on-demand services, which are originated from distributed servers in the network and deliver interactive and pay-per view content. Furthermore heterogeneous delivery networks that include fixed and mobile internet infrastructures require that adaptive video streaming should be carried out at network boundaries based on such protocols as HTTP Live Streaming (HLS). Distributed processing of nonlinear video services in cloud environments is addressed in the present work by defining *Distributed Acyclic Graphs* (DAG) models for multimedia processes executed by a set of non-locally confined virtual machines. A novel discrete multivalue *Particle Swarm Optimization* (PSO) algorithm is proposed in order to optimize task scheduling and workflow. Numerical simulations regarding such measures as *Schedule-Length-Ratio* (SLR) and *Speedup* are given for novel fat-tree cloud architectures.

Keywords: cloud networks, particle swarm optimization (PSO), fat-trees, DAG scheduling, non-linear multimedia services, H.264, MPEG-DASH, multithreading.

L. Iliadis, H. Papadopoulos, and C. Jayne (Eds.): EANN 2013, Part II, CCIS 384, pp. 257–268, 2013.
© Springer-Verlag Berlin Heidelberg 2013

1   Introduction – Cloud Architectures and Standards

Cloud computing delivers computing services from large, highly virtualized network environments to many independent users, using shared applications and pooled resources. One may distinguish between *Software-as-a-Service* (SaaS), in which case software is offered on-demand through the internet by the provider and it is parameterized remotely (like for example on-line word processors, spreadsheets, Google Docs and others), *Platform-as-a-Service* (PaaS), in which case customers are allowed to create new applications that are remotely managed and parameterized and the platform offers tools for development and computer interface restructuring (like for example Force, Google App Engine and Microsoft Azure) and *Infrastructure-as-a-Service* (IaaS), in which case virtual machines, computers and operating systems may be controlled and parameterized remotely (like for example Amazon EC2 and S3, Terremark Enterprise Cloud, Windows Live Skydrive, Rackspace Cloud, GoGrid, Joyent, AppNexus and others). Cloud computing may be divided into *public cloud*, where everyone may register and use the services, *private cloud* - that is accessible through a private network - and *partner cloud* - that offers services to specific partners/users. A hybrid cloud is a combination of private/internal and external cloud resources that enables outsourcing of noncritical functions whilst keeping the remainder services internal. The key functionality is the ability to use and release resources from public clouds as and when required. This is used to handle sudden demand surges ("flash crowds") and is known as "cloud-bursting". Cloud computing is an on-demand service whose size depends upon users' needs and should feature scale flexibility. It is built upon such network elements as switches supporting novel communication protocols, specific servers based on *Virtual Machine* (VM) technology and dynamic resource management as well as *Network-Attached-Storage* (NAS). Specific software platforms may be used for service orchestration in cloud environments (like for example OpenStack, OpenCloud and Eucalyptus).

Cloud computing and virtualized environments have placed many new and unique requirements on network protocols and infrastructure. Conventional architectures such as *Spanning Tree Protocol* (STP) may limit the scale, latency, throughput and virtual machine mobility for large cloud networks. STP protocol, which is standarized as IEEE 802.1D, ensures a loop-free topology for any bridged Ethernet local area and aggregation network. State-of-the-art topologies use the IEEE 802.1s Multiple Spanning Tree Protocol for load balancing and service differentiated QoS networking [1] (see Fig. 1). A Layer 2 Ethernet based fat-tree architecture (flattened Layer 2 network) on the other hand features full bisection bandwidth at each level, uses commodity low latency switches and scales to 10 GigE at the edge. The topology supports native multicast. Such novel protocols as *Transparent Interconnections of Lots of Links* (TRILL), that uses *routing bridges* and a new form of MAC-in-MAC encapsulation and OA&M, shortest path bridging SPB VLAN/MAC-in-MAC [2] and OpenFlow [3] may be used in order to provide rooting and loop free forwarding in a layering hierarchy which is based upon a multi-rooted tree.

Aggregation in a fat tree topology is carried out in edge pods. All switches are identical and feature the same number of ports. Size is defined by a factor *k*, i.e. the number of ports per identical switch in the network. A 3-level hierarchy is illustrated in Fig. 2. Each core switch uses all k ports to connect to k switches in the first layer of the pod level. The "pod" level consists of k pods featuring k switches/pod. Each pod has two internal layers with $k/2$ switches. The upper level switches connect $k/2$ of their ports to core level switches and $k/2$ of their ports to lower pod level switches. The lower pod level switches connect to upper pod level switches and a total of $\dfrac{k^3}{4}$ end hosts/DSLAMs.

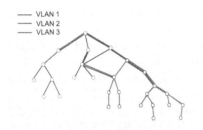

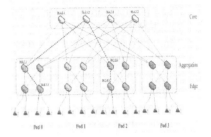

Fig. 1. Traffic engineering using multiple spanning trees (the IEEE 802.1s Multiple Spanning Tree Protocol)

Fig. 2. A 3-level fat tree Layer 2 architecture of size $k=4$

A Layer 2 fat tree architecture is proposed within the context of the present work for the implementation of multimedia cloud services. Workflow scheduling and load balancing is carried at pod level using the proposed PSO algorithm. Task-Resource scheduling basics are outlined in *Section 2* whereas proposed processes for video-to-video distributed cloud applications are described in *Section 3*. *Section 4* presents numerical results for abstract symmetric and asymmetric distributed processes. The proposed algorithm is the main contribution of this work. It is illustrated in *Subsection 2.3* and in Fig. 4.

2   Task-Resource Scheduling in Clouds

2.1   Background and Definitions of Workflow Scheduling

An application workflow is represented as a *Directed Acyclic Graph* (DAG), which is denoted as $G = (V, E)$ with $V = \{T_1, \ldots T_n\}$ being the set of tasks and E representing the data dependencies between these tasks. The data produced by task T_j and consumed by task T_k are denoted as $f_{j,k} = (T_j, T_k) \in E$. We have a set of storage sites $S = \{1, \ldots, k\}$, a set of compute sites $PC = \{1, \ldots, m\}$ and a set of tasks $T = \{1, \ldots, n\}$. The workflow scheduling problem reads: "Find a task-resource mapping instance M, such that when estimating the total cost incurred using each compute resource PC_j, the highest cost among all the compute resources is minimized". The mapping of jobs to the compute-resources is an NP-complete problem in the general form [4]. *Particle Swarm Optimization* (PSO) is a self-adaptive global search based optimization algorithm introduced by Kennedy and Eberhart [5]. PSO has become popular due to its simplicity and its effectiveness in a wide range of applications with low computational cost. It has been applied to solve NP-hard problems like scheduling [6, 7, 8] and task allocation [9, 10].

Cloud computing evolved from grid computing, service oriented computing and virtualization paradigms. The features that distinguish the scheduling algorithms are the following [11]:

- The target system, which is the system for which the scheduling algorithm was developed.
- The optimization criterion. Metrics as *makespan* and *cost* are frequently used for clouds.
- Multicore awareness and virtual machines (VM), which should be considered by scheduling algorithms in resource selection.
- On-demand resources treated as "single expense" during the execution of the workflow and long term leased resources
- The reserved resources that are used in a long term.
- The levels in a service level agreement (SLA) that the scheduling algorithm may consider implementing in a hierarchical fashion. Multiple levels require that the scheduling algorithm should run an intermediate facility between the IaaS cloud provider and the final client.

Efficient application scheduling based upon DAG theory is possible by such novel algorithms as the Heterogeneous Earliest-Finish-Time (HEFT) algorithm, the Critical-Path-on-a-Processor (CPOP) algorithm, the Hybrid Cloud Optimized Cost Scheduling algorithms (HCOC) and others [12, 13, 14].

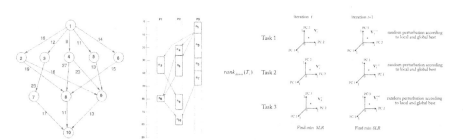

Fig. 3. Assignment of a sample composite task consisting of ten subtasks (as illustrated by the directed acyclic graph - DAG- on the left) to three processor units

Fig. 4. Outline of the proposed algorithm (probability velocities per task) – Each task corresponds to a line of the TP matrix

2.2   Particle Swarm Optimization and Task Scheduling

The particle swarm algorithm (PSO) is one of the novel evolutionary techniques that adjusts the trajectories of a population of "particles" through a problem space on the basis of information about each particle's previous best performance and the best previous performance of its neighbors. The approach uses the concept of population and a measure of performance similar to the fitness value used in evolutionary algorithms. The adjustments of individuals are analogous to the use of a crossover operator [15]. Some of the applications that have used PSO are data mining [16], pattern recognition [17] and reliability optimization in communication networks [18]. The general algorithm of PSO is outlined as follows [19]:

```
I - For i=1 to M (where M equals the size of the
population):
        - Initialize P[i] randomly (where P is the
          population-position of particles)
        - Initialize V[i]=0 (where V is the speed of each
          particle)
        - Evaluate P[i] and set GBEST = Best particle found
          in P[i]
II - End For
III- For i=1 to M (initialize the "memory" of each
particle)
        - PBESTS[i]=P[i]
IV - End For
V  - Repeat
     For i=1 to M
        - Calculate the speed of each particle
          V[i] = wt X V[i] + c₁ X r₁ X (PBESTS[i]-P[i]) +
          + c₂ X r₂ X (PBESTS[GBEST]-P[i])
          where wt stands for the inertia weight, c₁ and
          c₂ are positive constants and r₁ and r₂ are
          random numbers in the range [0…1]
        - Evaluate new P[i]=P[i]+V[i]
          If a particle gets outside the pre-defined
          hypercube then it is reintegrated to its
          boundaries, reevaluate P[i]
        - If new position is better then PBESTS[i]=P[i]
        - Set GBEST = Best particle found in P[i]
     End For
VI - Until stopping condition is reached (according to
some fitness function or the maximum number of
iterations)
```

PSO is applied to *Multiobjective Optimization* (MO) problems as well. Its ability to detect *Pareto Optimal* points and capture the shape of the *Pareto Front* is investigated in [20]. An aggregating function for multi-objective PSO is adopted, i.e. conventional linear aggregation, dynamic aggregation and bang-bang weighted aggregation.

Conventional versions of the particle swarm optimization operate in continuous space, where trajectories are defined as changes in position on some numbers of dimensions. A binary version of the PSO algorithm is proposed in [21]. Trajectories in the discrete PSO model are probabilistic whereas velocity on a single dimension is the probability that a bit will change. Should one assume that the probability of a bit being one equals $S(V[i])$ and its probability of being zero equals $1 - S(V[i])$ - where the function $S(.)$ is a sigmoid limiting transformation – he gets,

$$V_d^{t+1}[i] = V_d^t[i] + \varphi\left(PBESTS_d[i] - P_d^t[i]\right) + \varphi\left(PBESTS_d[GBEST] - P_d^t[i]\right) \text{ and}$$
$$if \quad (rand() < S(V_d^{t+1}[i]) \quad then \quad P_d^{t+1}[i] = 1 \quad else \quad P_d^{t+1}[i] = 0 \ . \tag{1}$$

Variables $PBESTS_d[i]$ and $P_d[i]$ are integers in $\{0, 1\}$, $\varphi(.)$ is a normalized random positive number and subscript d is a dimension index. Typical discrete optimization problems require ordering or arranging of discrete elements as in scheduling and routing problems. The velocity of the particle overall may be described by the number of bits changed by iteration or the Hamming distance between the particle at times t and $t+1$.

2.3   A Discrete Multivalue PSO Model for Scheduling Distributed Multimedia Processes in Clouds

A discrete multivalue version of the particle swarm PSO algorithm is proposed in the context of the described methodology. The total number of tasks of a set $V = \{T_1, \ ... \ T_n\}$ is defined as the dimensionality of the search space of optimization denoted as d. For compute sites featuring equal compute capacities PC_m the computation cost (execution time) of task T_j is defined as

$$w_j = \frac{instructions \ of \ T_j}{PC_k} \tag{2}$$

The communication cost pertaining to edge (j, i) of a DAG, which is the cost for transferring data from task T_j (scheduled on PC_k) to task T_i (scheduled on PC_l), is defined by

$$c_{j,i} = L_k + \frac{f_{j,i}}{B_{k,l}} \tag{3}$$

where L_k is the communication startup time associated with compute site PC_k and $B_{k,l}$ is the available bandwidth between compute sites PC_k and PC_l. Tasks are numbered by scheduling priority based on downward ranking. Downward rank of a task T_i is recursively defined by

$$rank_{down}(T_i) = \max_{T_j \in pred(T_i)} \{rank_{down}(T_j) + w_j + c_{j,i}\} \tag{4}$$

where $pred(T_i)$ is the set of immediate predecessors of task T_i according to the DAG. The downward ranks are computed recursively by traversing the task graph downward starting from the entry task of the graph. For the entry task T_{entry}, the downward rank value is equal to zero. Basically, $rank_{down}(T_i)$ is the longest distance from entry task to task T_i, excluding the computation cost of the task itself. Numbering the tasks of a set is such that $rank_{down}(T_1) \leq rank_{down}(T_2) \cdots \leq rank_{down}(T_d)$.

The proposed discrete multivalue version of PSO is a direct extension of the discrete binary version. It is assumed that vector $\mathbf{V}_j[i]$ is a m-value vector that holds the probabilities that task T_j is processed by the set of distributed compute sites $PC = \{1, \ldots, m\}$ for particle (schedule plan) indexed by i. Velocities are updated according to the following relationship in such a case (see Fig. 4),

$$\begin{pmatrix} V_{j,1}^{t+1}[i] \\ V_{j,2}^{t+1}[i] \\ \vdots \\ V_{j,m}^{t+1}[i] \end{pmatrix} = w \begin{pmatrix} V_{j,1}^{t}[i] \\ V_{j,2}^{t}[i] \\ \vdots \\ V_{j,m}^{t}[i] \end{pmatrix} + c_1 \begin{pmatrix} \varphi\left(PBESTS_{j,1}[i] - P_{j,1}^{t}[i]\right) \\ \varphi\left(PBESTS_{j,2}[i] - P_{j,2}^{t}[i]\right) \\ \vdots \\ \varphi\left(PBESTS_{j,m}[i] - P_{j,m}^{t}[i]\right) \end{pmatrix} + c_2 \begin{pmatrix} \varphi\left(PBESTS_{j,1}[GBEST] - P_{j,1}^{t}[i]\right) \\ \varphi\left(PBESTS_{j,2}[GBEST] - P_{j,2}^{t}[i]\right) \\ \vdots \\ \varphi\left(PBESTS_{j,m}[GBEST] - P_{j,m}^{t}[i]\right) \end{pmatrix} \quad (5)$$

and $\mathbf{P}_j^{t+1}[i]$ is a m-value zero vector except for $P_{j,\arg\{\max S(\mathbf{V}_j^{t+1}[i])\}}^{t+1}$ that equals one.

Each individual (particle) of the population is associated with an overall cost (as expressed by a fitness or objective function). The schedule length is known as the *makespan*. The assignment of tasks of a given application to processors should be such that its schedule length is minimized. *Schedule Length Ratio* (SLR) normalizes schedule length to a lower bound. It is defined as

$$SLR = \frac{makespan}{\sum_{T_j \in V} \min_{all\ compute\ sites} \{w_j\}} \quad (6)$$

The denominator is the summation of the minimum computation costs of tasks on some compute site. The task-scheduling algorithm that gives the lowest *SLR* of a graph is the best algorithm with respect to performance. An alternative criterion is the speedup value for a given graph that is computed by dividing the sequential execution time (i.e. cumulative computation costs of the tasks in the graph) by the parallel execution time (i.e. the *makespan* of the output schedule). The sequential execution time is computed by assigning all tasks to a single processor that minimizes the cumulative of the computation costs.

$$Speedup = \frac{\min_{all\ compute\ sites} \sum_{T_j \in V} w_j}{makespan} \quad (7)$$

Efficiency, i.e. the ratio of the speedup value to the number of processors used, is another comparison metric used for application graphs of real world problems. The optimal task-process matrix (TP-*matrix*) that holds the best schedule plan is given by matrix **PBEST** [*GBEST*] after completion of the proposed algorithm.

3   DAG Models for Nonlinear Multimedia Processes

Several open platforms may be used for the development of cloud applications[1]. Distributed processes in multimedia clouds include multiple GPU (*Graphics Processor Unit*) clusters that execute multithreading application code using the MPI network communication protocol [22], non-linear video interactive content [23] and advertisements [24] and HTTP Live Streaming (HLS) [25, 26]. GPU clusters (see Fig. 5.a) use multithreading application code over MPI network communication [27] (like for example the CUDA code). Setting an MPI Cluster on *OpenNebula* is a possible such implementation. One may use the OpenML[2] interface as well for distributed computing along with MPI and CUDA code. Rendering high definition graphics allows for distributed encoding schemes. Non-linear video interactive content and

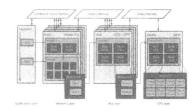

Fig. 5a Distributed processes in multimedia

advertisements are based upon distributed *Digital Program Insertion* (DPI) and ad-splicing of MPEG streams using input content originated from different local sites (the positions of advertising are signaled using for example SCTE protocols whereas such languages as the *Binary Format for Scenes*-BIFS may be used for dynamically describing locally generated contribution content). HTTP *Live Streaming* is HTTP-based media streaming that can distribute both live and on-demand files. It enables high quality streaming of media content over the Internet delivered from conventional HTTP web servers. It breaks the content into a sequence of small HTTP-based file segments. Each segments contains a short interval of playback time of a content that is potentially many hours in duration like a movie or the live broadcast of a sports event. The content is made available at a variety of different bit rates. Tasks may be outlined as follows (see Fig. 5.b): *Media Stream Segmenter, Media File Segmenter, Variant Playlist Creator, Metadata Tag Generator* and *Media Stream Validator*.

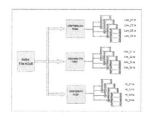

Fig. 5b Distributed processes in multimedia clouds (HLS)

4   Numerical Simulations

Numerical simulations are carried out for two different DAG models for multimedia processes as illustrated in Fig. 6, a non-symmetric one and a symmetric one. It is

---

[1] See for example the *Openstack* cloud project in http://www.openstack.org/ and the Opennebula cloud project in http://www.opennebula.org/

[2] See http://openmp.org/wp/. OpenMP uses the fork-join model of parallel execution.

assumed that each DAG requires input files for participating users as indicated. Clusters are accessed through a fat-tree architecture featuring 1Gbps links. Three distinct pod sizes (total number of 1 Gbps Ethernet ports per switch) are considered, i.e. four (4), eight (8) and sixteen (16). Core switches are connected in a full mesh topology using 1 Gbps links in all cases. It is assumed that each core switch features storage and sixteen (16) Virtual Machines (VMs) capable of executing 2.5×10^9 instructions/sec each.

Non-symmetric distributed process   Symmetric distributed process

Fig. 6. DAGs for non-symmetric and symmetric distributed processes used in numerical experiments (the numbers in nodes refer to sequences of instructions whereas the numbers at edges indicate output file size in bits)

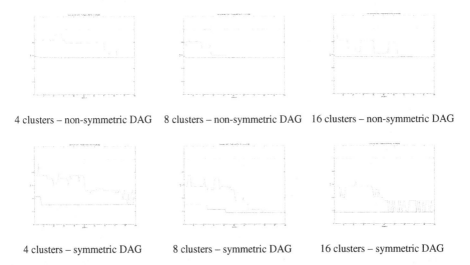

4 clusters – non-symmetric DAG   8 clusters – non-symmetric DAG   16 clusters – non-symmetric DAG

4 clusters – symmetric DAG   8 clusters – symmetric DAG   16 clusters – symmetric DAG

Fig. 7. Inverse Speedup during training for four (4), eight (8) and sixteen (16) clusters (solid line best value in the population, dashed line highest best value in the population)

The proposed PSO algorithm as described in Eq. 5 is applied for a population of fifteen (15) individuals for a total of 100 iterations. The inertia parameter equals 1 whereas the normalization parameters c_1 and c_2 equal 0.05. Optimal inverse Speedup values are obtained in all cases with more than one individual in the population converging to the optimal schedule as indicated in Fig. 7. Inverse Speedup values in the population range between the solid line (lower limit) and the dashed line (upper limit). The optimal *Task-Processor matrices* for eight (8) clusters are given in Table 1 both for the non-symmetric and the symmetric case. Inverse Speedup values are presented in Table 2.

Table 1. Task-Processor optimal schedule matrices for eight clusters

non-symmetric DAG									symmetric DAG							
	PC1	PC2	PC3	PC4	PC5	PC6	PC7	PC8		PC1	PC2	PC3	PC4	PC5	PC6	PC7	PC8
T1	0	0	0	0	1	0	0	0	**T1**	0	0	0	0	0	1	0	0
T2	0	0	0	0	1	0	0	0	**T2**	0	0	0	0	1	0	0	0
T3	0	0	0	0	0	0	0	1	**T3**	0	0	1	0	0	0	0	0
T4	0	0	0	0	0	1	0	0	**T4**	0	0	0	0	0	0	1	0
T5	1	0	0	0	0	0	0	0	**T5**	0	0	0	0	0	0	0	1
T6	0	0	0	0	0	0	1	0	**T6**	0	0	0	0	0	1	0	0
T7	0	0	0	0	0	0	1	0	**T7**	0	0	0	1	0	0	0	0
T8	0	1	0	0	0	0	0	0	**T8**	0	1	0	0	0	0	0	0
T9	0	0	0	1	0	0	0	0	**T9**	1	0	0	0	0	0	0	0
T10	0	0	0	0	1	0	0	0	**T10**	0	0	0	0	0	0	1	0

Table 2. Values of inverse Speedup and the corresponding execution time after 100 iterations of the proposed PSO algorithm

non-symmetric DAG (0.2218 sec) - **symmetric DAG** (0.2001 sec)

	Inverse Speedup	Execution time	Inverse Speedup	Execution time
4 clusters	0.5684	0.1261 sec	0.3027	0.0606 sec
8 clusters	0.5639	0.1251 sec	0.1765	0.0353 sec
16 clusters	0.5638	0.1250 sec	0.1758	0.0351 sec

5   Conclusion

A novel discrete multivalue *Particle Swarm Optimization* (PSO) algorithm is proposed for scheduling tasks of composite DAG of multimedia distributed processes.

The proposed method exhibits good convergence characteristics with more than one individual in the populations converging to the optimal solution, i.e. *Task-Processor* (TP) schedule matrix. The percentage of the individuals that have converged to the optimal solution to the total number of individuals in the population may be used as a termination criterion of the proposed algorithm. Several distributed multimedia processes that are implemented in cloud networks are presented. The proposed scheme allows for user input to distributed multimedia processes differentiating, thus, itself, from standard task scheduling algorithms like the HEFT and the CPOP algorithms. Fat-tree cloud architectures provide scalable, low cost and efficient access platforms to next-generation broadband Ethernet networks that feature low concentration ratios. Communication protocols that support such architectures are already implemented by several providers.

Acknowledgments. The present work has been inspired by the scope of the original *LiveCity* European Research Project and supported by the Commission of the EC.

References

1. He, X., Zhu, M., Chu, Q.: Traffic Engineering for Metro Ethernet Based on Multiple Spanning Trees. In: International Conference on Networking/International Conference on Systems/International Conference on Mobile Communications and Learning Technologies (2006), 10.1109/ICNICONSMCL.2006.216
2. DeCusatis, C.J.S., Carranza, A., DeCusatis, C.M.: Communication Within Clouds: Open Standards and Proprietary Protocols for Data Center Networking. IEEE Communications Magazine, 26–33 (September 2012)
3. ONF OpenFlow Switch Specication, Version 1.3.0 (Wire Protocol 0x04) (June 25, 2012), http://www.opennetworking.org
4. Ullman, J.D.: NP-complete Scheduling Problems. J. Comput. Syst. Sci. 10(3) (1975)
5. Kennedy, J., Eberhart, R.: Particle Swarm Optimization. In: IEEE International Conference on Neural Networks, vol. 4, pp. 1942–1948 (1995)
6. Pandey, S., Wu, L., Guru, S.M., Buyya, R.: A Particle Swarm Optimization-Based Heuristic for Scheduling Workflow Applications in Cloud Computing Environments. In: 24th IEEE AINA, pp. 400–407 (April 2010)
7. Yu, B., Yuan, X., Wang, J.: Short-term hydro-thermal scheduling using particle swarm optimization method. Energy Conversion and Management 48(7), 1902–1908 (2007)
8. Veeramachaneni, K., Osadciw, L.A.: Optimal Scheduling in Sensor Networks Using Swarm Intelligence. In: Proceedings of 38th Annual Conference on Information Systems and Sciences, pp. 17–19 (2004)
9. Yin, P.-Y., Yu, S.-S., Wang, Y.-T.: A hybrid particle swarm optimization algorithm for optimal task assignment in distributed systems. Computer Standards and Interfaces 28(4), 441–450 (2006)
10. Zavala, A.E., Aguirre, A.H., Villa Diharce, E.R., Rionda, S.B.: Constrained optimization with an improved particle swarm optimization algorithm. Intl. Journal of Intelligent Computing and Cybernetics 1(3), 425–453 (2008)
11. Bittencourt, L.F., Madeira, E.R.M., da Fonseca, L.S.: Scheduling in Hybrid Clouds. IEEE Communications Magazine, 42–47 (September 2012)
12. Topcuoglu, H., Hariri, S., Wu, M.-Y.: Performance-Effective and Low-Complexity Task Scheduling for Heterogeneous Computing. IEEE Trans. on Parallel and Distributed Systems 13(3), 260–274 (2002)
13. Bittencourt, L.F., Madeira, E.R.M.: HCOC: A Cost Optimization Algorithm for Workflow Scheduling in Hybrid Clouds. J. Internet Svcs. and Apps. 2(3), 207–227 (2011)
14. Yu, J., Buyya, R., Tham, C.K.: Cost-based Scheduling of Scientific Workflow Applications on Utility Grids. In: Int'l Conf. e-Science and Grid Computing, pp. 140–147 (July 2005)
15. Coello Coello, A.C., Lamont, G.B., Van Veldhuizen, D.A.: Evolutionary Algorithms for Solving Multi-Objective Problems, 2nd edn. Springer (2007)
16. Sousa, T., Silva, A., Neves, A.: Particle swarm based data mining algorithms for classification tasks. Parallel Computing 30(5-6), 767–783 (2004)
17. Louchet, J., Guyon, M., Lesot, M.J., Boumaza, A.: Dynamic flies: a new pattern recognition tool applied to stereo sequence processing. Pattern Recognition Letter 23(1-3), 335–345 (2002)
18. Bakare, G.A., Chiroma, I.N., Venayagamoorthy, G.K.: Comparison of PSO and GA for *K*-Node Set Reliability Optimization of a Distributed System. In: IEEE Swarm Intelligence Symposium, Indianapolis, IN, USA, May 12-14 (2006)

19. Shi, Y., Eberhart, R.C.: Parameter Selection in Particle Swarm Optimization. In: Porto, V.W., Waagen, D. (eds.) EP 1998. LNCS, vol. 1447, pp. 591–600. Springer, Heidelberg (1998)
20. Parsopoulos, K.E., Vrahatis, M.N.: Particle Swarm Optimization Method in Multiobjective Problems. In: SAC 2002, Madrid, Spain (2002)
21. Kennedy, J., Eberhart, R.C.: A Discrete Binary Version of the Particle Swarm Algorithm. In: Conference on Systems, Man, and Cybernetics, pp. 4104–4109. IEEE Service Center, Piscataway (1997)
22. Strengert, M.: Parallel Visualization and Compute Environments for Graphics Clusters. PH. D. Thesis, Institut für Visualisierung und Interaktive Systeme der Universität Stuttgart (2010)
23. http://bittubes.com/
24. Interactive Advertising Bureau. Digital Video In-Stream Ad Format Guidelines and Best Practices, http://www.iab.net/ (2008) and In-Stream Video Advertising, http://www.iab.net/media/file/IAB-Video-Ad-Format-Standards.pdf
25. MPEG-DASH. ISO/IEC DIS 23009-1.2 Dynamic adaptive streaming over HTTP (DASH)
26. http://Encoding.com
27. High Performance Computing Center Stuttgart (HLRS). MPI: A Message-Passing Interface Standard, Version 3.0

Intelligent and Adaptive Pervasive Future Internet: Smart Cities for the Citizens

George Caridakis, Georgios Siolas, Phivos Mylonas, Stefanos Kollias,
and Andreas Stafylopatis

Intelligent Systems, Content and Interaction Laboratory
National Technical University of Athens, Greece
{gcari,fmylonas}@image.ntua.gr,
{gsiolas}@islab.ntua.gr,
{stefanos,andreas}@cs.ntua.gr
http://www.image.ntua.gr

Abstract. Current article discusses the human centered perspective adopted in the European project SandS within the Internet of Things (IoT) framework. SandS is a complete ecosystem of users within a social network developing a collective intelligence and adapting its operation through appropriately processed feedback. In the research work discussed in this paper we will investigate SandS from the user perspective and how users can be modeled through a number of fuzzy knowledge formalism through stereotypical user profiles. Additionally, context modeling in pervasive computing systems and especially in the SandS smart home paradigm is examined through appropriate representation of context cues during overall interaction.

Keywords: Smart Cities, Smart Homes, Intelligent Systems, User Modeling, Context Aware Services, Future Internet.

1 Introduction

Thanks to pervasive computing practices, the IoT framework supports and enhances the cooperation between humans and devices in terms of: 1) facilitating communication between the (Internet of) Things and people, and among Things through a collective network intelligence driven by users in the SandS context, 2) people's ability to exploit the benefits of this communication with the increasing familiarity with ICT technologies, 3) a mashup vision where in certain respects people and things are homogeneous agents endowed with fixed computational tools. However, the ways of deploying the IoT paradigm may differ significantly, from the logistics-driven idea where individual consumer items are being tracked to the co-creative design approach where the user participates in a proactive manner in all stages of the product or system creation process.

Following a pragmatic approach, the FP7 European Project and FIRE framework Social & Smart (SandS)[1] aims to highlight the potential of IoT technologies

---

[1] http://www.sandsproject.eu

L. Iliadis, H. Papadopoulos, and C. Jayne (Eds.): EANN 2013, Part II, CCIS 384, pp. 269–281, 2013.
© Springer-Verlag Berlin Heidelberg 2013

in a concrete user-centric framework. The aim is for the user to collectively, via the SNS (Social Network Service), and intelligently, via the adaptive network intelligence, interface with and finally control his household appliances. The overall interface is orchestrated through a domestic infrastructure. The central role of the user is reflected on all aspects of the ecosystem, from the family of Things which are socially governed to the household appliances that affect our everyday life. This entire procedure is devised so as to optimally carry out usual housekeeping tasks with a minimal low level intervention from the part of the user.

By giving the means to the eahooker to intelligently control his domestic appliances and by placing him inside the ESN (Eahookers Social Network), SandS follows clearly a human and user centric approach. More precisely, User Modeling (UM) emerges as an important research direction inside the project. More precisely, UM not in a general sense, but relatively to the users activity inside the ESN and with respect to the task on hand, the efficient orchestration of his household appliances (context). We are considering in particular a context-aware UM of eahookers, that is taking into account all the contextual information that could characterize the situation and condition of the systems entities. In SandS case this could be context information about the eahooker (distance to his house, communication device used, time of the day, weather, etc.), usage information (recipes used, feedback provided by user, frequency of use,..), information about the homes (geolocalisation, proximity to other homes, surface area, number of rooms, etc.), about the appliances (location inside house, energy consumption levels etc) and information specific to the social network itself (friendship statements, content exchanged between users, graph structure, communities formation, etc.). As soon as the eahookers activity will start producing this data, Computational Intelligence algorithms will extract knowledge about groups of similar users and construct for these groups stereotypical users (or Personas). Ultimately, we will investigate how each individual eahooker could be modeled with a simple user model, consisting of a fuzzy combination of the extracted Personas.

2   Internet of People

A user model [23] is a computational representation of the information existent in a user's cognitive system, along with the processes acting on this information to produce an observable behavior. Concretely, a model receives inputs in a similar capacity to a person, performs mental and cognitive operations and outputs a response. Within the extended Human Computer Interaction framework, User Modeling serves to make systems more usable, useful, and to provide users with experiences fitting their specific background, knowledge and objectives. User profiling, on the other hand, is achieved by understanding user individual characteristics, including information related to age, gender, skills, education, experience, and cultural level or higher representations of these characteristics. User features could also include online usage log statistics or patterns.

Construction of user stereotypes or personas is quite common due to its correlation with the actors and roles used in software engineering systems, its flexibility, extensibility, reusability and applicability. A persona is an archetypal user that is derived from specific profile data to create a representative user containing general characteristics about the users and user groups and is used as a powerful complement to other usability optimisation methods. The use of personas is a growing popular way to customize, incorporate and share research about the users [21]. The personas technique fulfills the need of mapping and grouping a huge number of users based on the profile data, aims and behavior which can be collected both during design and run time, users and usage design respectively.

User modeling utilizes also Artificial Intelligence, Machine Learning (ML) and Data Mining techniques. In [2] the authors propose a user modeling framework addressing the issue of cost-intensiveness by integrating supervised and unsupervised machine learning. The application domain for the framework is learning during interaction with the Adaptive Coach for Exploration (ACE) environment using both interface and eye-tracking data. An unsupervised learning (K-means clustering) algorithm using vectors derived from offline and online interaction as input to form groups according to their similarity can be considered. Clustering is a very popular and effective approach for user modeling based on ML and many standard ML techniques are prime candidates for straightforward application to user modeling.

3   Future Internet in Context

Emerging ubiquitous or pervasive computing technologies offer "anytime, anywhere, anyone" computing by decoupling users from devices. To provide adequate service for the users, applications and services should be aware of their context and automatically adapt to their changing context, known as context-awareness. Context is any information that can be used to characterize the situation of an entity. An entity is a person, place, or object that is considered relevant to the interaction between a user and an application, including location, time, activities, and the preferences of each entity. Context-awareness means that one is able to use con-text information. A system is context-aware if it can extract, interpret and use context information and adapt its functionality to the current context of use.

Context models are used for representing, storing and exchanging contextual information [4]. A growing body of research investigates different approaches of context modeling and additionally reasoning techniques for context information [5]. The existence of well-designed context information models facilitates the development and deployment of future applications. Moreover, a formal representation of context data within a model is necessary for consistency checking, as well as to ensure that sound reasoning is performed on context data.Most of the context-aware systems [20] focus on the external context, called physical context. External context means context data collected by physical sensors. It involves context data of the physical environment, location data, distance, function on to other objects, temperature, sound, air pressure, time, lighting levels surrounding users, and so on. However, a few authors have addressed utilizing the cognitive elements of a user's context.

4   User Modeling via Fuzzy Personas

Before going further into the details of our approach, the motivation and development of our knowledge modeling notion and methods are grounded on a set of problems, assumptions, views and design decisions, which are stated next. We consider following settings: A set of users $\mathcal{U}$ interact with information objects, typically (though not mandatorily) containing a fair amount of unstructured or semi-structured content, e.g. text and/or multimedia objects and/or documents. The information objects are annotated with metadata, consisting of concepts, properties and values defined according to a domain ontology $\mathcal{O}$, and stored in a Knowledge Base (KB). At this point it should be made clear that the latter constitutes a clear assumption of this work and ontology matching or semantic similarity issues are not tackled herein.

Following the above common view, we define P as a set of *meanings* that can be found or referred to in items. Beyond raw keywords and multimedia descriptors, which are commonly used as semantic representation bricks for user needs, ontologies are being investigated in the field as enablers of qualitatively higher expressivity and precision in such descriptions [8], [16], [24], [30]. In our approach, P is described as a set of semantic entities that the user has interest for to varying degrees. This provides a fairly precise, expressive, and unified representational grounding, in which both user interests and content meaning are represented in the same space, in which they can be conveniently compared [9].

It is rather true that in the seek of an efficient user model representation formalism, ontologies ([3], [17]), present a number of advantages. In the context of this work, ontologies are suitable for expressing user modeling semantics in a formal, machine-processable representation. As an ontology is considered to be "a formal specification of a shared understanding of a domain", this formal specification is usually carried out using a subclass hierarchy with relationships among classes, where one can define complex class descriptions (e.g. in Description Logics (DLs) [3] or in Web Ontology Language (OWL) [32]).

4.1   Mathematical Background of Fuzzy Personas

Given a universe $\mathcal{V}$ of users $\mathcal{U}$, a crisp (i.e., non fuzzy) set S of concepts on $\mathcal{V}$ is described by a membership function $\mu_S : \mathcal{V} \rightarrow \{0,1\}$. The crisp set S may be defined as $S = \{s_i\}$, $i = 1, .., N$. A *fuzzy* set F on S may be described by a membership function $\mu_F : S \rightarrow [0,1]$. We may describe the fuzzy set F using the well-known sum notation for fuzzy sets [25] as:

$$F = \sum_i s_i/w_i = \{s_1/w_1, s_2/w_2, \ldots, s_n/w_n\} \tag{1}$$

where:

- $i \in N_n$, $n = |S|$ is the cardinality of the crisp set S,
- $w_i = \mu_F(s_i)$ or, more simply $w_i = F(s_i)$, is the membership degree of concept $s_i \in S$.

Consequently, equation (1) for a concept $s \in S$ can be written equivalently as:

$$F = \sum_{s \in S} s/\mu_F(s) = \sum_{s \in S} s/F(s) \qquad (2)$$

Let now R be the crisp set of fuzzy relations defined as:

$$R = \{R_i\}, R_i : S \times S \to [0,1], \quad i = 1, .., M \qquad (3)$$

Then the proposed fuzzy ontology contains concepts and relations and may be formalized as follows:

$$\mathcal{O} = \{S, R\} \qquad (4)$$

In equation (4), $\mathcal{O}$ is a fuzzy ontology, S is the crisp set of concepts described by the ontology and R is the crisp set of fuzzy semantic relations amongst these concepts.

Given the set of all fuzzy sets on S, $\mathcal{F}_S$, then $F \in \mathcal{F}_S$. Let $\mathcal{U}$ be the set of all users $\hat{u}$ in our framework, i.e. a user $\hat{u} \in \mathcal{U}$. Let $\mathcal{P}$ be the set of all user meanings and $\mathcal{P}_\mathcal{O}$ be the set of all user meanings on $\mathcal{O}$. Then $\mathcal{P}_\mathcal{O} \subset \mathcal{F}_S$ and $\mathcal{P}_\mathcal{O} = \mathcal{F}_Z \subset \mathcal{F}_S$, whereas $P_{\hat{u}} \in \mathcal{P}_\mathcal{O}$ depicts a specific user meaning.

4.2 Definition of Fuzzy Relations

In order to define, extract and use both a set of concepts, we rely on the semantics of their fuzzy semantic relations. As discussed in the previous subsection, a *fuzzy binary relation* on S is defined as a function $R_i : S \times S \to [0,1], i = 1, .., M$. The inverse relation of $R_i(x, y)$, $x, y \in S$ is defined as $R_i^{-1}(x, y) = R_i(y, x)$. We use the prefix notation $R_i(x, y)$ for fuzzy relations, rather than the infix notation xR_iy, since the former is considered to be more convenient for the reader. The *intersection, union* and *sup-t composition* of any two fuzzy relations R_1 and R_2 defined on the same set of concepts S are given by:

$$(R_1 \cap R_2)(x, y) = t(R_1(x, y), R_2(x, y)) \qquad (5)$$

$$(R_1 \cup R_2)(x, y) = u(R_1(x, y), R_2(x, y)) \qquad (6)$$

$$(R_1 \circ R_2)(x, y) = \sup_{w \in S} t(R_1(x, w), R_2(w, y)) \qquad (7)$$

where t and u are a fuzzy t-norm and a fuzzy t-conorm, respectively. The standard t-norm and t-conorm are the *min* and *max* functions, respectively, but others may be used if appropriate. The operation of the union of fuzzy relations can be generalized to M relations. If $R_1, R_2, ..., R_M$ are fuzzy relations in $S \times S$ then their union R^u is a relation defined in $S \times S$ such that for all $(x, y) \in S \times S$, $R^u(x, y) = u(R_i(x, y))$. A transitive closure of a relation R_i is the smallest transitive relation that contains the original relation and has the fewest possible members. In general, the closure of a relation is the smallest extension of the relation that has a certain specific property such as the reflexivity, symmetry or transitivity, as the latter are defined in [22]. The sup-t transitive closure $Tr^t(R_i)$ of a fuzzy relation R_i is formally given by:

$$Tr^t(R_i) = \bigcup_{j=1}^{\infty} R_i^{(j)} \tag{8}$$

where $R_i^{(j)} = R_i \circ R_i^{(j-1)}$ and $R_i^{(1)} = R_i$. It is proved that if R_i is reflexive, then its transitive closure is given by $Tr^t(R_i) = R_i^{(n-1)}$, where $n = |S|$ [22].

Based on the relations R_i we first construct the following combined relation T utilized in the definition of the taxonomic context C:

$$T = Tr^t(\bigcup_i R_i^{p_i}), \quad p_i \in \{-1, 0, 1\}, \quad i = 1 \ldots M \tag{9}$$

where the value of p_i is determined by the semantics of each relation R_i used in the construction of T. More specifically:

- $p_i = 1$, if the semantics of R_i imply it should be considered as is
- $p_i = -1$, if the semantics of R_i imply its inverse should be considered
- $p_i = 0$, if the semantics of R_i do not allow its participation in the construction of the combined relation T.

The transitive closure in equation (9) is required in order for T to be taxonomic, as the union of transitive relations is not necessarily transitive, independently of the fuzzy t-conorm used. In the above context, a fuzzy semantic relation defines, for each element $s \in S$, the fuzzy set of its ancestors and its descendants. For instance, if our knowledge states that "American Civil War" is before "WWI" and "WWI" is before "WWII", it is not certain that it also states that "American Civil War" is before "WWII". A transitive closure would correct this inconsistency. Similarly, by performing the respective closures on relations that correlate pair of concepts of the same set, we enforce their consistency.

Similarly, based on a different subset of relations R_i, we construct the combined relation $\widehat{T}$ for use in the determination of the runtime context $\widehat{C}$:

$$\widehat{T} = \bigcup_i (R_i^{\widehat{p}_i}), \quad \widehat{p}_i \in \{0, 1\}, \quad i = 1 \ldots \widehat{M} \tag{10}$$

For the purpose of analyzing textual descriptions, relation T has been generated with the use of a small set of fuzzy taxonomic relations, whose semantics are derived primarily both from the MPEG-7 standard and specific user requirements and are summarized in Table 1. On the other hand, relation $\widehat{T}$ has been constructed with the use of the entire set of relations available in the knowledge base. This approach is ideal for the user modeling interpretation followed herein; initially, when dealing with generic user information, focus is given on the semantics of high level abstract concepts, whereas additional precision and a more specific view is required as the runtime user modeling expansion comes into play. The latter demands the use of all available information in the KB. Of course,

Table 1. Semantic relations used for generation of combined relation T

Name	Inverse	Symbol	Meaning	Example	
				a	b
Belongs	Owns	$Bel(a,b)$	b belongs to a	house	device
Manufactured by	Constructs	$Made(a,b)$	b is manufactured by a	Siemens	fridge
Friend	NotRelated	$Fr(a,b)$	b is a friend of a	George	Bruno
Execute	ExecutedBy	$Exec(a,b)$	b is executed by a, or	user	recipe
			b undergoes the action of a		
Triggers	TriggeredBy	$Trig(a,b)$	b is triggered by a	rule	recipe

as the construction of relation $\widehat{T}$ implies, an intermediate step of removing its possible cycles, that are present due to the utilization of all relations and their inverses, is necessary before the application of the taxonomy-based expansion process.

The aforementioned relations are traditionally defined as crisp relations. However, in this work we consider them to be fuzzy, where fuzziness has the following meaning: high values of $Bel(a,b)$, for instance, imply that the meaning of b approaches the meaning of a, while as $Bel(a,b)$ decreases, the meaning of b becomes narrower than the meaning of a. A similar meaning is given to fuzziness of the rest semantic relations of Table 1, as well. Based on the fuzzy roles and semantic interpretations of R_i, it is easy to see that both aforementioned relations (9) and (10), combine them in a straightforward and meaningful way, utilizing inverse functionality where it is semantically appropriate. More specifically, relation T utilizes the following subset of relations:

$$T = Tr^t(Bel \cup Made^{-1} \cup Fr \cup Exec \cup Trig^{-1}) \tag{11}$$

Relation T is of great importance, as it allows us to define, extract and use contextual aspects of a set of concepts. All relations used for its generation are partial taxonomic relations, thus abandoning properties like synonymity. Still, this does not entail that their union is also antisymmetric. Quite the contrary, T may vary from being a partial taxonomic to being an equivalence relation. This is an important observation, as true semantic relations also fit in this range (total symmetricity, as well as total antisymmetricity often have to be abandoned when modeling real-life relationships). Still, the taxonomic assumption and the semantics of the used individual relations, as well as our experiments, indicate that T is "almost" antisymmetric and we may refer to it as "almost" taxonomic. Relying on its semantics, one may define the crisp taxonomic context $C^{'}$ of a single concept $s \in S$ as the set of its antecedents provided by relation T in the ontology. Considering the semantics of the T relation, it is easy to realize that when the concepts in a set are highly related to a common meaning, the context will have high degrees of membership for the concepts that represent this common meaning. Understanding the great importance of the latter observation, we plan to integrate such contextual aspects of user models in our future work.

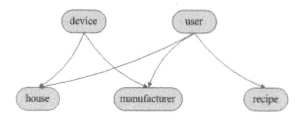

Fig. 1. Concepts and relations example

As observed in Figure 1, concepts *device* and *user* are the antecedents of concepts *house* and *manufacturer* in relation T, whereas concept *user* is the only antecedent of concept *recipe*.

5 Context Modeling

Existing approaches to context modeling differ in the ease with which they capture real world concepts, in their expressive power, in the support they provide for reasoning and in the computational performance of the reasoning. Early approaches to context modeling include key-value models and markup scheme models. Key-value models use simple key-value pairs to define the list of attributes and their values to describe context information. Markup based context information models use a variety of markup languages including XML. Fact-based context modeling like approaches that are based in database conceptual modeling and support query processing and reasoning. A special case of fact-based models are spatial context models that organise their context information by physical location, geometric or symbolic. Ontology based models of context information consider context as a specific kind of knowledge and use Web Ontology Language and Description Logics (OWL-DL) to augment the models' expressiveness and the complexity of reasoning.

Context modeling in SandS low level may seem to have strong similarities with context modeling in pervasive computing systems [18] because of the wirelessly interconnected appliances but considering the entire SandS architecture, many more modeling aspects become apparent. By giving the means to the eahouker, people who rule household appliances, to intelligently control his domestic appliances and by placing him inside the ESN, SandS follows clearly a human and user centric approach. We are considering in particular a context-aware user modeling of eahookers, that is taking into account all the contextual information that could characterize the situation and condition of the systems entities. In SandS case this could be context information about the eahooker (distance to his house, communication device used, time of the day, weather), usage information (recipes used, feedback provided by user, frequency of use), information about the homes (geolocalisation, proximity to other homes, surface area, number of rooms) and about the appliances (location inside house, energy consumption levels, etc.)

In order to be able to model these various aspects of context information inside SandS, several context models can be considered (see for example [6]). Of particular interest is the Context Broker Architecture (CoBrA) [10] which introduces a broker agent that maintains a shared model of the context for all computing entities in the space through a common ontology defined by using Semantic Web languages. Context Management Systems (CMS), an extension of the context server described in section 3, could also be considered in SandS case. As defined in [14], the role of a CMS is to acquire information coming from various sources, such as physical sensors, user activities, and applications in process or internet applications and to subsequently combine or abstract these pieces of information into context information to be provided to context aware services. The concept of CMS has also been used in the very relevant to SandS Amigo ("Ambient intelligence for the networked home environment") project[2]. From the projects description: "The Amigo project develops middleware that dynamically integrates heterogeneous systems to achieve interoperability between services and devices. For example, home appliances (heating systems, lighting systems, washing machines, refrigerators), multimedia players and renderers (that communicate by means of UPnP) and personal devices (mobile phones, PDAs) are connected in the home network to work in an interoperable way. This interoperability across different application domains can also be extended across different homes and locations.". Hence, the context server approach extended to a full CMS seems, at this project's early vision stage, as an adequate solution for context-aware user modeling in SandS.

6   Discussion on Related Datasets

In order to experimentally validate the proposed modeling and formalization architecture discussed in this paper as well as the adaptation mechanisms described as future work user, context and usage datasets are required. Datasets could be acquired in two ways: derived from SandS system integrated application or provided by a related project. Within SandS a small scale mockup and large scale experiments and validation will take place within WP7 and WP8 respectively. Related projects capable of providing a related dataset have been researched within the FIRE or related frameworks such as European Network of Living Labs (ENoLL).

A small scale mockup that will replicate the overall SandS system will be composed of a physical site located in Cartif[3] and seven virtual sites each one located in a server for each partner. Such a setting will allow validation of the entire system and its functionalities aiming to highlight problems while still on a manageable scale. On the other hand such a restricted application scale will also affect the completeness of the collected data. Additionally, such data, even if appropriate in terms of breadth and volume, will not be available during the SandS design and implementation stage constituting them usable only for a purely research objective.

---

[2] http://www.extra.research.philips.com/amigochallenge/

[3] http://www.cartif.com/en/

Large scale validation of the SandS system will include deploying the on Crew (w-iLab.t Ghent), OpenLab and SmartSantander [4] FIRE facilities to stress different aspects of applicability. Namely the respective aspects are, robustness and deliverability, large scale cooperation and communication of the different layers and finally, embedding the system in real life. Although the nature of the collected data, especially the ones corresponding to the SmartSantander application, will be appropriate, the timing of their availability almost excludes the integration of the research based on them within the SandS context. Analysis of the collected data in terms of stereotype personas and usage/context analysis is required in order to construct the knowledge base and fuzzify the ontology definition. This analysis will be followed by integration of its analysis to the adaptation mechanisms into the SandS system. Since large scale validation ends at M24 the integration and most importantly its validation is unfeasible constituting adoption of datasets from previous or ongoing (but more mature) research within related projects the only viable solution.

Attempting to pursuit this only viable solution we initially research bibliography in the related research area in order to detect a dataset close to the requirements set during the design of the user and context modeling process described in the respective sections of the current article. Our research yielded some results that although relevant could not be placed within the scope of our research goals and could not meet the set requirements. These results include:

- Home Activity and Sensors Datasets [7] that were collected and reported also related to the CHI 2009 Workshop on Developing Shared Home Behavior Datasets to Advance HCI and Ubiquitous Computing Research. This collection of home activity datasets includes mainly instrumented living environments recorded data that are somewhat irrelevant to our research aims since we are not dealing with in house user behavior.
- the Ambient Intelligence Datasets [1] which contains links to smart home datasets, as well as data gathered from wearable sensors.
- the Smart* Data Set [31] that deals with energy consumption and continuously gathers a wide variety of data in three real homes, including electrical (usage and generation), environmental (e.g., temperature and humidity), and operational (e.g., wall switch events).
- the HomeData [19] which is a collection of publicly available datasets recorded from different homes for use in research on Load Disaggregation, Smart Homes, and Ambient Assisted Living.
- a common repository [11] for context recognition data sets initiated during Pervasive 2004 Workshop on Benchmarks and a Database for Context Recognition.
- the ContextPhone [12] dataset on mobile context and communication.
- Nodobo [26] containing data gathered during a study of the mobile phone usage of high-school students, from September 2010 to February 2011.

---

[4] http://www.smartsantander.eu/ SmartSantander is a city-scale experimental research facility in support of typical applications and services for smart cities.

Finally, we contacted various recipients within the FIRE or related frameworks aiming to disseminate our quest for publicly available datasets that would prove useful to user and context modeling as well as usage information that could simulate the respective adaptation process. Initially, we contacted the FIRE facilities that will be utilized during SandS large scale validation. Crew [13], OpenLab [28] and SmartSantander were contacted but this communication did not produce any results although a number of contact redirections were performed. In the following we widened the scope of our research and contacted the European Network of Living Labs [15], OneLab [27] and PlanetLab [29] both centrally as well as individually. None of our attempts proved successful but important conclusions were drawn. Initially, we feel that on the one hand FIRE facilities neither monitor closely the projects implemented on them nor they followup on the obtained results and on the other hand frameworks and networks of labs such as ENoLL do not centrally manage interchangeable information that would be of common interest to the participating institutions and to the research community as a whole similar to the one encountered in the areas discussed above. We feel that the user aspect within this research area is largely ignored although individual institutions have highlighted this need. We could not summarize this better that to use a quote from one of our interlocutors "We'd love to know who our users are but actually have no idea!".

7   Conclusions and Future Work

In this paper the pervasive Future Internet, and namely the SandS paradigm, has been discussed from the human centered perspective. The research questions investigated were: a) how users can be modeled through a number of fuzzy knowledge formalism based on stereotypical user profiles and b) how context can be modeled and integrated in modeling in computing systems and especially in the SandS smart home paradigm. User stereotypes or personas on the one hand provide flexibility, extensibility, reusability and applicability and on the other hand knowledge management is incorporated as an efficient user model representation formalism. In addition, this formal, machine-processable representation is used in order to define, extract and use both a set of concepts and their fuzzy semantic relations. Finally, moving forward from existing, conventional context modeling approaches relying on data of the physical environment, we address the issue of incorporating contextual information characterizing the entire system's entities state and interaction, usage information and social activity derived information.

Ongoing, the progress relying heavily on the issues raised in section 6, and future work includes incorporation of user, usage and context information, through a unified semantic representation, driving an adaptation mechanism aiming to provide a personalized service and optimizing the user experience.

Acknowledgments. Current research work has been funded by FP7, Future Internet Research and Experimentation - FIRE project Social & Smart, Social housekeeping through intercommunicating appliances and shared recipes merged in a pervasive web-services infrastructure.

References

1. Ambient intelligence datasets,
   `http://www.cise.ufl.edu/~prashidi/Datasets/ambientIntelligence.html`
2. Amershi, S., Conati, C.: Unsupervised and supervised machine learning in user modeling for intelligent learning environments. In: Proceedings of the 12th International Conference on Intelligent user Interfaces, pp. 72–81. ACM (2007)
3. Baader, F., Calvanese, D., McGuinness, D.L., Nardi, D., Patel-Schneider, P.F.: The Description Logic Hand-book: Theory, Implementation and Application. Cambridge University Press (2002)
4. Baldauf, M., Dustdar, S., Rosenberg, F.: A survey on context-aware systems. Int. J. Ad Hoc Ubiquitous Comput. 2(4), 263–277 (2007), `http://dx.doi.org/10.1504/IJAHUC.2007.014070`, doi:10.1504/IJAHUC.2007.014070
5. Bettini, C., Brdiczka, O., Henricksen, K., Indulska, J., Nicklas, D., Ranganathan, A., Riboni, D.: A survey of context modelling and reasoning techniques. Pervasive and Mobile Computing 6(2), 161–180 (2010), `http://www.sciencedirect.com/science/article/pii/S1574119209000510`, doi:10.1016/j.pmcj.2009.06.002; Context Modelling, Reasoning and Management
6. Bolchini, C., Curino, C.A., Quintarelli, E., Schreiber, F.A., Tanca, L.: A data-oriented survey of context models. SIGMOD Rec. 36(4), 19–26 (2007), `http://doi.acm.org/10.1145/1361348.1361353`, doi:10.1145/1361348.1361353
7. Boxlab wiki page, `http://boxlab.wikispaces.com/List+of+Home+Datasets`
8. Castells, P., Fernandez, M., Vallet, D.: An Adaptation of the Vector-Space Model for Ontology-Based Information Retrieval. IEEE Transactions on Knowledge and Data Engineering 19(2) (February 2007); Special issue on Knowledge and Data Engineering in the Semantic Web Era
9. Castells, P., Fernández, M., Vallet, D., Mylonas, P., Avrithis, Y.: Self-tuning Personalized Information Retrieval in an Ontology-Based Framework. In: Meersman, R., Tari, Z. (eds.) OTM 2005 Workshops. LNCS, vol. 3762, pp. 977–986. Springer, Heidelberg (2005)
10. Chen, H.: An Intelligent Broker Architecture for Pervasive Context-Aware Systems. PhD thesis, University of Maryland, Baltimore County (2004)
11. Context database, `http://www.pervasive.jku.at/Research/Context_Database/`
12. Contextphone, `http://www.cs.helsinki.fi/group/context/#data`
13. Crew cognitive radio experimentation world, `http://www.crew-project.eu/`
14. Dey, A.K.: Understanding and using context. Personal and Ubiquitous Computing 5, 4–7 (2001)
15. European network of living labs, `http://www.openlivinglabs.eu/news/enoll-mou-partners`
16. Gauch, S., Chaffee, J., Pretschner, A.: Ontology-Based Personalized Search and Browsing. Web Intelligence and Agent Systems 1(3-4), 219–234 (2004)
17. Gruber, T.R.: A Translation Approach to Portable Ontology Specification. Knowledge Acquisition 5, 199–220 (1993)
18. Henricksen, K., Indulska, J., Rakotonirainy, A.: Modeling context information in pervasive computing systems. In: Mattern, F., Naghshineh, M. (eds.) PERVASIVE 2002. LNCS, vol. 2414, p. 167. Springer, Heidelberg (2002), `http://dx.doi.org/10.1007/3-540-45866-2_14`
19. Homedata, `https://github.com/smakonin/HomeData`

20. Hong, J., Suh, E., Kim, S.: Context-aware systems: A literature review and classification. Expert Systems with Applications 36(4), 8509–8522 (2009)
21. Junior, P., Filgueiras, L.: User modeling with personas. In: Proceedings of the 2005 Latin American Conference on Human-Computer Interaction, pp. 277–282. ACM (2005)
22. Klir, G., Bo, Y.: Fuzzy Sets and Fuzzy Logic, Theory and Applications. Prentice Hall, New Jersey (1995)
23. Kobsa, A.: Generic user modeling systems. User Modeling and User-Adapted Interaction 11(1), 49–63 (2001)
24. Kiryakov, A., Popov, B., Terziev, I., Manov, D., Ognyanoff, D.: Semantic Annotation, Indexing, and Retrieval. Journal of Web Sematics 2(1), 47–49 (2004)
25. Miyamoto, S.: Fuzzy Sets in Information Retrieval and Cluster Analysis. Kluwer Academic Publishers, Dordrecht (1990)
26. Nodobo dataset, http://nodobo.com/release.html
27. Onelab, https://www.onelab.eu/
28. Openlab, http://www.ict-openlab.eu/
29. Planetlab europe, http://www.planet-lab.eu/
30. Popov, B., Kiryakov, A., Ognyanoff, D., Manov, D., Kirilov, A.: KIM - A Semantic Platform for Information Extraction and Retrieval. Journal of Natural Language Engineering 10(3-4), 375–392 (2004)
31. Smart* data set, http://traces.cs.umass.edu/index.php/Smart/Smart
32. W3C Recommendation, OWL Web Ontology Language Reference (February 10, 2004), http://www.w3.org/TR/owl-ref/

Creative Rings for Smart Cities

Simon Delaere[1], Pieter Ballon[1], Peter Mechant[1], Giorgio Parladori[2],
Dirk Osstyn[3], Merce Lopez[4], Fabio Antonelli[5], Sven Maltha[6], Makis Stamatelatos[7],
Ana Garcia[8], and Artur Serra[9]

[1] iMinds, Gaston Crommenlaan 8/102, Gent, 9050, Belgium
[2] Alcatel-Lucent Italia S.P.A., Piazzale Biancamano 8, Milano 20124, Italy
[3] Alcatel-Lucent Bell NV, Copernicuslaan 50, Antwerpen 2018, Belgium
[4] i2cat Foundation
[5] Center for Research and Telecommunication Experimentation for Networked Communities,
via Alla Cascata 56/D, Trento 38123, Italy
[6] Dialogic Innovatie & Interactie BV, Hooghiemstraplein 33-36, Utrecht 3514AX, Netherlands
[7] National and Kapodistrian University of Athens, Department of Informatic and
Telecommunications, Panepistimiopolis, Ilissia, Athens, Greece
[8] European Network of Living Labs, Pleinlaan 9, Brussels, 1050, Belgium
[9] Fundacio Privada I2CAT, Internet i Innovacio Digital a Catalunya, Calle Gran Capita 2-4,
Edifici Nexus I, Barcelona, 08034, Spain
simon.delaere@vub.ac.be, pieter.ballon@iminds.be,
peter.mechant@ugent.be, giorgio.parladori@alcatel-lucent.com,
dirk.osstyn@alcatel-lucent.com, merce.lopez@i2cat.net,
fabio.antonelli@create-net.org, maltha@dialogic.nl,
makiss@di.uoa.gr, ana.garcia@enoll.org, artur.serra@i2cat.net

Abstract. In order to make an impact on citizens' lives, projects within the framework of *Smart Cities* shall address - along with safety and transportation issues – shared services based on novel as well as creative applications that exploit *Future Internet* platforms and relevant network infrastructures. It is individual creativity, skills and talents, which lie at the crossroads between arts, business and technology, that provide a strong competitive advantage in novel applications aiming at the production and the commercialization of creative content. The vision of the SPECIFI project for European Creative Rings is presented in this paper. Creative communities are in most cases isolated from their counterparts in other cities having no access to *Future Internet* technologies and solutions. SPECIFI proposes the use of *Creative Rings* as a means of sharing creative and innovative content and enabling internet activities all over Europe. Thus *Creative Rings* intend to bring together infrastructure solutions that facilitate the use of future internet systems and applications. *Creative Industries* may experiment and deploy these systems and take advantage of the distribution of innovative content. *Creative Rings* are presented herein in terms of scenarios, infrastructures and applications.

Keywords: Smart Cities, Future Internet, Creative Industries, Creative Rings.

1 Introduction

Smart Cities have to advance safety, transportation and public data infrastructures beyond their current status. As Richard Florida [1] famously demonstrated, the

L. Iliadis, H. Papadopoulos, and C. Jayne (Eds.): EANN 2013, Part II, CCIS 384, pp. 282–291, 2013.
© Springer-Verlag Berlin Heidelberg 2013

leading force of economic growth worldwide are cities that foster a "Super-Creative Core", i.e. the combination of high-tech industries with media, design and arts. It is the guiding principle of the ICT PSP SPECIFI project [2] that Smart Cities can be a success story should they become Smart Creative Cities as well by employing open platforms and Future Internet infrastructures that transform them into thriving regional, European and worldwide centres of media, arts, leisure and urban discovery activities.

In this paper, we present a roadmap towards **European Creative Ring** in terms of concepts, scenarios infrastructures and applications, which are developed within the ICT PSP SPECIFI project. The concept of the Creative Rings is introduced in the next section. The three SPECIFI Creative Rings are subsequently presented in terms of selected use cases and applications. *Section* 4 elaborates upon the European Creative Ring providing insights regarding initial concept, architecture, targeted deployment and infrastructures together with their interconnection plan. *Section* 5 reports on the project's work-plan. Section 6 concludes the paper and sums up results.

2   Creative Rings

A new market-based definition of the term "*Creative Industries*" emerged recently. It describes *Creative Industries* as economic activities that involve the creation and maintenance of social networks as well as the generation of value through production and consumption of network-valorised choices [3].

Indeed, a **new generation of entrepreneurs and creative individuals** is emerging in our cities. Consisting of a diverse set of urban "tribes" (wikipedians, digital artists, local media producers, fablabbers, arduins, new designers…), they are creating new jobs and opportunities in difficult times, not just by creating and improving products and services on their own but also enriching and revitalizing existing economic activities. This informal community is discovered by scientific and cultural institutions such as universities, cultural centres and academia as well as by corporations and local governments that start to accept this system for creating innovation and try to create bridges.

Creative Industries are based upon individual creativity, skill and talent and have a potential for wealth creation through the generation and exploitation of intellectual property linked to film and video content, music, performing arts, printing and publishing on TV and radio [4]. *Creative Industries* lie at the crossroads between arts, business and technology and evolve rapidly driven by ICT technologies.

In Europe, *Creative Industries* consist typically of SMEs, micro-enterprises or semi-professionals strongly concentrated in cities that have to face the lack of advanced and interoperable tools as well as insufficient data links to surrounding regions and between cities across national borders.

Creative Industries can enable Smart Cities to become *Smart Creative Cities* in the aforementioned sense by employing open platforms and Future Internet infrastructures in order to promote centres of media arts, leisure and urban discovery activities regionally, European-wide and worldwide.

Urban creative communities that are isolated from their counterparts in other cities and do not have access to Future Internet technologies and solutions (both in terms of high-speed wired and wireless infrastructure and in applications), may benefit from the **creation and the exchange of innovative content and applications**. The immense potential of European Creative Industries for growth and diversity is yet to be fulfilled.

Creative Ring is proposed as a new experimental community facility and platform for sharing creative and innovative content and activities all over Europe in order to address this issue through advanced internet technologies. It is based on an open collaboration between local artists and creative industries (comprising cultural, media arts, advertising people and others) with universities, local authorities and ICT companies in each city and region. Creative Ring intends to bring together solutions for infrastructure, proven Future Internet systems and applications that *Creative Industries* may use to produce and distribute innovative content.

European Creative Rings will emerge through scaling up the rich but fragmented creative assets in Europe by linking Creative Cores with other cities and regions utilizing ultra fast network infrastructures (FTTH) and platforms.

3   European Creative Ring

3.1   The Architecture

The set-up of the European Creative Ring Architecture allows for the reuse of different components of the pilots and for secondary content dissemination as dictated by the specified functionalities, abstractions and requirements per case within the framework of the project. The procedure will be based on:

- Identifying the daily organisational and business processes and practices of the national Creative Industries for each Smart Pilot City and analyse them in order to prepare integration processes between the local Creative Rings and the broader European Creative Ring Architecture; this may require, for example, certain adaptation capabilities,
- Assisting the three pilot cities to integrate a component of their Creative Ring into a broader European Creative Ring. An overall scenario will be developed for this purpose, whose certain parts could be addressed by different local Rings as detailed in the sequel of this paper ,
- Evaluating integration activities into this broader European Creative Ring based on certain feedback loops and KPIs related to provided QoE to Rings' users. A white book will be issued detailing guidelines for the European Creative Ring architecture.

The different steps towards this objective encompass: (*i*) elaborating on a global "European Creative Ring" scenario which will be realized by the European Creative Ring Architecture based on the different Use Cases, (ii) deriving and specializing corresponding requirements in order to outline the functionalities that enable interconnection between Creative Rings, and (ii) analysing each Use Case in terms of available infrastructure in order to capture diverse infrastructures and related capabilities. Certain adaptation requirements will be taken into account in order to efficiently address any heterogeneity along the different routes when interconnecting the different pilots.

3.2   The Infrastructure

The European Creative Ring will provide an infrastructure for connecting the three pilot cities and regions. Using this infrastructure, the pilots will assess the possibility for reuse of different components and content of the Kortrijk, Barcelona and Trento Creative Ring. The networks of the three Creative Rings are already interconnected with each other through a high speed fiber internet connection underpinning that the SPECIFI Consortium can rely on existing, proven and monitored setting. As shown in Fig. 1 the three Creative Rings are connected to the European research network Geant2 [5]. Kortrijk and Barcelona through 1Gb/s, Trento through 155Mb/s but with a plan to upgrade to 1Gb/s.

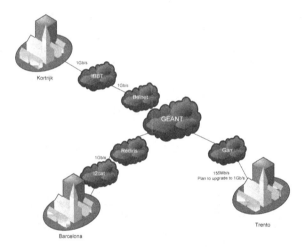

Fig. 1. Infrastructure of SPECIFI Creative Ring

The NRENS (National research networks) Belnet, Rediris and Garr are connected with 10Gb/s to Geant2 which has European-wide multiple 10Gb/s wavelengths. High bandwidth connectivity between the three sites over the research networks is foreseen. Hence nothing has to be installed extra or paid extra.

The European Creative Ring is not merely a supporting high-performance infrastructure; it provides a European platform of loosely coupled technical components for Creative Industries defining a market and networking place for the stakeholders. It addresses the need to overcome technological limitations, geographical fragmentation and the lack of sustainable business models that limit the ability of European cities to fully leverage "Super-Creative Cores" of media, advertising activities, design arts and ICT activities. SPECIFI aims at pioneering these fields by involving both the public and private sector to set-up and test Future Internet platforms and services in real urban environments, rather than focusing on preparing for an infrastructure roll-out. SPECIFI will interlink and open up a combination of state-of-the-art technologies and tools in the Future Internet context so as to provide new opportunities to existing urban innovation ecosystems in order to enhance their creative processes and businesses.

Summing up, the European Creative Ring is planned to combine three things in one: (i) an infrastructure to connect, (ii) a set of ICT components aimed at boosting Creative Industries, and (iii) a network of creative industries and ICT players that can interchange experiences, content and technologies.

4   Use Cases within SPECIFI Creative Rings

SPECIFI will demonstrate the positive impacts of a European Creative Ring of Smart Cities and Regions, with real-life Future Internet infrastructures and platforms, with real-life users and producers, and in five (5) real-life (ENoLL-certified) Living Labs active in three (3) smart cities (Kortrijk, Barcelona and Trento as in Fig. 2) and their surrounding regions. The project will foster the creation, delivery, promotion and participation of creativity and culture (media, arts, leisure and urban discovery services), yielding concrete and measurable effects at the city, regional and European levels. To this aim, the SPECIFI pilot will combine (i) connected smart areas – locally, regionally, and internationally, (ii) open ultrafast Internet platforms and infrastructures, and (iii) innovative arts, media, leisure and urban discovery services from the Creative Industries. The next sub-sections present the three SPECIFI pilots focusing on certain use cases and related platforms and applications.

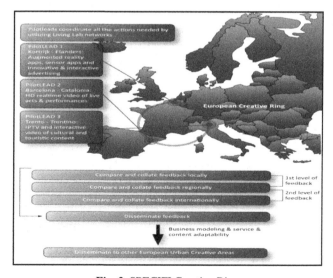

Fig. 2. SPECIFI Creative Ring

Indicative scenarios will be described, within the scope of each of the three SPECIFI pilot cities in the following sub-sections. A use case visualisation is also provided based on the Logical Framework Approach (LFA) approach [8]. It can be traced as follows (see Fig. 3, Fig. 4, and Fig. 5):

Key resources (technical components, physical equipments, organisation and financing) feed use case activities (application development, testing, marketing, dissemination, and collaborations) towards use case outputs (applications, services, IPR/Licences and reputation) and revenues (users, revenue models, pricing models). All these converge to use case purpose and global impact).

4.1   Kortrijk Creative Ring

The pilot city Kortrijk aims at creating an environment and tool-set for providing various expressive affordances for the Creative Industries in the context of Internet-of-Things and Future Internet. Activities include the creation of a "Captation Box", a capturing device (indoors/outdoors) specifically aimed at capturing Creative Industries events without any hassle which not only captures the actual event but also maps environmental and context parameters.

In "Captation Box" use case, city managers of stakeholders in the Creative Industries are provided with a "Captation Box", an easy to use, easy to set-up, robust capturing device that not only captures the actual event (audiovisual data-stream) but also maps environmental and context parameters (e.g. noise levels). Ideally, these raw data-streams can be staged and edited. The data-stream generated by (a cluster of) capitation box(-es) is released as open data and can be used as a means of feedback or to build new apps or services on top of it (e.g. a dynamic map showing the most busy 'Creative Industries' spots.

Integration of different data streams such as multimedia content and sensory data is envisaged in a second use case for the city of Kortrijk. This allows supporting a wide range of touristic, cultural, social and other initiatives and experiences. This use case will bring social network affordances to cultural experiences and the Creative Industries. It will use various media channels such as smart phones, tablets or public displays to visualize and enrich experiences in the city (like attending creative performances and others).

The goal of both use case is to enhance visibility of Creative Industries in Kortrijk and promote their activities, as well as gather data and feedback on (Creative Industries) public events (e.g. in terms of number of audience) whilst the purpose: Capture (in- and outdoor) Creative Industries events to the fullest extent possible and disseminate this stream via a variety of tools and devices. Gain insight in the nature and number of target groups and customer/citizen segments reached via Creative Industries events. Fig. 3 visualizes the "Captation Box" use case of the Kortrijk pilot city.

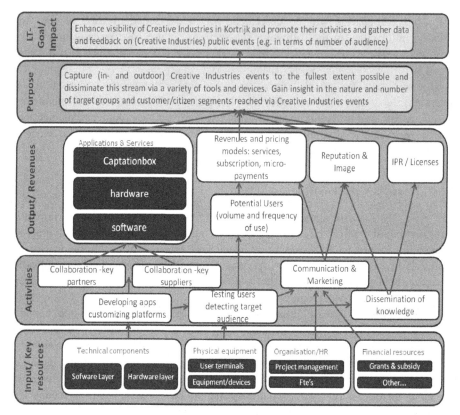

Fig. 3. Visualisation of Kortrijk "Captation Box" use case

4.2   Barcelona Creative Ring

Two existing pilots - which are related to cultural content creation - will be extended upon introducing real time components over a distributed platform for the Spanish pilot city Barcelona. Specifically, the existing use cases of "Anella Cultural" [6] and "Arts en viu a la llar" [7] will be adjusted towards a distributed platform for real-time culture co-creation and sharing. Creative industry will be invited to take part in this pilot exploring new ways of use of the technologies deployed in the platform, involving end-users and all the other actors identified as potential stakeholders. At least one feasible scenario is subject of the pilot within the Barcelona Creative Ring, (see Fig. 4).

One of the Barcelona use cases makes use of advanced e-Infrastructures, throughout a distributed platform, to foster the active cooperation and co-creation among Creative Industries, Public Bodies and End-Users. Activities aim at testing and demonstrating innovative applications, activities or uses of multimedia technologies, with the objective of stimulating participation of end-users in the co-creation process jointly with creative industries and exploring new and innovative ways that brings together culture and technology, while making both more accessible to a wider range of audience, creating and making them join, new target groups.

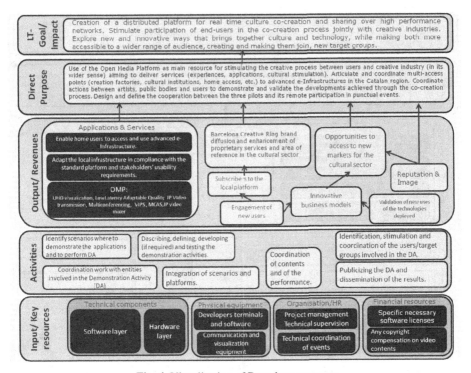

Fig. 4. Visualisation of Barcelona use case

4.3   Trento Creative Ring

The pilot city Trento aims at setting up an ICT platform offering thematic channels supporting communication and promotion services for the tourism sector, targeting both public and private customers, and integrating connected TV services to enable the delivery of services via TV screen. At the same time, this ICT platform will offer the opportunity to local creative industries and all actors working in the leisure market value chain to monetize their products using the delivered visitors services as promotion and advertising channels. The Trento use case is visualised in Fig. 5.

For example, tourists can get informed during their visit, when staying in their apartment, on the area and on the offered city highlights (such as sport areas, cultural events, museums, shopping, restaurants, etc...) by accessing various thematic TV Channels, some of them (co-)created and participated by citizens, local administrations, and cultural associations.

The Trento Creative Ring intends foster the availability of high bandwidth networks on the local territory as an opportunity to offer distinctive and attractive content and services, giving at the same time the opportunity to access these services also in mobility, in order to fully support tourists during their visits and support a video-based augmented reality representation of Trento city and vicinity while personalizing the themes and material of interest in order to know better the Trentino area (see Fig. 5 for a visualization of the IPTV use case).

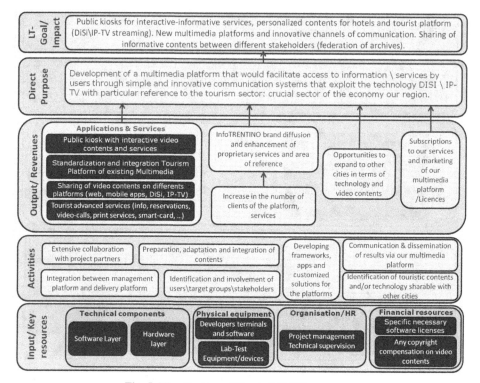

Fig. 5. Visualisation of Trento "IPTV" use case

5   Conclusion

In this paper, we presented the European Creative Rings concepts, use cases and infrastructures as elaborated within the SPECIFI project. Three pilot cities have been identified each of which with a couple of specific use cases involving Future Internet networking, platforms and applications as well as local creative industries. The main target of SPECIFI, the European Creative Ring has been presented together with the SPECIFI project structure was then presented as a reference framework ensuring the project success. The next immediate steps are focusing on defining user and technical requirements as well as setting the pilot scenarios.

Acknowledgment. This work has been performed in the framework of the ICT- PSP 325094 SPECIFI Project funded by the European Commission.

References

1. Florida, R.: The Rise of the Creative Class: And How it's transforming work, leisure, community and everyday life. Perseus Book Group, New York (2005)
2. ICT PSP SPECIFI project, http://www.specifi.eu (last access date: May 31, 2013) http://ec.europa.eu/information_society/apps/projects/factshe et/index.cfm?project_ref=325094 (last access date: May 31, 2013)

3. O'Connor, J.: Creative industries: a new direction? International Journal of Cultural Policy 15(4), 387–402 (2009)
4. Understanding Creative Industries. Cultural statistics for public-policy making, http://portal.unesco.org/culture/en/files/30297/11942616973cultural_stat_EN.pdf/cultural_stat_EN.pdf (last access date: May 31, 2013)
5. The GÉANT pan-European research and education network, http://www.geant2.net (last access date: May 31, 2013)
6. The Anella Cultural Event, http://www.barcelonafestivalofsong.com/find-an-event/details/46
7. The Arts en viu a la llar event, http://www.iglor.es/content/?page_id=6 (last access date: May 31, 2013)
8. The Logical Framework Approach. Handbook for objectives-oriented planning, http://www.norad.no/en/tools-and-publications/publications/publication/_attachment/106231?_download=true&_ts=11eb62dcb2d (last access date: May 31, 2013)

Energy Efficient E-Band Transceiver
for Future Networking

Evangelia M. Georgiadou[1], Mario Giovanni Frecassetti[2], Ioannis P. Chochliouros[1],
Evangelos Sfakianakis[1], and Ioannis M. Stephanakis[1]

[1] Hellenic Telecommunications Organization (OTE) S.A.,
99 Kifissias Avenue, GR-151 24, Athens, Greece
{egeorgiadou,ichochliouros,esfak}@oteresearch.gr, stephan@ote.gr
[2] Alcatel-Lucent,
via Trento 30, Vimercate (MB), Italy
mario_giovanni.frecassetti@alcatel-lucent.com

Abstract. Rapidly changing network requirements and traffic growth, call for
new deployment architectures and utilization of emerging technologies in
higher operational frequencies, such as the E-band. The E3NETWORK project
aims at designing an E-band transceiver capable of fulfilling the enhanced
backhaul challenges set by future networking. In this paper, the unique
characteristics of E-band systems are highlighted. A regulatory overview of the
E-band allocations is presented. The main potential applications of E-band
technology are investigated. Among others, focus is given to comparative
advantages of E-band backhauling against conventional backhaul solutions.

Keywords: E-band technology, light-licensing, millimeter wave backhaul.

1 Introduction

The 2020 Digital Agenda for Europe previsions to speed up the roll-out of high speed
internet and reap the benefits of a digital single market for households and firms [1].
Indeed, as users' requested experiences grow more challenging and demanding, the
need for ubiquitous, broadband infrastructures across Europe becomes crucial.
Consumers tend to seek seamless service provisioning in their homes, their offices,
and on the go. Advanced multimedia applications of high data rates and peaky
bandwidth are pursued. Interoperability of fixed and mobile network topologies is
therefore becoming a prerequisite to fulfill Europe's digital vision. Utilization of the
E-band (70/80GHz) is expected to serve the convergence of future evolving
heterogeneous networks, due to congestion of conventional lower frequency bands
and constantly increasing requirements on bandwidth resources, network capacity,
coverage and performance.

The E3NETWORK project targets the backhauling segment of future networks,
which is responsible for connecting the core network or backbone to the small subnets
at the edge of the entire hierarchical network, and which is turning into a bottleneck
for the development of novel infrastructure solutions. All different backhaul

L. Iliadis, H. Papadopoulos, and C. Jayne (Eds.): EANN 2013, Part II, CCIS 384, pp. 292–301, 2013.
© Springer-Verlag Berlin Heidelberg 2013

technologies available in the literature, i.e. copper, fiber and microwave, are characterized by their separate advantages and drawbacks. Copper, while already present in many regions of the world, presents poor capacity and excessive latency to address the advanced requirements of future networks. This, along with the high maintenance cost of existing copper infrastructures, is expected to lead to their gradual obsolescence. On the other hand, fiber appears a very attractive candidate from the capacity, latency and OPEX point of view. Capability to serve complex cell site configurations is unquestionable; nevertheless the very high CAPEX of optical infrastructure renders fiber deployment for extended use prohibitively expensive. Microwave backhauling is proved to be the optimum solution in terms of CAPEX and OPEX. Microwaves are able to provide latency performance close to that of fiber; they are characterized by rapid deployment, low cost, comparatively high performance and growing capacity as higher frequencies are employed. The microwave E-band transceiver that will be developed within the project aims at enhancing the data transmission capacity of backhaul microwave links to 10Gbps, providing communication over 1km with availability of 99.995%. Since the power consumption of a microwave system is practically independent of the data transmission capacity, the E3NETWORK prototype is expected to be a "best-in-class solution" in terms of energy efficiency. In this paper, an effort to point out all the potential diverse applications of such an E-band transceiver is made, covering different scenarios and network topologies.

2   E-Band Regulatory Overview

The ITU has approved the worldwide utilization of the 71-76 / 81-86 GHz band for ultra high capacity point-to-point communications [2]. The E-band frequencies were specifically opened up to enable ultra-high capacity links that require high operational bandwidths, not available at microwave frequencies and below. At millimetre wavelengths, spectrum is not deemed scarce, since only few services currently exist, so a highly efficient re-use of the spectrum can allow implementation of multiple services and applications. Systems operating in the E-band have several unique propagation characteristics, not experienced by conventional lower frequency radio systems. More precisely, the possibility to employ very directional antennas enables highly focused pencil beam transmissions, and thus greatly reduces interference concerns. Moreover, propagation impairments, particularly rain fading, limit high frequency links to relatively short range distances of a few kilometres, thus promoting frequency reuse patterns.

For all these reasons, systems operators utilizing the E-band are capable of realising networks with minimal frequency coordination, high degree of frequency reuse, and close link configurations with minimal interference concerns. Furthermore, many countries have configured the E-band frequency allocation as a single pair of channels, namely 71-76 GHz and 81-86 GHz, each with a 5GHz bandwidth. Since these bands are not channelized, traditional frequency coordination does not need to be considered. Consequently, the licensing process of such frequencies can be significantly simplified, through 'light licensing' procedures that implement streamlined interference analyses and automatic link approval. These techniques still award the link operator with 'first come first served' registration rights and full interference protection benefits of a traditional link license, whereas administration burden and reflected costs are dramatically reduced. Many countries worldwide have set up light licensing schemes, also adopting new fee structures that reflect the reduction in administration associated with the light license procedure, thus encouraging the adoption of competitive high data rate services at the E-band frequencies [3].

3   E-Band Potential Applications

Figure 1 depicts the forecast of E-band technology market penetration in the coming years, foreseeing wide adoption of E-band solutions. The range of applications for services using the 71-76 and 81-86 GHz bands (E-band) is wide and evolving quickly [4]. It includes voice, data and entertainment services of many kinds. Each user may require a different mix of services. Traffic flow may be unidirectional, asymmetrical or symmetrical, and this balance will change with time and application. In the following, some of the numerous cases where the E3NETWORK microwave transceiver shall be applicable are presented.

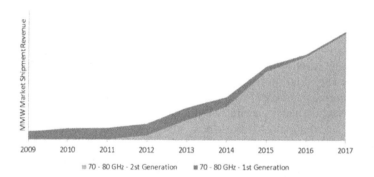

Fig. 1. E-band forecast curve [5]

3.1   Fixed and Mobile Backhaul

Huge increases in traffic on mobile and fixed networks means that service providers are faced with significant upgrades on their backhaul networks. Video streaming, content downloading, gaming and other high-bandwidth, data-intensive multimedia applications are accelerating data traffic growth and pushing networks to saturation. When these networks are located in areas that lack backhaul fiber infrastructure, the costs associated with building this new capacity is often prohibitive relative to what can be charged to the end user. Cost and time effective delivery of such backhaul services can be achieved through wireless broadband E-band technology, with its possibility of very high data rates, as shown in Fig. 2. Some of the network technologies that can reap the benefits of E-band backhaul are IEEE 802.11, third generation (3G) cellular, LTE and WiMAX in dense urban environments. Specifically, this technology can be incorporated to establish communication between the Base Transceiver Stations (BTS) and Base Station Controllers (BSC). Additionally, portable and temporary links for high definition video or high definition television (HDTV) transport can be built by applying this technology.

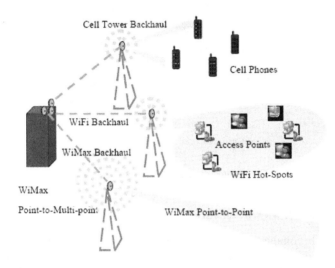

Fig. 2. Wireless backhaul [4]

For the case of LTE networks, mobile broadband operators are under increasing pressure to deploy more backhaul transmission capacity, both in the form of more links for new radio sites (including heterogeneous networks) and as more capacity to support new, high-bandwidth services on existing radio sites. E-band indeed offers more capacity per backhaul link, supporting LTE as a high-capacity layer on top of 2G/3G networks. It also offers more high-capacity backhaul links in urban areas to meet broadband demand and network densification. Finally it serves lowering the cost per unit bandwidth to improve mobile network operator profitability.

The 300Mbit/s rule-of-thumb value for LTE backhaul transmission requirement to and from a single three-sector LTE base station is already in the borderline for typical single bitstream capabilities of packet microwave radios operating in the traditional bands. If we add to this the need to transport multiple backhaul traffic streams aggregating traffic from several LTE sites e.g. in ring topologies, the choice of deploying E-band backhaul to 3G/LTE base station sites is easily explained. Figure 3 depicts the evolution of backhaul capacity needs when mobile broadband services evolve from 3G/HSPA+ to LTE and LTE-A point-to-point links [6].

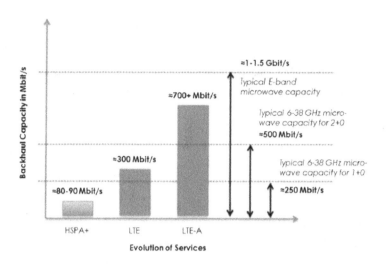

Fig. 3. The need for backhaul capacity from 3G to LTE-A [6]

3.2   Fiber Extensions / Metro Networking

With the economy becoming more information dependent, the bandwidth needs of large and small corporations continue to grow apparently without bound. However a large majority of corporate buildings are still being served only by archaic copper wires barely able to deliver a few megabits per second bandwidth. What is even more outstanding is that while many commercial buildings are 'out of the loop', literally the fiber-loop of the metro rings, a large majority of these buildings are within a small distance of a high bandwidth metro ring. What has been missing is the practical ability to extend the metro network services from an existing metro ring to the commercial buildings not touched by the ring. E-band technology creates an opportunity to fill these gaps in a cost effective manner. As illustrated in Figure 4a), a single millimetre wave link can be used to connect a commercial building with a metro ring. With the bandwidth of the millimetre wave link being comparable to that of the metro core itself, this single wireless link would be sufficient to serve a large-occupancy building with high bandwidth demands. That way, service providers can extend their fibre communication network by transmitting from one of their buildings to nearby customer buildings using millimetre wave technologies (such as the broadband E-band transceiver to be developed within E3NETWORK). Wireless links can be quite attractive in cases where fibre connections are not physically possible or financially feasible [7].

3.3   Enhancing Network Paths

Organisations can enhance their network integrity by connecting to backbone networks using technically and physically diverse techniques such as Fibre Optic and Wireless Broadband E-band technologies. This idea is illustrated in Figure 4b).

It is often necessary or more efficient to enhance network coverage by a system of distributed antennas that are basically extensions of the antenna of base stations. These are often used to provide cellular coverage in spots that are shadowed by large structures, such as buildings, from base station antennas, or to provide coverage in areas where it is not efficient to install a base station. For example, an area behind a large commercial building may be covered better by installing a remote antenna behind the building and transmitting the radio signal back to the nearest base station. In another scenario, for a corporate building with a large subscriber base, it may be desirable to distribute antennas throughout the building and transport the signal to the base station over several wireless paths. These signals often need to be digitized thus generating high digital data throughput. While fiber optic cables are usually utilized, millimetre wave technology is the ideal solution when signals of this kind need to be transported wirelessly [7].

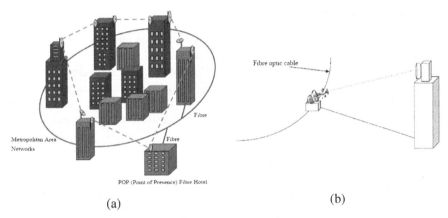

(a)   (b)

Fig. 4. a) Typical deployment of Fibre extension using a wireless link [4].
b) Enhancing network connections using a wireless link [8].

3.4   Temporary Service Restoration and Disaster Recovery

For applications requiring high end-to-end bandwidth, broadband connectivity by means of fiber optic cables is often the technology of choice when access to fiber optic cables is readily available. However, cases abound where fiber connections have been broken by accident, for instance during trenching operations, often bringing down mission critical networks for a substantial period of time. Therefore, it is highly desirable to design such mission critical networks with redundancies that minimize probability of such failures.

A millimetre wave wireless link is very well suited to provide such redundancy. As an example, a data center connected to a network service provider's point-of-presence (PoP) by means of a fiber optic network may also be connected to the PoP by means of a high capacity millimetre wave wireless link. In the event that a failure is detected in the fiber optic network, the data traffic could be routed through the millimetre wave link without impacting the availability or performance of the network [7]. In the case of fibre breakage, the temporary service restoration could be delivered more rapidly by the means of a radio backup. The major advantage of this technique is its shorter set-up time in comparison with the time needed to restore the original fibre link. By adding 'virtual fiber links' between two or more networks that are already connected via traditional fiber links, E-band technology provides users with an economical and resilient means of obtaining fiber optic backbone access diversity and disaster recovery. Figure 5a) shows an example of how a broadband E-band link, such as the one that will be developed in E3NETWORK, can be used for network temporary recovery. E-band is indeed an ideal back-up solution to existing connections, offering optical fiber-like speeds and different failure mechanisms, e.g. different entry/exit points to the building. Since it is most likely to enter the building via a roof or high window, a wireless link is impervious to the most common causes of fiber outages, such as flooding, construction work, earthquakes and other ground-based disturbances.

3.5   LAN-to-LAN Connections (LAN Bridges)

Network administrators operating LAN networks in campus environments often face the challenge of establishing a private, high speed network connection between their different buildings. This could be for simple Ethernet LAN extensions, for offsite back-up of files or transfer of customer data, billing, images or other large data files. Instead of laying a fiber optic cable or leasing an expensive lower-speed network connection, the operator can use an E-band radio to establish network connectivity between the remote locations [9].

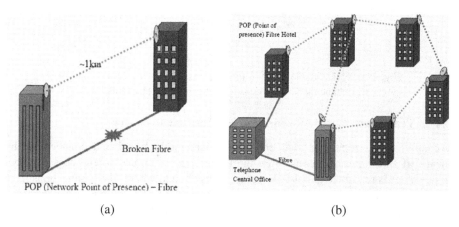

(a)   (b)

Fig. 5. a) Network recovery using a wireless link [4].
b) Use of a wireless link in a campus LAN [4].

Utilization of the E-band frequencies in campus LAN connectivity provides users with a flexible, fast and safe way to establish the network and fiber optic backbone access. As depicted in Figure 5b), buildings in a campus environment can be connected together with wireless broadband E-band links that offer high bandwidth, excellent data security and high availability. Using wireless broadband E-band technology makes it possible to extend fibre-optic communication networks with gigabit wireless LANs and private networks. Furthermore, high-speed (gigabit) access will be maintained within the wireless part of the network as well as in the rest of fibre-optic communication network, but without problems and expenses connected with fibre installation.

3.6   Centralized Base Band Unit (BBU) Architectures

For the near future, a common approach is to use a distributed base station architecture where the baseband processing of the base station (BS) is separated from the radio unit (Remote Radio Head, RRH). Communication between the two entities happens by means of a digital interface. This has multiple advantages. Firstly, the RRH can be mounted directly on the antenna which reduces the power loss and the RF cable between the BS and the antenna. Secondly, it is possible to have one BS baseband controlling several RRHs. Figure 6 illustrates the evolution of the mobile backhaul network. The last generation of mobile backhaul networks employs centralized baseband processing. For the communication between the baseband and the RRH, the industry has agreed upon defining a common digital interface protocol standard, like e.g. the common public radio interface (CPRI). In fact, the last generation of mobile backhaul networks presented in Figure 6 is based on CPRI interconnection and requires a link capacity in the order of 5-10 Gbps.

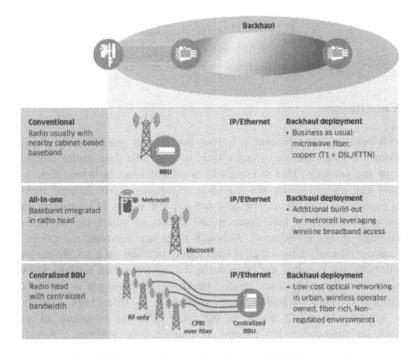

Fig. 6. Evolution of the mobile backhaul deployment [10]

Figure 7 compares the Conventional Base Station arrangement with the Centralized Base Stations proposed in the last evolution of the backhaul links. This centralized baseband processing enables both infrastructure suppliers and network operators to reduce their cost and to optimize their solutions. As explained before, the most popular technology solution for Backhaul is based on a microwave Point-to-Point link. However, the current microwave links have some strong limitations for this CPRI approach. The main concern is related to the huge capacity required by CPRI. Indeed current CPRI requires extremely high bandwidths, ranging from 1.25 to 10 Gbit/s. Microwave backhaul solutions are traditionally working in frequency bands from 2 to 40 GHz frequency bands, using up to 56 MHz channel, with capacity up to 600Mbps, which is not enough for the CPRI interconnection. As E-band radios in some ways are similar in performance to fiber, E-band can be applicable to such approaches.

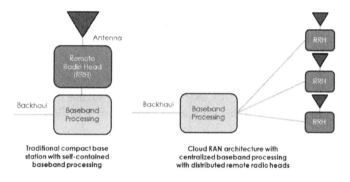

Fig. 7. Current and future base station arrangements [11]

4   Conclusions

Bandwidth bottleneck has become a serious problem, as more broadband intensive applications are rolled out and rapidly increasing mobile data traffic threatens the stability of the network. This is especially a critical issue in urban centers where subscribers and mobile traffic are the most concentrated. The E-band technology provides users with high speed, low cost and scalable means of increasing data and voice traffic to and from hard-to-reach cell towers or where the deployment of terrestrial fiber is not economical. Although there are licensing exceptions, these frequencies are typically licensed or at least lightly licensed on a worldwide scale. E-band millimetre wave radios are expected to be used in urban settings for various applications, including: Fixed and mobile backhaul, last mile gap between the fiber backbone and commercial buildings not accessible by fiber, high capacity links between main transmission nodes in the radio network, provision of network alternatives or redundancy for existing communications systems, connection of multiple buildings in a campus setting using a dedicated network, and CPRI interconnect. The E-band transceiver to be designed within the E3NETWORK project will be a promising solution towards the realization of enhanced wireless backhaul and other applications of future networks.

Acknowledgments. The present work has been performed in the scope of the *E3NETWORK ("Energy Efficient E-band Transceiver for Backhaul of the Future Networks")* European Research Project and has been supported by the Commission of the European Communities (FP7-ICT-2011-8, Grant Agreement No.317957).

References

1. Commission of the European Communities: Digital Agenda: Commission proposed over €9 billion for broadband investment, MEMO/11,709 (October 19, 2011)
2. International Telecommunication Union (ITU): Recommendation ITU-R F.2006: Radio-frequency channel and block arrangements for fixed wireless systems operating in the 71-76 and 81-86 GHz bands (March 2012)
3. Wells, J.: Licensing and License Fee Consideration for E-band 71-76 GHz and 81-86 GHz Wireless Systems, Light Licensing, E-Band Communications Corporation (2010)
4. Australian Communications and Media Authority: Planning of the 71-76 GHz and 81-86 GHz Bands for Millimetre Wave High Capacity Fixed Link Technology, Spectrum Planning Discussion Paper SPP (December 2006)
5. Johnson, E.: Mobile Data Backhaul: The Need for E-Band, Sky Light Research, Mobile World Congress (2013)
6. Ayvazian, B., Hetting, C.: Second-Generation E-Band Solutions: Opportunities for Carrier-Class LTE Backhaul, Heavy Reading white paper (February 2013)
7. Adhikari, P.: Understanding Millimeter Wave Wireless Communication, Loea White Paper (2008)
8. Australian Communications and Media Authority – Australian Government: 60 GHz Band Millimetre Wave Technology (December 2004)
9. Huang, K.-C., Edwards, D.J.: Millimetre Wave Antennas for Gigabit Wireless Communications: A Practical Guide to Design and Analysis in a System Context. Wiley (2008)
10. Dicke, D., Cameirao, P.: LightRadio Baseband Processing and Backhauling,
   http://www2.alcatel-lucent.com/techzine/lightradio-baseband-processing-and-backhauling/2011
11. Green, R.: Sizing Cloud RAN for Dense 3G and LTE Populations (2012),
   http://www.billingworld.com/blogs/vpisystems/2012/09/~/media/A1795B0F68EF4BA492313EB0DBF1C1C8.ashx?w=500&h=233&as=1

Social and Smart: Towards an Instance of Subconscious Social Intelligence

M. Graña[3], B. Apolloni[1], M. Fiasché[1], G. Galliani[1], C. Zizzo[1], G. Caridakis[2], G. Siolas[2], S. Kollias[2], F. Barrientos[4], and S. San Jose[4]

[1] Dept. of Computer Science, University of Milano, Milano, Italy
apolloni@di.unimi.it
[2] Intelligent Systems Laboratory, NTUA, Athens, Greece
gcari@image.ece.ntua.gr
[3] Dept. of Computer Science and Artificial Intelligence,
Universidad del Pais Vasco, San Sebastian, Spain
manuel.grana@ehu.es
[4] Fundation Cartif, Valladolid, Spain
frabar@cartif.es

Abstract. The Social and Smart (SandS) project aims to lay the foundations for a social network of home appliance users endowed with a layer of intelligent systems that must be able to produce new solutions to new problems on the basis of the knowledge accumulated by the social network players. The system is not a simple recollection of tested appliance use recipes, but it will have the ability generate new or refine existing recipes trying to satisfy user demands, and to perform fine tuning of recipes on the basis of user satisfaction by a hidden reinforcement learning process. This paper aims to advance on the specification of diverse aspects and roles of the system architecture, to get a clearer picture of module interactions and duties, along with data transfer and transformation paths.

1 Introduction

Social networks can be seen as a repository of information and knowledge that can be queried when needed to solve problems or to learn procedures. This is well known in the social sciences, where social networks have been useful to spread educational innovations in health care [1], manage product development programs [2], engagement in agricultural innovations by farmers [3]. A definition of *social computing* as "intra-group social and business actions practiced through group consensus, group cooperation, and group authority, where such actions are made possible through the mediation of information technologies, and where group interaction causes members to conform and influences others to join the group" has been proposed by Vannoy and Palvia [4] in the study of social models for technology adoption. This definition, and the fact that somebody publishing in a well reputed and high impact journal, is rather interesting for the purposes of this paper, where we deal with the novel idea of building social computing

L. Iliadis, H. Papadopoulos, and C. Jayne (Eds.): EANN 2013, Part II, CCIS 384, pp. 302–311, 2013.

systems where some of the computing is hidden from the social network player, hence it is a form of *subconscious* computing[1].

For a given social network mediated by a social software implemented as a web service we can distinguish between the conscious and the subconscious computing. The former is defined by the decisions and actions performed by the social players (aka the users) on the basis of the information provided by the social service. Conscious computing is quite close to Vannoy and Palvia's definition. The latter is defined by the data processing performed automatically and autonomously by the web service in order to search or produce the information offered to the social players, or for other purposes, such as data mining for the advertising industry. Social subconscious computing can be termed *intelligent* when new solutions to new or old problems are generated when posed to it. Let us clarify by proposing a short taxonomy of tasks that can be done by the social subconscious computing:

- Crowd-sourcing: the social players explicitly cooperate to build a knowledge object following some explicit and acknowledged rules. The foremost example: wikipedia.
- Information gathering: The social player asks for a specific data, i.e. the restaurant closest to a specific landmark, and the social framework searches for it. In the internet of things framework [5], this search may mean that "things" are chatting between them sharing bits of information that lead to the appropriate information source.
- Solution recommendation: the social player asks for the solution of a problem, i.e. the best dating place for a first date, and the social framework broadcasts the question searching for answers in the form of recommendations by other social players. Answers can be tagged by trust values.
- Solution generation: the social player asks for the solution of a problem, i.e. how to cook a 5 kg turkey?, and the social framework provides solutions based on previous reported experience from other social players. This experience can be explicitly or implicitly provided by the social players, but the hidden intelligence layer effectively provides the best fit solution, even if nobody in the social network has ever cooked a turkey.

Social and Smart (SandS) project is proposed as an instance of the Internet of Things [6] at the boundary with the Internet of People. SandS aims to lay the foundations of social subconscious intelligent systems in the realm of household appliances and domestic services. The term "eahouker" is introduced in this context, meaning "easy household worker", to denote the household appliance user empowered by the social network and social intelligence discussed in this paper. The contents of the paper are as follows: Section 2 gives an intuitive definition of Sands Network. Section 3 gives a primitive attempt to formalize the ideas supporting SandS subconscious intelligence. Section 4 further comments on SandS architecture. Section 5 gives a detailed discussion on the user language versus the recipes.

---

[1] As far as we know, this idea is novel: there is no published reference where it is identified and defined. Claiming novel ideas is risky, and needs careful spelling.

2   Intuitive Definition of SandS

Figure 1 gives an intuitive representation of the architecture and interactions between the system elements. The SandS Social Network mediates the interaction between a population of users (called eahoukers in SandS [7]) each with its own set of appliances. The SandS Social Network has a repository of tasks that have been posed by the eahoukers and a repository of recipes for the use of appliances. These two repositories are related by a map between (to and from) tasks and recipes. This map needs not to be one-to-one. Blue dashed arrows correspond to the path followed by the eahouker queries, which are used to interrogate the database of known/solved tasks. If the task is already known, then the corresponding recipe can be returned to the eahouker appliance (solid black arrows). The eahouker can express its satisfaction with the results (dashed blue arrows). When the queried task is unknown and unsolved then the social network will request a solution from the SandS Networked Intelligence that will consists in a new recipe deduced from the past knowledge stored in the recipe repository. This new solution will be generated by intelligent system reasoning. The eahouker would appreciate some explanation of the sources and how it has been reasoned to be generated, therefore explicative systems may be of interest for this application.

The repository of recipes solving specific tasks can be loaded by the eahoukers while commenting among themselves specific cases, or by the appliance manufacturing companies as an example of use to foster their sales by appealing their clients with additional services. These situations correspond to the conscious computing done on the social web service by human agents. The role of the Networked Intelligence is to provide the subconscious computing that generates innovation without eahouker involvement.

3   An Attempt to a Formal Specification of SandS

We can advance an abstract definition of the SandS SNS' collection of eahoukers $S = \{U_i\}_{i=1}^{N}$, where each eahouker defined by a personal profile P_i, a collection of appliances $A_i = \left\{A_i^1, A_i^2, \ldots, A_i^{nA_i}\right\}$ governed by the local middleware, and a collection of recipes $R_i = \left\{R_i^1, R_i^2, \ldots, R_i^{nR_i}\right\}$, i.e. $U_i = \{P_i, R_i, A_i\}$. In general, the number of eahoukers as well as the eahouker's information will be time varying, but for the descriptive contents of this section we remove the time parameter.

The Networked Intelligence will need to perform searches in the space of recipes in order to provide best matching solutions or to generate new solutions to the tasks queried by the eahoukers. In an attempt to formalize these ideas, we say that there is a similarity measure between appliances $d_A : \mathcal{A} \times \mathcal{A} \to \mathbb{R}$, such that $d_A(A, A^*) = 0$ iff $A = A^*$, $d_A(A, A^*) \geq 0$ iff $A \neq A^*$, and other charactistic axioms of the definition of a distance may also hold. Similarly we have a similarity measure between recipes $d_R : \mathcal{R} \times \mathcal{R} \to \mathbb{R}$, where $\mathcal{A}$ and $\mathcal{R}$ are the spaces of

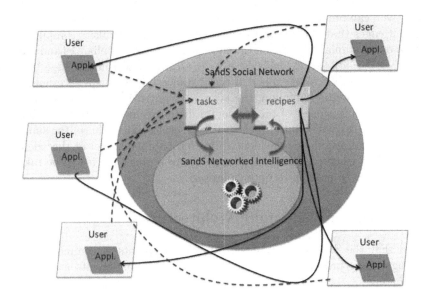

Fig. 1. Social and Smart system prototypical architecture

appliances and recipe objects, respectively, which are defined by the specification provided by the computational intelligence versant. Notice that appliances may be similar if they perform similar processes. Appliance similarity is, therefore, a function that may be highly parameterized. The personal profile similarity $d_P : \mathcal{P} \times \mathcal{P} \to \mathbb{R}$ will be rather transparent for the Networked Intelligence.

Advancing in the formalization, we may say that recipes have two components, the task to be achieved and the specification of the appliance's settings and sequential steps accomplishing it, i.e. the recipe. Additionally a degree of satisfaction can be given by the eahouker. This degree can be used also by the search engine to give a personal value to the "expertise" of the eahouker solving the task. Therefore a recipe can be formalized as $R = (\tau, \Gamma, \alpha)$, where the components are the task τ, the appliance setting Γ, and the degree of satisfaction α. The relation between tasks and recipes need not be bijective, as many tasks can be achieved with the same pattern of actions performed on/by the appliance. The eahouker's set of recipes may include a several "solutions" to the same task, with similar or different degrees of satisfaction. Notice that the ultimate semantic of the objects in this formalization will be provided by the technological specifications provided by the manufacturers.

Linking Eahoukers. The existence of a link between eahoukers can be specified explicitly by the eahoukers including each other in their profiles, like in conventional social network services. Then we may say that $d_p (P_i, P_j) = 0$ if the eahoukers are explicitly connected, and there is a corresponding edge in the social graph. As eahouker profiles may be complex structures, the distance among them can be small even when they are not explicitly linked. Implicit links are latent edges that depend on the distance between eahoukers computed as a function of the similarity between the eahoukers' profiles, appliances and recipes: $d_U (U_i, U_j) = \mathcal{F} [d_P (P_i, P_j), d_A (A_i, A_j), d_R (R_i, R_j)]$, so that an edge between the eahoukers can become explicit if the distance falls below a threshold. This functional may be the minimum, the average or any other function, it can even be dependent on a subset of the elemental distances and can be dependent on the process being carried out in the social web service. The value of the similarity between two eahoukers is equivalent to the probability model of the adequacy of an eahouker when providing some response to a query [8], it can be taken as a trust value when making recommendations.

4   More about the SandS Architecture

We find that the SandS social network has many points in common with the searching engine proposed in [8], so that its main components are somehow corresponding.

1. The indexer component that manages the eahouker identifications and labels. This component deals with the interaction with the middleware layer of the household eahouker's environment.
2. Eahouker Interface and Query analyzer component that allows the eahouker to perform queries, extracting the information from the eahouker near natural language specification. Queries are in fact task specifications.
3. Ranking function allowing to guide the search and select the best response. It manages the information embedded in the latent links between eahoukers, so that it can forward the query to the most appropriate eahouker.

When a eahouker joins the system, s/he fills his/her profile as in a conventional SNS, including explicit links to friends already in the system. The system will also gather appliance information and status from the middleware layer of the household. This information may include some standard recipes for standard tasks. The evolution of the eahouker in the system will be increasing the number of recipes and/or appliances. As in [8] the system may gather the opinions of the eahouker's friends to obtain an additional modulation of the expertise of the eahouker solving a task. Opinions can have the same structure (task, settings, satisfaction) as the recipe plus some ID tag, but the satisfaction corresponds to the opinion value. The SandS system benefits from its focused nature so that (1) semantic analysis is simplified, (2) search for expertise is bounded to the knowledge stored in the system.

4.1   Main Search Process

The main service of the SandS SNS besides providing conventional social network interaction among the eahoukers will be the automated search for recipes solving a task. The process starts when the eahouker introduces a query for a specific task, the eahouker interface and query analyzer tries to identify the task and solve specific parameters. The task specification may require some feedback to the eahouker, showing similar tasks known to the system in case the analysis fail or for confirmation previous to launching the search. The appropriate appliance is selected, if not specified. The task specification is compared with already stored recipes, searching for the ones with maximum similarity against the query and maximum eahouker satisfaction for eahoukers whose appliances are the most similar to the ones own by query's originator. The ranking function will combine the eahouker's recipe satisfaction of the owner and the friends that have tried it. The result appliance setting is communicated to the eahouker. The search process can explain the answer in terms of the distances between the diverse components of the distance guiding the search.

4.2 Consensus and Statistics

The SNS may automatically explore the stored recipes to reach some consensus or compute statistics on the information contained in the system:

- average degree of satisfaction of the recipes solving a task by the appliances provided by some specific brand, this is equivalent to compute the average of the $\{\alpha_i; i \in \mathbb{A}_\tau\}$ where the set $\mathbb{A}_\tau$ includes all recipes R_j^k such that the distance between the task τ_j^k in the recipe and the target task is below a threshold, and the appliances involved are of the same brand.
- consensus for some task on the optimal settings for an specific appliance allowing optimal experience to the eahoukers, which can be reached by voting on the settings Γ_i^k of all recipes solving task τ with appliance A^* weighted by the satisfaction α_j^k.
- consensus on the general settings for an specific task across appliances or brands to provide starting values for the initialization of new appliances, which are equivalent to extending the average of settings Γ_i^k of all recipes solving task τ to all appliances in the global knowledge database.
- communities can be built on the basis of the similarities between the appliances owned by the eahoukers, the tasks performed and the preferred appliance settings. These communities can be based on one these similarities or on a combination of them, so that communities can overlap and coexist. They can be made explicit or latent, guiding the search and recommendation processes.
- Average appliance satisfaction: may be easily computed on the satisfaction associated with the recipes that have been tested on it. This average can be across all eahouker.
- Usefulness of appliances: some statistical measure of diversity on the tasks performed over an appliance may be used as a measure of the versatility of the appliance. Frequency of use can also be readily measured.

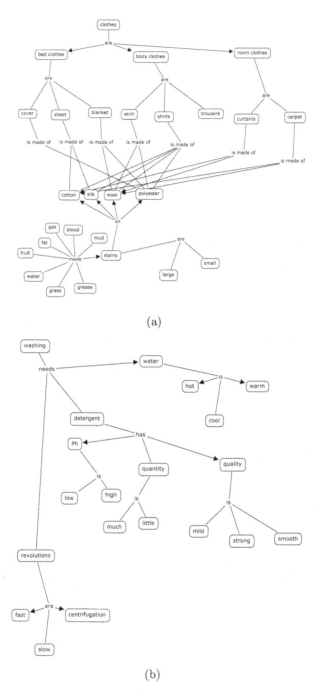

Fig. 2. Example of concept graphs for washing task. (a) Eahouker language for task, (b) appliance language for recipes.

5   Eahouker Language and Appliance Recipes

Eahoukers will specify their needs and the tasks they require to be fulfilled by
the use of a language which is different from the language required to talk to
the appliances. In other words, the language of appliance users and the language
of recipes will be different. Figure 2 shows examples of the concept maps for
the washing activity from the point of view of the user (Figure 2(a)) and the
appliance (Figure 2(b)). The user needs to specify the kind of cloth that needs
washing, the kind of stain and some parameters that may be relevant to the
appliance, such as the weight, may be implicit in the kind of cloth considered.
The eahouker wants to specify in natural language its needs, i.e. "my son's jeans
got some mud and blood from playing rough at school", while the appliance needs
an specification as a process involving the kind of detergent, the temperature of
the water and the rotation speed of the washing drum, such as the one provided

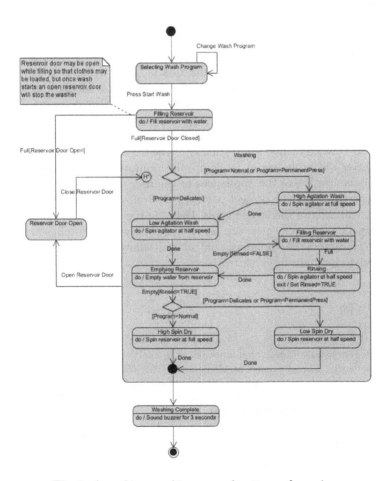

Fig. 3. A washing machine general pattern of a recipe

in Figure 3. This differentiation is needed because all learning processes will be relaying on the feedback given by the eahouker, which is the only agent entitled to say if the outcome of the appliance process is as desired or not. Then, learning will try to adapt the appliance recipes to reach the satisfaction of the user. Users may be presented with the recipes, and they may discuss and share them explicitly, hence they can specify their satisfaction with the recipe, but the ground truth value for learning needs to be the final result of the recipe.

From the point of view of language specification, the eahouker may specify the task in natural language, while the recipes need to be specified in some regular language, with a specific formal syntax. The task specification consists in some keywords, while the recipe is a formal construct. The Network Intelligence needs to devise mappings between changes in the task specification and the changes in the recipe specification, so that it may be possible to perform the inverse transformation to obtain the desired recipe changes from the user inputs.

6 Conclusion

The SandS project is an ambitious endeavor aiming to lay the grounds and propose prototypical instances of the realization of social systems with a subconscious intelligent computing layer. Though the project is focused on household appliances, its philosophy may be exported to many other domains. The main difference with other approaches is that the system will autonomously elaborate on the knowledge provided by the social players to innovate and obtain solutions to new problems, and to increase the satisfaction of the eahouker by solving better old problems by underground reinforcement learning, obtaining thus a personalization of the appliances to the user and its conditions.

In its current state, the project partners are consolidating the computational tools that will achieve the stated goals. Defining the interfaces between the system layers, i.e. the social network service, the networked intelligence, the domotic interface and the appliance control, will allow a smooth development of the project in its main research tracks.

References

1. Jippes, E., Achterkamp, M.C., Brand, P.L., Kiewiet, D.J., Pols, J., van Engelen, J.M.: Disseminating educational innovations in health care practice: Training versus social networks. Social Science & Medicine 70(10), 1509–1517 (2010)
2. Kratzer, J., Leenders, R.T., van Engelen, J.M.: A social network perspective on the management of product development programs. The Journal of High Technology Management Research 20(2), 169–181 (2009)
3. Oreszczyn, S., Lane, A., Carr, S.: The role of networks of practice and webs of influencers on farmers' engagement with and learning about agricultural innovations. Journal of Rural Studies 26(4), 404–417 (2010)
4. Vannoy, S.A., Palvia, P.: The social influence model of technology adoption. Commun. ACM 53(6), 149–153 (2010)

5. Atzori, L., Iera, A., Morabito, G., Nitti, M.: The social internet of things (siot) – when social networks meet the internet of things: Concept, architecture and network characterization. Computer Networks 56, 3594–3608 (2012)
6. Uckelmann, D., Harrison, M. (eds.): F.M.: Architecting the Internet of Things. Springer (2011)
7. Apolloni, B., Fiasche, M., Galliani, G., Zizzo, C., Caridakis, G., Siolas, G., Kollias, S., Grana Romay, M., Barriento, F., San Jose, S.: Social things - the sands instantiation. In: IoT-SoS 2013. IEEE (2013)
8. Horowitz, D., Kamvar, S.D.: Searching the village: models and methods for social search. Commun. ACM 55(4), 111–118 (2012)

Living Labs in Smart Cities as Critical Enablers for Making Real the Modern Future Internet

Ioannis P. Chochliouros[1], Anastasia S. Spiliopoulou[2], Evangelos Sfakianakis[1], Evangelia M. Georgiadou[1], and Eleni Rethimiotaki[3]

[1] Research Programs Section,
Hellenic Telecommunications Organization (OTE) S.A.,
99 Kifissias Avenue, GR-151 24, Athens, Greece
{ichochliouros,esfak,egeorgiadou}@oteresearch.gr
[2] Hellenic Telecommunications Organization (OTE) S.A.,
99 Kifissias Avenue, GR-151 24, Athens, Greece
aspiliopoul@ote.gr
[3] National and Kapodistrian University of Athens,
Faculty of Law, Athens, Greece
erethemn@law.uoa.gr

Abstract. The present position paper discusses several challenges originating from the fast growth of the Future Internet (FI) in the context of ICT, mainly within urban environments where a multiplicity of digital-based opportunities may appear for the benefit of various categories of citizens (corporate and residential) as well as of several legal entities. In particular, we focus upon the concept of actual European Living Labs that can be identified as real "enablers" for the fast growth and the proper adoption and/or implementation of modern FI-based facilities and of related services-products in a variety of domains as well as for a diversity of applications, also covering many innovative ones. Living-Labs can provide a diversity of benefits for the market, the technological evolution and the society, thus enhancing progress and development.

Keywords: Future Internet (FI), (open) innovation, interoperability, Living Lab, smart city, urbanization.

1   Introduction

The *Digital Agenda for Europe* [1] aims to help Europe's citizens and businesses to get the most out of digital technologies. The overall aim is *to deliver sustainable economic and social benefits from a digital single market based on fast and ultra-fast internet and interoperable applications*. The financial crisis has wiped out years of continuous economic and social progress and has exposed several structural weaknesses in Europe's economy. To overpass the actual state of difficulty and uncertainty, Europe's primary goal today should be "*to get Europe back on track*". To achieve an effective and sustainable future offering adequate opportunities for growth, it is essential, however, "to look beyond the short term". Faced with demographic

L. Iliadis, H. Papadopoulos, and C. Jayne (Eds.): EANN 2013, Part II, CCIS 384, pp. 312–321, 2013.

ageing and global competition, the European economy seems to take into account three essential options: work harder, work longer or work smarter, with the aim of enhancing standards of life for the European citizens. In order to guarantee a proper action plan for smart, sustainable and inclusive growth -as an immediate response to the previous challenge- the European Commission has structured the *Europe 2020 Strategy* [2] to exit the crisis and prepare the European Union's (EU) economy for the challenges of the next decade. The *Digital Agenda for Europe* is one of the seven flagship initiatives of the *Europe 2020 Strategy*, set out to define the key enabling role that the use of Information and Communication Technologies (ICT) will have to play if Europe wants to succeed in its ambitions for 2020. Among the core priorities for maximizing the social and economic potential of ICT should be the proper development and the dispersion of a variety of modern infrastructures and/or corresponding facilities, composing the so called *"Future Internet" (FI)* [3]. Internet is and will continue to be the means for doing business, working, playing, communicating and expressing ourselves, affecting both citizens and businesses.

2   Future Internet as a "Tool" for *Making Cities Smarter*

During the last decade and following to the fast technological growth that characterizes most -*if not all*- sectors and/or processes of modern digital and ICT-based societies[1], it became evident that Internet has been transformed to a more "complex" and a more "sophisticated" conceptual "cluster entity" than it was originally considered to be in past years. Although Internet actually comprises of numerous essential parts/components of modern networks-platforms-infrastructures and of related (usually modern and innovative) services-applications-facilities together with multi-generated "content" and a variety of connected equipment and/or devices, it is actually much more than simply a "communication system"; on the contrary, it is the undoubted essential "core" of our modern world towards creating a real knowledge-based society with a variety of businesses and simultaneous investments in a fully competitive context, thus implicating numerous challenges for development and for progress [4].

The swift deployment on a large scale of modern digital-based technologies in the European society is a prerequisite for the transition to a more inclusive society of growth and also a key strategic objective of the EU. A necessary condition for the timely market roll-out of various Internet-based technologies [5] that should also be able to support competition and openness to innovative business models is the acceleration of their development and demonstration. This is also catalysed by the related EU initiatives and by other specific plans for development through the streamlining and amplifying of the European human and financial resources dedicated to technology innovation. Innovation activities [6] are all the scientific, technological,

---

[1] ICT-driven innovation refers to the use of Information and Communication Technologies for the creation-implementation of new processes/products/services and methods of delivery which result in significant improvements in the efficiency, effectiveness and quality of related services offered in the marketplace.

organisational, financial and commercial steps that essentially lead, *or are intended to lead,* to the implementation of technologically new or improved products, processes, services etc. "Innovation" can also be defined as the process by which value is created for customers (or end-users) through the transformation of new knowledge and technologies into profitable products and commercial services. This also includes technology deployment, market assessment and the other typical practices leading to the diffusion of innovations across markets and societies.

In this scope, new and unexpected applications and/or services are nowadays emerging from cutting-edge technological developments that "shape" requirements for future progress, while this dynamic and continuous evolution makes the entire context of reference "more fascinating" [7]. In particular, as Internet is evolving very rapidly, this also implicates several parallel effects on socio-economic, environmental and cultural sectors as well, not purely dependent on technology factors. The future (Internet-based) networks' research is expected to "deliver" the next generation of network technologies and to enable *smart connectivity for all, anywhere, at any time at the highest speed and efficiency* so as to meet a concentration overwhelming demand by today's modern societies. The present Internet with its services and the related social networks has been converted to a critical part of our daily life, while Internet-based services are at the core of our society and economy. Simultaneously, new and possibly disruptive Internet technologies are continuing to emerge (such as location-based technologies, Internet of Things, new trust and security platforms, multimodal user interfaces, 3D content, simulation technologies, *to name just a few*). Consequently, this leads to the FI consideration which is expected to be the basis for a new wave of Internet-based services, broadly dispersed.

On the other hand, the rapid urbanization which is also a feature of our today's life, has resulted in migration -*or "concentration"*- of people in urban areas and spatial expansion of relevant urban infrastructures; this has resulted to a "diversity" of challenges-opportunities but to several undesirable "malaises" as well [8]. Worldwide, cities now perform an ever-more important role in social and economic development. Already today, 50% of the global population lives in cities and by the year 2050 the urban population will almost double from 3.3 billion to 6.4 billion. The top 100 urban agglomerations account for 25% of global gross domestic product (GDP). Rapid urbanisation generates vast stresses in developing countries, giving rise to mega-cities (those with more 10m people) across the developing world. Developed countries too have underinvested in cities and are feeling the strain in terms of congestion, pollution and land use. Although they occupy only 2% of the world's land area, they account for 75% of greenhouse gas emissions, putting cities at the forefront of efforts to cut carbon emissions and increase use of renewable resources. But this high density of challenges, users and opportunities also makes cities the ideal platform for new digital applications and services – and to drive these through Living Labs [9]. ICT and especially new Internet-enabled technologies and services will be critical for cities in meeting the challenges of sustainable economic, social and environmental development. ICT can also help to fulfil the increasing request for improved quality of life within urban settings.

To deal with this emerging priority, there were various policies and applied initiatives that have been proposed -*and occasionally become effective*- within the framework of modern cities; among the most promising ones, it was the idea of "*making cities smarter*", for the benefit of all citizens, via the implementation and/or usage of modern Internet-based technologies-applications, with the Future Internet being a vital component for any innovative process. More specifically, under this consideration the concept of "*Smart Cities*" has lately become a critical objective in the framework of urban developmental policies in the EU, affecting both policy initiatives as well as applied programmes with the aim of creating more "inclusive", sustainable and "competitive" cities [10], at least regarding business aspects and capital expenditures that could lead to further modernization and growth. (In broad terms, a "*smart city*" [11] is understood to imply a city that makes a conscious effort to innovatively employ ICT to support a more inclusive, diverse and sustainable urban environment). Smartness can include several distinct "*dimensions*", namely: (i) Smart economy (i.e.: competitiveness); (ii) smart people (thus implying social and human capital); (iii) smart governance (thus implying options for broader participation); (iv) smart mobility (affecting transport via suitable usage of ICT); (v) smart environments (also including natural resources), and; (iv) smart living (which mainly refers to the quality of life). Thus, a Smart City could be assessed as a city performing well in a forward-looking way in the previously mentioned dimensions, built on the "smart" combination of endowments and activities of self-decisive, independent and aware citizens. "Making cities smarter" means improving them in these identified dimensions. New Internet-enabled services and applications are "key-parts" in this process, increasing the efficiency, accuracy and effectiveness of operation of the city's complex ecosystem. The EU, *in particular,* has devoted constant efforts to devising a strategy for achieving urban growth in a "smart" sense for its metropolitan areas [12]. In that sense, smart cities can be considered as "*platforms*" for designing, developing and testing new Internet-based services. The next generation of such services has the potential to radically "transform" our lives, society and business both in the "smart" city context and in general. As microcosms at the heart of social and economic life, properly selected "city-based" environments can thus offer important platforms for the development, testing and benchmarking of such modern Internet-enabled services. In order to accelerate take-up and certify everybody is able to benefit from this transformation, it is important that these new Internet-based services are relying on common open platforms, accessible to (numerous) users with no particular terms and/or constraints.

Although the original related approaches were focused upon the potential role of the corresponding ICT infrastructures, further research has also identified other important drivers for realizing a proper urban growth, such as the role of human capital/education, social implications [13], environmental interest, etc. Simultaneously, the role of the FI as an "enabler" of city services has been well "identified" as a major factor for a proper urban development within environments where huge amounts of digital information and of "data" (also including pictures and video) are created and exchanged, on a daily basis. Without any doubt, cities are increasingly realizing a critical role as "*drivers*" -or "*enablers*"- of innovation in diverse areas, including health, culture, social inclusion, security aspects, environment and businesses. This is the reason why we do need to enable a very fast Future Internet affecting a multiplicity of sectors and also to ensure that the EU citizens and businesses should be able to "access and use" the most advanced content and services they call for, especially in their urban environments. Following to the previous concept, cities and urban spaces are anticipated to develop -*and/or to gradually "evolve"*- into smarter frameworks and to be better connected, exploiting the FI to organise, optimise, and provide facilities to citizens, *under a multi-contextual aspect.* The Future Internet will increasingly be a decisive "tool" and/or a critical infrastructure for the public sector, the citizens and the business of variable sizes and, under this sense, it is expected to "perceive" an Internet less directed upon technological evolution but, *instead,* an Internet of content, things and people. Realizing universal, systemic and sustainable innovation for the future society [14] and making future cities a "good place to live-work" simply necessitates wider, more rigorous and novel approaches, under new schemes for development.

3   Living Labs: Enabler for Supporting FI in Urban Environments

The research and development (R&D) of *Future Internet systems* covers an extensive variety of technology areas. In particular, an end-user seldom experiences a *single* technology, but rather, a "complex orchestration" of many technologies (i.e., routers, protocols, services, application control, devices and so forth) all leveraging each other to produce a single experience. Accordingly, whilst experimentation, and hence experimental facilities, often deal with a single technology, major complexity occurs in measuring the overall impact of a particular complex advance [15]. Many cities in Europe -and across the world- have meaningfully invested in common platforms for Internet-based services cutting across application domains. The present situation is fragmented, *however,* with many developments appearing as "islands" within their own city environments or as "limited networks". Neither the individual application areas nor individual cities have the resources -or the potential- to deal with this challenge alone; thus, it is important that cities connect to "share" best practices and explore together, in becoming scouts for this new wave of technologies and services. Open innovation [16] has the prospective to become the key enabler for this change. As a consequence, user-driven open innovation methodologies or ecosystems, such as *Living Labs* [17], have massive potential in linking the innovation gap between technology development and the rapid active use of new Internet-based services. On the other hand, openness facilitates interoperability, integration and user-friendliness, and has become a key focus in the development of FI-enabled services. Until today, user-driven open innovation methodologies have confirmed that they can deeply advance the effectiveness of the modernization process by combining R&D aspects with real market requirements. The concept of *"Living Labs"* is a specific example of such open innovation environments in real-life settings, in which user-driven innovation is fully integrated within the co-creation process of new services, products, as well as societal infrastructures. Under the consideration of detailed existing user-driven innovative processes in Europe, the concrete challenge is to realize a broader implementation and performance of open platforms for the expansion and the offering of Future Internet-enabled services in the cities, with the "inclusion" of as many citizens as possible in the related "interactive" procedures. These actions can so lead to "innovative Internet-based ecosystems" and will support the "transition" towards "smarter cities" by providing opportunities for novel facilities and services. Different approaches have been proposed in order to explain -or to assess- the huge expansion of *"Living Labs"-like* initiatives in Europe, a perspective that has certainly resulted to a strong impact upon the European innovation effort over the past few years. Several among the involved -*and quite distinct regarding their nature*- actors, such as researchers and science experts, industry and market representatives and policy makers seem to "share" the perception that the movement still under way covers a broad collection of activities, making difficult an overall explicit assessment. However, this does not reflect the real sense of the issue as the present European Living Labs constitute a successful combination of various features (such as, *inter-alia*: ICT-based collaborative and mutually dependent environments, open innovation platforms, user centred product/service development methods, and Public Private Partnerships (PPPs) contexts) and do exercise disruptive and medium- to long-term transformational effects on the European industry, markets, regional economies and societal landscapes.

Under a purely *institutional perspective*, a Living Lab can be conceived as *"a system based on a business citizens-government partnership allowing users to actively participate in the research, development and innovation process"*. In this sense, products and services can be developed in a real-life environment and in a human-centric and a co-creative way [18], based on continuous feedback mechanisms between the developers and the users. Functioning as open innovation platforms, European Living Labs aim *"at creating a user environment where users are confronted with ideation and prototypes or demonstrators of technology from the early stages of the research, development and innovation process, not only at the end"*, as it is the case *"in more classical field trials or product testing approaches"*. In this consideration, Living Labs do work as a *"tool"* to motivate innovation [19] in the interest of all players in the involved value chain: that is, from technology providers to service providers and up to different levels of involved users [20]. Thus, they can be considered as "similar" -*somehow*- to technology transfer centres, technology transfer nodes, innovation centres and technology take-up initiatives. The major attribute of Living Labs is their human-centric and user-driven perspective; they are often structured by application domains/user groups rather than in terms of technologies applied. Under a purely *business perspective*, a Living Lab can be seen as a *service provider* in the topic of research, development and innovation, based on the *"co-creation"* concept, focusing on people in their daily living environments as active -*if not decisive*- contributors to products and services design, development and testing. In any case, the creation and the proper functioning of a Living Lab require the orchestration of both the financial and operational efforts of a multitude of stakeholders, *as already pointed out*.

The *Living Labs phenomenon* [21], *unfolded in the late 1990's in Europe*, has resulted to the identification of several distinct cases of scientific, business and social interest. Viability has effectively been displayed via an operational interaction among "key pillars" of our society (i.e.: citizens and users communities; industries and businesses; public administrations, *and*; the academic and research institutes) addressed to human centred, open innovation – particularly in the early stages of ideation and concept design along the product/service development process [22]. Despite the fact that the original concept came in from the US, the way Living Labs have been established and are actually being developed complements the core character of a truly original *"European model of innovation"* that can, *in turn*, supplement the known boundaries-limitations in market size by influencing local communities and by promoting diversity (of actors, topics, requirements, etc....) for the provision of co-creative, user centred innovation to industrial players and especially to SMEs. In this sense, Living Labs can quite significantly improve the take-up ratio of patents and, *more generally*, a variety of R&D results, leading to unexpected market opportunities [23]. To this aim, they have already created a perceptible effect in certain industrial domains and businesses (especially in ICT and communication), and also have the potential to provide cross domain services in parallel areas (such as energy, environment, transportation, rural, health and social care, leisure and culture, etc) mostly where the involvement of citizens is assumed as fundamental for a more complete definition of product / service features and functionalities. Universities and research institutes play a key role in leading and operating European Living Labs, which already see the participation of several large / multinational enterprises, while a more intensive involvement of SMEs and venture capitalists should be more intensively pursued in the near future. In addition, the fact that Living Labs are also used by regional and local authorities, can contribute to the provision of extra mechanisms for enhancing the performance of territorial innovation systems; in that respect, they deserve a further appreciation and dissemination as complementary policy making instruments to the more traditional ones, like business incubators and technology transfer centres [24]. Living Labs have now been established at a local scale and can be considered as: (i) environments for real-life testing and experimentation of new services, products and systems with communities of real users; (ii) allowing early feedback and co-design by end-users; (iii) following a collaborative, iterative and stochastic process; (iv) focused on sustainable, social innovation (i.e.: not just purchase decision, but behavioural patterns and changes are the central concern), and; (v) offering an open and neutral platform where all involved "stakeholders" (technology suppliers, service providers, business customers, institutions, policy makers and regulators, end-users) can interact and co-innovate.

Current Living Labs in the European environment constitute a fresh, vibrant, vivacious and growing community, also collecting an immense variety in applications, domains, methodologies, concepts, approaches, action plans, as well as many other relevant attributes [25]. The involvement of people into product and service creation in order to get better market acceptance is of vital importance for any investment activity and can so influence further growth. Real users have to be captured during their daily life activities, to join the design, development and validation process since the very beginning and in all its phases and act as co-producers or co-creators, so that a service facility or a technical solution would be directly applicable in the market and would be meaningful and efficient. This fact can be further expanded by taking into account potential federation aspects on European scale, such as via the *European Network of Living Labs* (*ENoLL*) which is a sort of federation of Living Labs conforming to a number of general benchmark criteria or via independent European projects addressing exchange of best practices and methodologies for individual labs. In this concept, among the main challenges can be considered options such as: (i) To leverage local implantation and to overcome local limitations; (ii) to perform joint testing in cross-border "living lab" projects, via the establishment of a common ecosystem approach, common research benchmark and/or priorities, common platform guidelines and common integration frameworks, and; (iii) to offer opportunities for innovators (i.e. SMEs) to innovate and scale up internationally much faster, in particular by testing and entering new markets.

At a close check, the composition features -or elements- of most among the current European Living Labs are the following: (i) An *academic or research institute* can usually play a "leading" role, to safeguard a methodical and coherent application of fundamental underlying methodologies; (ii) One or more *local/global industries* are involved and participated as "technology providers", usually interested in designing, testing and/or validating their prototype products and services; (iii) The *Open Innovation concept*[2] is presumed, implicating that there are more opportunities for market actors-companies in "sharing and dispersing" instead of "storing and protecting" the digital knowledge context created by their internal and/or external sources (for example, employees, customers, suppliers, etc.); (iv) A *real-life testing environment* is properly and effectively established, where a variety of users' feedback on innovation is gathered and combined while it is mainly originating from the continuous and spontaneous interaction between humans and technologies; this, in fact, becomes the most reliable implication of the Living Labs context, which is critically differentiating if compared to the traditional, laboratory-based, prototype testing environment. (v) A *community of variable users* who are iteratively asked to become integrated in certain phases of the design/development/validation and marketing process, and whose response is assembled via a variety of socio-ethnographic research methods (i.e.: from focus groups to surveys, from TV-recorded debates to web-based interviews and polls and many more).

Living Labs support the creation of novel environments for territorial marketing or business promotion, adding to the framework of local and regional innovation systems tools available to policy making [26]. As purely innovation and testing environments, they cover a great number of technology domains or business applications. The key-dimensions of a Living-Lab include operation, interoperability, impact analysis and supportability. *Operation* means how a Living Lab works, the services it provides, its business model, the techniques used for community building and management, and the level of success in doing all of the above. A high level of variability and *heterogeneity* may be found across Europe, due to a number of factors. *Interoperability* means the harmonisation and/or integration perspectives in terms of methods, tools, infrastructures, applications, etc. - both among European Living Labs and in relation with FI related research projects and experimental facilities. *Impact analysis* of Living Labs' activities, services and results on the local/regional and/or European innovation systems, stakeholders and communities is relevant to issues of measuring performance, efficacy, efficiency, and quality, as well as for evaluating the potential for benchmarking, standardisation, and accreditation of best practices to be used. *Supportability* is the potential policy design and orientation several levels, also in comparison with international, non-EU experiences, to promote and expand the

---

[2] "Open innovation" is the use of purposive inflows-outflows of knowledge to accelerate internal innovation, and expand the markets for external use of innovation, respectively [16]. This paradigm assumes that firms can -and should- use external ideas as well as internal ideas, and internal & external paths to market, as they look to advance their technology.

opportunity and scope to which Living Labs can support other important R&D- and innovation- oriented initiatives.

4   Synopsis

The experimental facilities required for Future Internet research are wide ranging, both in terms of technical scope and the type of experimentation required. A classification of FI research activities can be done along two main dimensions: (i) The type of technology used (developed or tested) in the related activity, and; (ii) the type of experimentation performed. These "dimensions" also specify the nature of the innovation set forth and therefore play a principal role in assessing the effectiveness of the corresponding "Living Lab"-like approach. From a technical perspective (vertical dimension), the scope of Future Internet research is grouped loosely into four layers, which cover significantly different relevant technologies and usage patterns, (i.e.: networks-Internet infrastructures; security-safety; services and data; multimedia/user interaction) plus a fifth one addressing the applications themselves. In the second dimension, the type of experimentation required is classified according to criteria related to the functionality, robustness, scalability, integration, usability and flexibility/adaptation. Three main *categories of benefits* can emerge as being a common basis to many Living Labs: (i) Firstly, the *generation of innovation* in its different forms and shapes (i.e.: creation of new services, improvement of services and development of new tools/approaches); (ii) Secondly, the *promotion of collaboration:* In this respect Living Labs may be considered as "catalysers of interaction" providing a meeting place for actors with complementary knowledge assets; (iii) The final type of benefit is a direct consequence of the previous one and has to do with a more fluid *circulation and dissemination of knowledge*, making the participation to Living Labs activities an important learning opportunity for all involved potential stakeholders.

Acknowledgments. The present work has been inspired by the scope of the original *LiveCity* ("*Live Video-to-Video Supporting Interactive City Infrastructure*") European Research Project, and supported by the Commission of the European Communities – *DG CONNECT* (FP7-ICT-PSP, Grant Agreement No.297291).

References

1. Commission of the European Communities: Communication on A Digital Agenda for Europe (COM (2010) 245 final, August 26, 2010), Brussels, Belgium (2010)
2. Commission of the European Communities: Communication on Europe 2020: A strategy for smart, sustainable and inclusive growth (COM (2010) 2020 final, March 03, 2010), Brussels, Belgium (2010)
3. Future Internet Assembly (FIA): Position Paper: Real World Internet (2009), http://rwi.future-internet.eu/index.php/Position_Paper
4. Kusiak, A.: Engineering Design: Products, Processes, and Systems. Academic Press, San Diego (1999)
5. Brockman, B., Morgan, R.: The role of existing knowledge in new product innovativeness and performance. Decision Sciences 34(2), 385–419 (2003)
6. Garcia, R., Calantone, R.: A Critical Look at Technological Innovation Typology and Innovativeness A Literature Review. Journal of Product Innovation Management 19(2), 110–132 (2002)

7. Adner, R., Levinthal, D.: Demand Heterogeneity and Technology Evolution: Implications for Product and Process Innovations. Management Science 47(5), 611–628 (2001)
8. Begg, I.: Cities and Competitiveness. Urban Studies 36(5-6), 795–810 (1999)
9. Mulvenna, M., et al.: Living Labs as Engagement Models for Innovation. In: Cunningham, P., Cunningham, M. (eds.) Proceedings of eChallenges-2010, Warsaw, Poland, October 27-29, pp. 1–11. IIMC International Information Management Corporation (2010)
10. Parkinson, M., Hutchins, M., Simmie, J., Clark, G., Verdonk, H.: Competitive European Cities: Where Do The Core Cities Stand?, European Institute for Urban Affairs, Office of the Deputy Prime Minister, London (2003),
    http://www.vrm.ca/documents/competitive.pdf
11. Curwell, S.: E-Cities: Intelligent cities. In: Proceedings of the 2006 International Conference on Digital Government Research (DG.O 2006), pp. 142–143. ACM Press (2006)
12. http://www.smart-cities.eu/
13. Mingers, J., Willcocks, L.: Social Theory and Philosophy for Information Systems. John Wiley & Sons Ltd., Chichester (2004)
14. Moss Kanter, R.: Innovation: The Classic Traps. Harvard Business Review 84(11), 73–83 (2006)
15. Ballon, P., Pearson, J., Delaere, S.: Test and Experimental Platforms for Broadband Innovation: Examining European Practice. In: Proceedings of the 16th European Regional Conference, Porto, Portugal, September 4-6 (2005)
16. Chesbrough, H.: Open Innovation: The New Imperative for Creating and Profiting from Technology. Harvard Business School, Boston (2003)
17. Kusiak, A.: Innovation: The Living Laboratory Perspective. Computer-Aided Design and Applications 4(6), 863–876 (2007)
18. Maguire, M.: Methods to Support Human-centred Design. International Journal of Human-Computer Studies 55, 587–634 (2001)
19. Bergvall-Kåreborn, Ihlström-Eriksson, C., Ståhlbröst, A., Svensson, J.: A Milieu for Innovation - Defining Living-Labs. In: Proceedings of the 2nd ISPIM Innovation Symposium, New York, December 6-9 (2009)
20. Sood, A., Tellis, G.: Technological Evolution and Radical Innovation. Journal of Marketing 69(3), 152–168 (2005)
21. Markopoulos, P., Rauterberg, G.W.M.: LivingLab: A White Paper. Technical University Eindhoven, IPO Annual Progress Report 35, 53–65 (2000),
    http://www.idemployee.id.tue.nl/g.w.m.rauterberg/
    publications/IPOapr35LL.PDF
22. Ulwick, A.W.: Turn Customer Input into Innovation. Harvard Business Review 80(1), 91–97 (2002)
23. Følstad, A.: Living Labs for Innovation and Development of Information and Communication Technology: A Literature Review. The Electronic Journal for Virtual Organizations and Networks 10 (Special Issue on Living Labs), 99–131 (2008)
24. Bergvall-Kåreborn, B., Ståhlbröst, A.: Living Lab: An Open and Citizen Centric Approach for Innovation. Int. Journal of Innovation and Reg. Development 1(4), 356–370 (2009)
25. Niitamo, V.-P., Kulkki, S., Eriksson, M., Hribernik, K.A.: State-of-the-Art and Good Practice in the Field of Living Labs. In: Proceedings of The 12th International Conference on Concurrent Enterprising (ICE 2006), Milan, Italy, June 26-28, pp. 349–357 (2006)
26. Pierson, J., Lievens, B., Ballon, P.: Living Labs for Broadband Innovation: Configuring User Involvement. In: Proceedings of BBEurope 2005 (BroadBand Europe), Bordeaux, France, December 12-15 (2005)

Author Index

Agathokleous, Marilena I-496
Alexakos, Christos II-70
Alexandrides, Theodore II-174
Alexandiris, Antonios K. I-12
Alonso, Serafín I-370
Amali, Ramin I-203
Anagnostopoulos, Ioannis II-20
Anagnostopoulos, Christos-Nikolaos II-20, II-193, II-203
Anastassopoulos, George I-253, II-203
Antonelli, Fabio II-282
Antonopoulou, Hera II-70
Apolloni, B. II-302
Arbab, Masood Ahmad I-282
Asvestas, Pantelis II-40
Avlonitis, Markos II-60

Bacauskiene, Marija I-396
Bakic, Predrag R. II-146
Ballon, Pieter II-282
Barrientos, F. II-302
Barrientos, Pablo I-370
Barton, Carl II-11
Battipede, Manuela I-313
Beligiannis, Grigorios II-222
Belo, João I-182
Bentivoglio Colturato, Adimara I-406
Bentivoglio Colturato, Danielle I-406
Benyettou, Abdelkader I-273
Benyettou, Mohamed I-424
Boroş, Tiberiu I-42
Burget, Radim I-380

Calvo-Rolle, Jose Luis I-350
Caridakis, George II-257, II-269, II-302
Cassaro, Mario I-313
Castelo Branco, Luiz Henrique I-406
Chatziioannou, Aristotelis A. II-249
Chaudhari, B.N. I-61
Chochliouros, Ioannis P. II-257, II-292, II-312
Cortés Ramírez, Jorge Armando I-263
Crespo-Ramos, Mario J. I-350
Crnojević, Vladimir I-388
Ćulibrk, Dubravko I-388

Daoudi, Rima I-273
Debakla, Mohammed I-424
Delaere, Simon II-282
del Canto, Carlos J. I-370
De Moor, Bart II-222, II-241
Dimitrakopoulos, Christos II-70, II-231
Djemal, Khalifa I-273, I-424
Dolenko, Sergey I-81
Domínguez, Manuel I-370
Dumitrescu, Stefan Daniel I-42

Economou, George-Peter K. II-156
Essig, Kai I-446

Fiannaca, Antonino II-212
Fiasché, M. II-302
Fitkov-Norris, Elena I-213
Folorunso, Sakinat Oluwabukonla I-213
Fountas, Nikos I-144
Frecassetti, Mario Giovanni II-292
Fuertes, Juan J. I-370
Furtado, Edson Luiz I-406

Galliani, G. II-302
Garcia, Ana II-282
Garofalakis, John II-50
Gartsov, Alexander II-185
Gelzinis, Adas I-396
Georgakopoulos, Spiros V. I-292
Georgiadis, Panagiotis II-249
Georgiadou, Evangelia M. II-292, II-312
Giannoukos, Ioannis II-193
Gili, Piero I-313
Giotopoulos, Konstantinos II-70
Gkantouna, Vassiliki II-119
Gomes Benjamin, André I-406
Gopych, Petro I-71
Graña, M. II-302
González Guerra, José Luis I-263

Habermann, Danilo I-112
Hájek, Petr I-302, II-1
Hata, Alberto I-112
Hatzilygeroudis, Ioannis II-30
Hsieh, Bernard I-102

Hughes, Bradley J.   I-203
Hunt, Doug   I-233

Iakovakis, Vassilis   I-144
Iliadis, Lazaros   I-132, I-485
Iliopoulos, Costas S.   II-11
Iliou, Theodoros   II-203
Ioannou, Zafeiria-Marina   II-174
Iosifidis, Alexandros   I-1
Isaev, Igor   I-81

Jadhav, Sadhana V.   I-61
Jesus, Adelaide P.   I-182

Kalampakas, Antonios   I-485
Kalita, Oksana   II-185
Kampouridis, Michael   I-12
Kanavos, Andreas   II-100
Kapsouras, Ioannis   I-172
Karagiannis, Stefanos   I-144
Karakasidis, Alexandros   II-164
Karanasiou, Irene   II-40
Karanikolos, Stylianos   I-172
Karapilafis, Georgios   I-132
Kardara, Mania   II-90
Karwowski, Jan   I-122
Karydis, Ioannis   II-60
Kasabov, Nikola   I-233
Kasampalis, Dimitrios   I-360
Katakis, Ioannis   I-496
Kateris, Dimitrios   I-360
Katsavounis, S.   I-132
Kechagias, John   I-144
Kłoszewska, Iwona   II-193
Kmet, Tibor   I-52
Kmetova, Maria   I-52
Knoll, Alois   I-330
Kokkinos, Yiannis   I-340
Kollias, Stefanos   II-257, II-269, II-302
Kompatsiaris, Ioannis   I-223
Korvesis, Panagiotis   II-138
Kossida, Sophia   I-165
Kouneli, Marianna   II-50
Krause, André Frank   I-446
Kuzmin, Vadim   I-466
Kyrtopoulos, Soterios   II-249

La Rosa, Massimo   II-212
Legierski, Jarosław   I-122
Leon, Bobrowski   I-456

Lerro, Angelo   I-313
Li, Xirong   I-32
Likothanassis, Spiros   II-70, II-110,
    II-231, II-174
Lopez, Merce   II-282
López-García, Hilario   I-350
Lovestone, Simon   II-193
Lucas Jaquie Castelo Branco, Kalinka
    Regina   I-406
Lucena, Rui   I-182

Machón-Gonzílez, Iván   I-350
Maglogiannis, Ilias   I-292
Mahmud, Sahibzada Ali   I-22, I-91, I-282
Makris, Christos   II-100
Malcangi, Mario   I-323
Maltha, Sven   II-282
Maragoudakis, Manolis   I-474
Margaritis, Konstantinos   I-340
Martins, Leonardo   I-182
Masek, Jan   I-380
Matsopoulos, George K.   II-40
Mavroudi, Seferina   II-231
Mechant, Peter   II-282
Mecocci, Patrizia   II-193
Megalooikonomou, Vasilis   II-138, II-146
Merekoulias, Georgios   II-80
Mitroulias Athanasios   II-110
Mora, Ben   I-192
Morán, Antonio   I-370
Moreau, Yves   II-222, II-241
Moschopoulos, Charalampos   II-222,
    II-241
Moshou, Dimitrios   I-360
Moumtzidou, Anastasia   I-223
Moutafi, Konstantina   II-70
Mporas, Iosif   II-138
Muhammad Khan, Gul   I-22, I-91, I-282
Mulder, Nicola   II-11
Mylonas, Phivos   II-20, II-269

Nanopoulos, Photis   II-185
Natekin, Alexey   I-330
Nayab, Durre   I-91
Nikolaidis, Nikolaos   I-172
Nikolakopoulos, Athanasios N.   II-50
Nogueira, Pedro A.   I-243
Nuzhnaya, Tatyana   II-146

Obornev, Eugeny   I-81
Oikonomou, Maria   I-414

Okulewicz, Michał   I-122
Olej, Vladimír   I-302, II-1
Oliveira, Eugénio   I-243
Osório, Fernando   I-112
Osstyn, Dirk   II-282

Panić, Marko   I-388
Pantazi, Xanthoula Eirini   I-360
Papadopoulos, Harris   I-253
Papaioannou, Vaios   II-156
Papaoikonomou, Thanos   II-90
Parladori, Giorgio   II-282
Parry, Dave   I-233
Paschali, Kallirroi   II-174
Pavlidis, Georgios   II-185
Pegkas, Andreas   II-231
Perikos, Isidoros   II-30
Persiantsev, Igor   I-81
Piefke, Martina   I-446
Pigatto, Daniel Fernando   I-406
Pimenidis, Elias   I-132
Pitas, Ioannis   I-1
Plagianakos, Vassilis P.   I-292
Plegas, Yannis   II-100
Plerou P., Antonia   I-433
Politopoulou, Vicky   I-474
Popovic, Dusan   II-222, II-241
Prada, Miguel A.   I-370

Quaresma, Cláudia   I-182

Ratcliff, Jay   I-102
Razis, Gerasimos   II-20
Rethimiotaki, Eleni   II-312
Roschildt Pinto, Alex Sandro   I-406
Richardson, Mark   I-192
Rizzo, Riccardo   II-212
Rodrigues, Rui   I-243

San Jose, S.   II-302
Santos, Marcelo   I-182
Salazar Mendiola, Juan Luis   I-263
Schack, Thomas   I-446
Schliebs, Stefan   I-233
Serra, Artur   II-282
Sfakianakis, Evangelos   II-292, II-312
Shimelevich, Mikhail   I-81
Shrestha, Durga Lal   I-466
Sifakis, Emmanouil G.   II-249
Sifrim, Alejandro   II-222, II-241

Simmons, Andrew   II-193
Siolas, Georgios   II-269, II-302
Sioutas, Spyros   II-119
Skoura, Angeliki   II-146
Sladojević, Srdjan   I-388
Soininen, Hikka   II-193
Solomatine, Dimitri   I-466
Sourla, Efrosini   II-80
Sourla, Georgia   II-119
Spartalis, Stefanos   I-132, I-485
Spenger, Christian   II-193
Spiliopoulou, Anastasia S.   II-312
Spiridonidou, Antonia   II-60
Spiros, Likothanassis   II-110
Stafylopatis, Andreas   II-269
Stamatelatos, Makis   II-282
Stamatopoulou, Konstantina-Maria   II-80
Stephanakis, Ioannis M.   II-203, II-257, II-292
Stoykova, Velislava   II-129
Syrimpeis, Vasileios   II-80
Szupiluk, Ryszard   I-154

Tasoulis, Sotiris K.   I-292
Tefas, Anastasios   I-1, I-172, I-414
Theodoridis, Evangelos   II-100
Theofilatos, Konstantinos   II-231
Tsakalidis, Athanasios   II-80, II-119, II-174
Tsakona, Anna   II-174
Tsamandas, Athanasios   II-174
Tsapatsoulis, Nicolas   I-496
Tserpes, Konstantinos   II-90
Tsiliki, Georgia   I-165
Tsitiridis, Aristeidis   I-192
Tsolaki, Magda   II-193
Tsolis, Dimitrios   II-174
Tsouvaltzis, Pavlos   I-360
Tzimas, Giannis   II-80, II-119

Uher, Vaclav   I-380
Ullah, Fahad   I-22
Urso, Alfonso   II-212

Vaiciukynas, Evaldas   I-396
Valavanis, Ioannis   II-249
Vargas Luna, José Luis   I-263
Varvarigou, Theodora   II-90

Vaxevanidis, Nikolaos   I-144
Vellas, Bruno   II-193
Ventouras, Errikos M.   II-40
Vergeti, Paraskevi   II-70
Verikas, Antanas   I-396
Verykios, Vassilios S.   II-164
Vieira, Pedro   I-182
Vlachakis, Dimitrios   I-165
Vlamos, Panayiotis M. I-433
Vrochidis, Stefanos   I-223

Watson, Bruce   II-11
Wolf, Denis   I-112
Wu, Shaohui   I-32

Xu, Jieping   I-32

Yang, Gang   I-32

Ząbkowski, Tomasz   I-154
Zacharaki, Evangelia I.   II-138
Zizzo, C.   II-302